Crop Plant: New Frontiers in Crop Science

Crop Plant: New Frontiers in Crop Science

Edited by **Corey Aiken**

New York

Published by Callisto Reference,
106 Park Avenue, Suite 200,
New York, NY 10016, USA
www.callistoreference.com

Crop Plant: New Frontiers in Crop Science
Edited by Corey Aiken

International Standard Book Number: 978-1-63239-132-2 (Hardback)

Contents

Preface

New frontiers in crop science are elucidated in this comprehensive book. It provides the readers with a comprehensive overview of Crop Plant with latest advances in research. This book deals with information regarding markers and gen-next sequencing technology and its application. It also describes how silicon can be utilized for drought tolerance. The major problem of rising CO_2 and O_3 causing environmental pollution has been addressed in this book along with detailed accounts on the phenomena of RNAi and its application in crop science and review for boron deficiency in soils and how to deal with it for better crops. Information regarding contemporary works underway in the field of crop science has also been presented in this book making it a wholesome read for those interested in crop management and agricultural science.

This book unites the global concepts and researches in an organized manner for a comprehensive understanding of the subject. It is a ripe text for all researchers, students, scientists or anyone else who is interested in acquiring a better knowledge of this dynamic field.

I extend my sincere thanks to the contributors for such eloquent research chapters. Finally, I thank my family for being a source of support and help.

Editor

Part 1

Biotechnology

1

Progression of DNA Marker and the Next Generation of Crop Development

Herry S. Utomo, Ida Wenefrida and Steve D. Linscombe
Louisiana State University Agricultural Center
USA

1. Introduction

Advancement in genomic technology has been the main thrust for the progression of DNA markers that is now approaching a critical point in providing a platform for the next generation of varietal development. Improving total yield production to meet the increasing need to feed the world population remains the major goal. However, achieving more sophistication in providing high quality crop products to meet the emerging demand for better nutritional values and food functionalities will increasingly become important goals. Progression in high throughput marker analyses, significant reduction in the cost per data point, sophistication in computational tools, and creation of customized sets of markers for specific breeding applications are continuing and expected to have direct implications for highly efficient crop development in the near future. An advanced DNA marker system can be used to accomplish breeding goals, as well as achieve various scientific goals. The goals encompass a wide array of targets from understanding the function of specific genes so detailed that the quality of gene output or products can be controlled to attaining a global view of genomic utility to improve crop development efficiency. The combination of molecular understanding at the individual gene levels and genetic manipulation at the genome levels may lead to a significant yield leap to meet global food challenges.

Historically, plant breeding always integrates the latest innovations to enhance crop improvement. Starting out with the prehistoric selection based on systematic visual observations leading to the first plant domestication (Harlan, 1992), crop development was further enhanced by employing Darwin's scientific principles of hybridization and selection, then applying Mendel's principles of association between genotype and phenotype, and now through DNA markers and genomics that will lead to the next generation of crop development. Recent progression in high throughput marker genotyping, genome scanning, sequencing and re-sequencing, molecular breeding and bio-informatics, software and algorithm, and precise phenotyping (Delseny et al. 2010; Edwards & Batley, 2010; Varshney et al., 2009; Mochida & Shinozak, 2010; Davey et al., 2011) is a conduit for the next generation of crop development that uses different views and avenues to approach the same goal. This chapter will discuss various aspects of DNA molecular markers associated with crop development. They include the progression of molecular marker technology, prospect and current limitations, applications in unraveling global genetic potential, and specific utilization for exploiting global genetic sources and re-purpose some of the traits to fulfill local demand in the given environmental conditions.

2. Advancement in DNA marker technology

2.1 DNA marker

Genetic markers are heritable polymorphisms found among individuals or populations that can be measured (Davey et al., 2011). They become the center point of modern genetics to answers many important questions in population genetics, ecological genetics and evolution. A genetic marker is a gene or DNA sequence with a known location on a chromosome that can be used to identify cells, individuals, and species, or traits of interest. It can be described as an observable variation from mutation or alteration in the genomic loci. A genetic marker can be a short DNA sequence, such as a sequence surrounding a single base-pair change (single nucleotide polymorphism, SNP), or a long one, like microsatellite or simple sequence repeats (SSRs). DNA markers are available in many forms. In addition to older types of markers, such as RFLP, RAPD, and AFLP, combination of molecular markers, simple sequence repeats (SSRs), indels, PCR-based SNPs, and CAPS (cleaved amplified polymorphic sites) markers, are markers that can be used for many purposes including genotyping.

DNA markers can be used to identify or verify a true identity of cultivars or breeding lines, the F_1 hybrids, seed purity, or intra-varietal variation. The analyses are relatively simpler, easier, and more accurate than phenotypic evaluation. Few well-selected markers are typically adequate for providing definitive discrimination of cultivars in question. One of the major marker contributions is better understanding of genetic diversity, population structures, genetic relationship among subspecies, genetic relationship within specific germplasm collections, and family relationship among breeding lines and cultivars. Molecular markers provide high quality genetic data that may not be possible to produce through other genetic methods. Genetic relationship and connectivity among available germplasm is invaluable knowledge to realize the overall genetic potential that can be managed to develop more productive cultivars. Gene survey to find specific target alleles from different groups of germplasm is very useful for a breeder to efficiently use and manage available germplasm. An example in rice includes the survey of *Pi-ta* blast-resistant genes among a large collection of rice lines (Wang, et al., 2007b).

2.2 Simple sequence repeats (SSRs)

SSR markers are co-dominant, reliable, randomly distributed throughout the genome, and highly polymorphic. Due to its superiority, SSRs have been used widely for mapping purposes. They are generally transferable between mapping populations. SSRs consist of tandemly repeated short nucleotide units 1–6 bp in length. The abundance of SSR markers in the genome is generated through SSR mutational rates that occur between 4×10^{-4} to 5×10^{-6} per allele and per generation (Primmer et al., 1996; Vigouroux et al., 2005). The mutation is caused predominantly through 'slipped strand mis-pairing' during DNA synthesis that will result in the gain or loss of one or more repeat units (Levinson and Gutman, 1987). The most widely distributed SSRs are di-, tri- and tetranucleotide repeated motifs, such as (CA)n, (AAT)n and (GATA)n (Tautz & Renz, 1984). Prior to whole genome sequence, SSR markers initially were developed from expressed sequence tags (ESTs) and bacterial artificial chromosome (BAC) end sequences. For example, prior to the completion of rice genome sequence, 2,240 SSR markers were identified using publicly available BAC and PAC clones (McCouch et al., 2002). The completion of whole genome sequence in many plant species has led to the identification of numerous SSR markers. A large number of SSR markers totaling 797,863 SSRs were identified in three monocots *Brachypodium*, sorghum and rice and three dicots *Arabidopsis*, *Medicago* and *Populus* using their whole genome sequence information (Sonah et al., 2011). Mono-nucleotide repeats were the most abundant repeats, and the frequency of repeats decreased with increase in motif length both in monocots and dicots. The frequency of SSRs was higher in dicots than in monocots both for nuclear and chloroplast genomes. Based on SSR analyses of these six species, GC-rich repeats were the dominant repeats in monocots, with the majority of them being present in the coding region that involved in different biological processes, predominantly binding activities. Their locations on the physical map can be accessed through various online databases. The SSR markers are available to the public providing marker density of approximately 51 SSR per Mb or less and are suitable for use in mapping and various applications of MAS.

2.3 Single nucleotide polymorphism (SNP)

Single nucleotide polymorphism (SNP) is an individual nucleotide base difference between two DNA sequences. SNP markers are the ultimate form of genetic polymorphism and, therefore, may become the predominant markers in the future. SNPs can be categorized according to nucleotide substitution, i.e. transitions (C/T or G/A) or transversions (C/G, A/T, C/A or T/G). In both human and plant, C/T transitions constitute 67% of the SNPs observed (Edwards et al., 2007). In human, about 90% of variation is attributed to SNPs, equating to approximately 1 SNP in every 100–300 bases. Based on partial genomic sequence information from barley, soybean, sugarbeet, maize, cassava, potato and other crops, typical SNP frequencies are also in the range of one SNP every 100–300 bp (Edwards et al., 2007; Hyten et al., 2010). Rice serves a crop model and, therefore, discovery and utilization of SNP markers in rice can be enhanced and perfected for other crop plants. A genome-wide analysis in rice cultivars Nipponbare (japonica subspecies) and 93-11 (indica subspecies) revealed 1,703,176 SNPs and 479,406 indels (Shen et al., 2004), giving a frequency of approximately 1 SNP per 268 bp. Using alignments of the improved whole-genome shotgun sequences for japonica and indica rice, the SNP frequencies vary from 3 SNPs/kb in coding sequences to 27.6 SNPs/kb in the transposable elements (Yu et al., 2005). Major re-sequencing effort based on hybridization based re-sequencing of 20 diverse *O. sativa* varieties by Perlegen BioSciences for the OryzaSNP project, a total of 159,879 SNPs were identified (McNally et al. 2009). The quality of the SNPs in this study was very good, but the SNP discovery pool covered only about 100Mb of the genome and had a low discovery rate (approximately 11% within the tiled 100 Mb region). To improve the SNP discovery pool, hundreds of diverse *O. sativa*, *O. rufipogon/O. nivara*, *O. glaberrima and O. barthii* accessions are currently being re-sequenced by groups in several countries (McCouch et al., 2010).

SNP detections can be done through a number of methods, including gel electrophoresis, fluorescence resonance energy transfer (FRET), fluorescence polarization, arrays or chips, luminescence, mass spectrophotometry, and chromatography. Fluorescence detection method currently is the most widely used in high-throughput genotyping. Fluorescence has been used in different detection applications, including plate readers, capillary electrophoresis and DNA arrays. Many types of fluorescent plate readers are available with the capability to detect fluorescence in a 96- or 384-well format by using a light source and narrow band-pass filters to select the excitation and emission wavelengths and enable semi-quantitative steady state fluorescence intensity readings to be made (Jenkins & Gibson, 2002). It also has been applied for genotyping with TaqMan, Invader and rolling-circle amplification. Fluorescence plate readers allow measurement of additional fluorescence parameters, including polarization, lifetime and time-resolved fluorescence, and fluorescence resonance energy transfer. In addition, mass spectrometry and light detection are also used for high throughput SNP genotyping.

DNA chip or gene chip is a SNP detection platform for high-throughput genotyping. It consists of a collection of microscopic DNA spots attached to a solid surface. This is one of the fastest research developing areas. More than 1.8 million markers (about 906,600 SNPs and 946,000 probes) are available from the Affymetrix® Genome-Wide Human SNP Array 6.0 for the detection of copy number variation. Luminex has developed a panel of 100 bead sets with unique fluorescent labels that can be processed by flow analyzer. Besides detecting SNPs, genotyping, or re-sequencing mutant genomes, DNA microarrays has been used to measure gene expression. SNP detection also can be done using mass spectrometry based on molecular weight difference of DNA bases. Variation of this technique includes MALDI-TOF (matrix-assisted laser desorption/ionization-time of flight) mass spectrometry that uses allele-specific incorporation of two alternative nucleotides into an oligonucleotide probe to allow measurement of the mass of the extended primers. This approach can also detect PEX products in multiplex very efficiently. Both DNA microarrays developed by Affymetrix (Santa Clara, USA) and a high-density biochip assay by Illumina Inc. (San Diego, USA) are two major chip-based high-throughput genotyping systems that offer different levels of multiplexes of several thousands (Yan et al., 2010). When an ultra-high density SNP map is used, QTL gain detection efficiency has improved considerably compared to using maps from traditional RFLP/SSR markers (Yu et al., 2011).

2.4 DArT (diversity array technology) and RAD (restriction site associated DNA)

Dramatic advancement of SSR and SNP marker technology and their applications have been achieved in important organisms, including humans and a number of model animals and crops. However, discovering sequence polymorphism in non-model species, especially 'orphan' crop and other crops that have complex, polyploid genomes, remains slow. DArT (diversity arrays technology) is a microarray hybridization-based marker system that can be used to overcome the problem, since it does not require prior knowledge of genetic or genomic sequence (Yang et al., 2011; Alves-Freitaset al., 2011; Jaccoud et al., 2001; Wenzl et al., 2004). It has relevant applications for species with complex genomes and especially for the 'orphan' crops important for Third World countries. In addition to its high throughputness, DArT is relatively quick, highly reproducible, and cost effective about tenfold lower than SSR markers per data point (Xia et al., 2005). It is designed for open use and not covered by exclusive patent rights. Users can freely specify the scope of genetic analyses and it can be expanded as needed.

Typical DArT analyses include 1) Constructing a reference library representing the genetic diversity of a species through extraction of total genomic DNA (metagenome) from a pool of individuals (i.e. a group of cultivated genotypes or to be combined together with their wild relatives, followed by complexity reduction to produce genomic representation, and cloning using suitable vector and *E. coli*; 2) Preparing "discovery array" containing individual clones; 3) Generating genomic representations of individual lines studied; 4) Hybridizing with genomic representations of all genomes in the metagenome library; 5) Identifying polymorphic clones and assembling polymorphic data into "genotyping array"; and 6) Genotyping analyses, including construction of linkage mapping or other type of analyses.

Beside its great potentials, DArT has inherent limitations. First, DArT markers are dominant markers (present or absent), which restrict its value in some applications. Second, it is a microarray-based technique that involves several steps, including preparation of genomic representation for the target species, cloning, and data management and analysis. These steps require expertise, additional cost, and also utilization of supporting software, such as DArTsoft, DArTdb, and DArtsoft 7. These may pose some limitation to its full utilization potential in the developing countries. Beside a slow start centered around the team that developed the system, an increasing number of independent research groups now have routinely utilized the methodology involving a broader range of species for various purposes, including linkage mapping (Yang et al., 2011), genotyping of closely related species (Alves-Frietas et al., 2011), genotyping very large and complex genomes such as wheat (Paux et al., 2008) and sugarcane (Wei et al., 2010).

More recently, a variety of microarrays (including tiling/cDNA/oligonucleotide arrays) also has been used to develop the so-called RAD markers for study of genomewide variations associated with restriction sites for individual restriction enzymes. For this purpose, first a genome-wide library of RAD tags is developed from genomic DNA, which is then used for hybridization on to the chosen microarray to detect all restriction site-associated variations in a single assay. The development of RAD tags involves the following steps: (i) digestion of genomic DNA with a specific restriction enzyme; (ii) ligation of biotinylated linkers to the digested DNA; (iii) random shearing of ligated DNA into fragments smaller than the average distance between restriction sites, leaving small fragments with restriction sites attached to the biotinylated linkers; (iv) immobilization of these fragments on streptavidin-coated beads; and (v) release of DNA tags from the beads by digestion at the original restriction sites. This process specifically isolates DNA tags directly flanking the restriction sites of a particular restriction enzyme throughout the genome. The RAD tags from each of a number of samples, when hybridized on to a microarray, allows high-throughput identification and/or typing of differential hybridization patterns. These markers have clear advantage over the existing marker systems (for example, restriction fragment length polymorphisms, AFLPs and DArT markers) that could assay only a subset of SNPs that disrupt restriction sites. RAD markers were successfully developed in a number of organisms, including fruit fly, zebrafish, threespine stickleback, and Neurospora (Lewis et al., 2007; Miller et al., 2007a, b) and will certainly find their way in most of the laboratories working on higher plants.

Another high throughput restriction-based marker is RAD (Restriction site Associated DNA) markers that can be used for genetic mapping. To generate RAD markers, RAD tags (the DNA sequences immediately flanking each instance of a particular restriction enzyme site throughout the genome) need to be isolated. This involves digesting DNA with a particular restriction enzyme, ligating biotinylated adapters to the overhangs, randomly shearing the DNA into much smaller fragments than the average distance between restriction sites, and isolating the biotinylated fragments using streptavidin beads (Miller et al., 2007b). Different RAD tag densities can be obtained by utilizing different restriction enzymes during the isolation process. Once RAD tags are isolated, they can be used for microarray analysis (Miller et al., 2007a; Lewis et al., 2007).

As an alternative, RAD analyses can be incorporated into high-throughput sequencing (i.e. on the Illumina platform; Baird et al., 2008). For that, the RAD tag isolation procedure will need to be modified. After the production of DNA fragments much smaller than the average distance between restriction sites by random shearing, it will be followed by preparation of the sheared ends and ligation of the second adapter, and amplification of specific fragments that contain both adapters using PCR. The first adapter contains a short DNA sequence barcode. Different DNA samples can be prepared with different barcodes to allow for sample tracking when multiple samples are sequenced in the same reaction (Hohenlohe et al., 2010; Baird et al., 2008). These RAD tags can then be subjected to high-throughput sequencing for more efficient RAD mapping. The sequencing approach produces higher genetic marker density than microarray methods.

2.5 Random, genic, and functional markers

DNA markers can be classified as 1) random markers (anonymous or neutral markers) when they are derived at random from polymorphic sites across the genome, 2) gene targeted or candidate gene markers when they are derived from polymorphisms within genes, and 3) functional markers when they are derived from polymorphic sites within genes that are causally associated with phenotypic trait variation (Andersen & Lübberstedt, 2003; Wei et al., 2009). Each marker type may be used for specific purposes. Random markers, for example, can be used as an effective tool for establishing a breeding system, studying a gene flow among natural populations, and determining a genetic structure of population or characterizing a GeneBank collection (Xu et al., 2005). Although the predictive value of a random marker depends on the known linkage phase between marker and target locus alleles (Lübberstedt et al., 1998), so far, a random marker remains the marker system of choice for marker-assisted breeding and QTL analyses in a wide variety of crop plants (Semagn et al., 2010; Xu, 2003b).

Both genic and functional markers are derived within the genes. Therefore, they are correlated well with gene function and have a high predictive value for the targeted gene in selection (Anderson & Lübberstedt, 2003; Wei et al., 2009). Because of that, they are most suited for use in marker-assisted breeding. The number of both genic and functional markers increases substantially in the recent years due to the availability of DNA sequence information from whole genome sequence projects that are available publically for a number of plant species, including rice, soybean, cassava, maize, barley, wheat, potato, and tomato (Mochida & Shinozak, 2010). Sequence data of fully characterized genes and full-length cDNA clones are also available for some plant species, including those described above. The sequence data for ESTs, genes, and cDNA clones can be downloaded from GeneBank and scanned for identification of markers, including SSRs which are typically referred to as EST-SSRs or genic microsatellites. Many gene-derived SSR markers for maize, for example, have been developed from genes using the available information in GeneBank and their primer sequences are available at www.maizeGDB.org.

Genic SSRs are more transferable across species than genomic markers, especially when the primers are designed from more conserved coding regions (Varshney et al., 2005). EST-SSR markers could, therefore, be used in related species where information on SSRs or ESTs is limited. These markers can also be used effectively for comparative mapping (Shirasawa et al., 2011; Yu et al., 2004; Varshney et al., 2005; Oliveira et al., 2009). EST-SSRs can be used to produce high quality markers, but they are often less polymorphic than genomic SSRs (Wang et al., 2011; Aggarwal et al, 2007; Eujayl et al., 2002; Thiel et al., 2003). The EST resources can be further mined for SNPs (Ramchiary et al., 2011; Li et al., 2009).

2.6 Physical and molecular genetic map

2.6.1 Physical map

A physical map provides information on the order of genetic components on the chromosomes in terms of physical distance units (base pairs). Deciphering actual biological functions of the physical map hold the key to unravel the overall genetic potential of organisms. However, a construction of a whole-genome physical map is crucial. It provides a solid blueprint for quantifying species evolution, revealing species-specific features, delineating ancestral biological functions shared by a certain group of plant species, predicting and interpreting regulatory signatures, and for practical purposes identifying candidate genes needed in crop improvement through sequences of functional or structural orthologs among closely related or model species. The construction of a physical map has been the critical component in numerous genome projects, including the Human Genome Project (HGP), the first genome project initiated in 1990 and completed in 13 years, to produce and integrate genetic, physical, gene and sequence maps. As of today, 25 published plant genome sequence (complete, publicly available, and can be used without restriction) is available, including for potato, grape, *Arabidopsis thaliana, A. lyrata*, Thellungiella, *Brassica rapa*, poplar, cucumber, cannabis, apple, strawberry, soybean, Pigeon pea, lotus, Medicago, Date palm, maize, sorghum, Brachypodium, rice, selaginella, and Physcomitrella (CoGePedia, 2011). Rice (*Oryza sativa*) genome sequence was the second (after *Arabidopsis*) to be published in plants, but it is the first monocot, grass, grain, and food crop genome. Its original published genome published in 2002 is consisted of a dual publication from two independent groups, using two subspecies of rice, japonica and indica. The current version of the rice genome contains ~370 megabases of sequence and 40,577 non-transposon related genes spread across 12 chromosomes (CoGePedia, 2011).

Physical mapping can be carried out using BAC-by-BAC or clone-by-clone strategy using two-step progression; First is the establishment of BAC clones (typically 100–150 kb) for the target genome/chromosome together with a set of overlapping clones representing a minimal tiling path (MTP) to be ordered along the chromosomes of the target genome. Shotgun sequencing is then applied to the individually mapped clones of the MTP. The DNA from each BAC clone is randomly fragmented into smaller pieces to be cloned into a plasmid and then subjected to Sanger sequencing (dideoxy sequencing or chain termination method) or sequenced directly using Next Generation Sequencing (NGS) technologies. The resulting sequence data are then aligned so that identical sequences overlap and contiguous sequences (contigs) are assembled into a finished sequence. Unlike the Sanger sequencing technology, NGS technologies are based on massive parallel sequencing, do not require bacterial cloning, and only rely on the amplification of single isolated DNA molecules. Tens of millions of single-stranded DNA molecules can be immobilized on a solid surface, such as a glass slide or on beads, and analyze them in a massively parallel way providing extremely rapid sequencing. Physical mapping can also be done through whole-genome shotgun (WGS) strategy involving the assembly of sequence reads generated in a random, genome-wide fashion. The entire target genome (chromosome) is fragmented into pieces of certain sizes that can either be subcloned into plasmid vectors or sequenced directly using NGS technologies. Highly redundant sequence coverage across the genome or chromosome can be generated through sequence reads from many subclones and using various computational methods, the sequences are assembled to produce a consensus sequence.

One of the most expected outcomes of the genome sequence is high-throughput development of molecular markers to assist genetic analysis, gene discovery and breeding programs (Fukuoka et al. 2010). Because of the genome sequence, rice, for example, is now rich in tools for mapping and breeding. It has high density SSRs of about 51 SSR per Mb, comprehensive SNPs (1,703,176 SNPs, approximately one SNP every 268 bp), insertion–deletion polymorphisms (IDPs) and custom designed (candidate gene) markers for marker-assisted breeding (Feuillet et al, 2011). Upon the completion of genome sequence, various efforts have been dedicated to tributary SNP markers identified from the sequence into breeder's chips where it can be used as a breeding tool (McCouch et al. 2010). A combination of low-, medium- and high resolution SNP assays are being developed for variety of purposes. The low density SNP chips, the 384-SNP OPAs are particularly attractive to the breeding and geneticists because of their reliability and require little technical adjustment once they are designed and optimized. Hundreds or thousands of individuals can be assayed within a short time window and are relatively inexpensive compared to the time, labor and bioinformatics requirements of other marker technologies.

2.6.2 Genetic map

A genetic map is produced by counting recombinant phenotypes revealing important genetic layouts of the organism. Marker based-genetic linkage map is generally constructed using the same principles for constructing classical genetic maps. The components of mapping include selection of markers, development of mapping populations from selected parental lines, genotyping and phenotyping each individual in the mapping population using molecular markers; and constructing linkage maps from the phenotypic and marker data. To define a recombination frequency between two linked genetic markers, genetic distance units known as centiMorgans (cM) or map units are used. Two markers are 1 cM apart if they are found to be separated in one of 100 progeny. However, 1 cM does not always correspond to the same length of DNA physical distance. The actual length of DNA per cM is referred to as the physical to genetic distance. In the genome areas where recombination occurs frequently (recombination hot spots), shorter length of DNA per cM - as low as 200 kb/cM, can be found. The characteristic of recombination hot spot is that the gene or genes where crossovers occurred are mostly located in very small genetic intervals, consisting mostly of 1-2 genes, and that those genes almost always harbor one or more single feature polymorphisms (Singer et al., 2006). In other parts where recombination may be suppressed, the physical to genetic distance can be 1500 kb per 1 cM. The lowest recombination rates typically occur at the centromeres due to heavily methylated heterochromatin (Haupt et al., 2001). The proportion of recombinant gametes depends on the rate of crossover during meiosis and is known as the recombination frequency (r). The maximum proportion of recombinant gametes is 50% when crossover between two genetic loci has occurred in all the cells. This is equivalent to non-linked genes where the two loci are inherited independently. The recombination frequency depends on the rate of crossovers which in turn depends on the linear distance between two genetic loci. Recombination frequencies, range from 0 (complete linkage) to 0.5 (complete independent inheritance). A measure of the likelihood that genes are linked is expressed as the logarithm of the odds (LOD). The LOD score (logarithm (base 10) of odds), developed by Newton E. Morton, is a statistical test to determine that two loci are linked. Positive LOD scores indicate the presence of linkage, whereas negative LOD scores indicate that linkage is less likely. By convention, a LOD score greater than 3.0 is considered evidence for linkage.

2.7 Mapping populations

There are various types of populations that can be used to create genetic maps, develop marker linked to target genes, and facilitate marker verification. The most common populations created for mapping purposes include F_2s, backcrosses (BCs), double haploids (DHs), recombinant inbreed lines (RILs), and near isogenic lines (NILs). In association mapping, natural populations are used. DHs are produced from chromosome doubling of haploids via in vivo and in vitro. They have several advantages over other diploid populations of F_2s, F_3s, or BCs, since no dominance or dominance-related epistasis effects involves in the genetic model. As a result, additive, additive-related epistasis, and linkage effects can be investigated properly. As a permanent population, DH lines can be replicated as many times as desired across different environments, seasons and laboratories, providing endless genetic material for phenotyping and genotyping and to evaluate the genotype-by-environment interaction (Forster & Thomas, 2004; Bordes et al., 2006). In DH populations, the additive component of genetic variance is larger than that of F_2 and BC populations. Detailed quantitative genetics associated with DH populations have been previously discussed, including detection of epistasis, estimation of genetic variance components, linkage test, estimation of gene numbers, genetic mapping of polygenes and tests of genetic models and hypotheses (Choo et al., 1985; Bordes et al., 2006).

Recombinant inbred lines or random inbred lines (RILs) can be produced through various inbreeding procedures. They include full-sib mating for open-pollinated plants and selfing for self-pollinated plants. In self-pollinated plants, RIL can be developed through a bulking method where hybrids are bulk planted and harvested until F_5 to F_8 before they are planted by families. RIL can also be produced through single seed descent (SSD) where one or several seeds are harvested from each F_2 plant and planted to produce the next generation until F_5 to F_8. Near-isogenic lines are the product of inbreeding through successive backcrossing.

2.8 Mapping software and tools

Almost all molecular maps on the first generation of molecular markers, such as RFLPs, were constructed using MAPMAKER/EXP (Table 1). For severe distortion of segregation, statistical modifications will be needed and MAPDISTO can be used to solve this problem. JOINMAP can be used for construction genetic linkage for BC_1, F_2, RIL, F_1- and F_2-derived DH and out-breeder full-sib families. It can combine ('join') data derived from several sources into an integrated map, with several other functions, including linkage group determination, automatic phase determination for out-breeder full-sib family, several diagnostics and map charts (van Ooijen & Voorrips, 2001). A software package CMAP, a web-based tool, allows users to view comparisons of genetic and physical maps. The package also includes tools for curating map data (Ware et al., 2002). There are many commercial or freely available software packages for establishing association between marker genotypes and trait phenotypes. The most commonly used are QTL CARTOGRAPHER, MAPQTL, PLABQTL and QGENE. All of these only handle bi-allelic populations, while MCQTL (Jourjon et al., 2005) can perform QTL mapping in multi-allelic situations, including bi-parental populations from segregating parents, or sets of biparental, bi-allelic populations. The most frequently used QTL software during the 1980s and 1990s was MAPMAKER/QTL. MAPL allows a user to get results on segregation ratio, linkage test, recombination value, group markers, and order of markers by metric multi-dimensional scaling, and to draw a QTL map through interval mapping and analysis of variance (ANOVA).

A currently widely used QTL mapping software is QTL CARTOGRAPHER (Table 1). PLABQTL uses composite interval mapping with many functions similar to QTL CARTOGRAPHER. QTL can be localized and characterized in populations derived from a biparental cross by selfing or production of DHs. Simple and composite interval mapping are performed using a fast multiple regression procedure and can be used for QTL × environment interaction analysis (Utz & Melchinger, 1996). Recently, QGENE has been rewritten in the Java language and can be used for analyses of trait and QTL permutation and simulation for populations and as well as traits. Several software packages can be used for constructing linkage maps in out-crossing plant species, using full-sib families derived from two outbreed (non-inbreeding) parent plants (Garcia et al., 2006). Bayesian QTL mapping has received a lot of attention in recent years. Several software packages have been developed; For example, BQTL can perform maximum likelihood estimation of multi-gene models, Bayesian estimation of multi-gene models using Laplace Approximations, and interval and composite interval mapping of genetic loci. BLADE was for Bayesian analysis of haplotypes for LD mapping. MULTIMAPPER is a Bayesian QTL mapping software for analyzing backcross, DH and F_2 data from designed crossing experiments of inbred lines (Martinez et al., 2005). MULTIMAPPER/OUTBRED for populations derived from out-bred lines. Several mapping software packages were developed for QTL mapping for some specific situations. MCQTL was developed for simultaneous QTL mapping in multiple crosses and populations (Jourjon et al., 2005), including diallel cross modeling of the QTL effects using multiple related families. MAPPOP was developed for selective and bin mapping by selecting samples from mapping populations and for locating new markers on pre-existing maps (Vision et al., 2000). In addition, QTLNETWORK was developed for mapping and visualizing the genetic architecture underlying complex traits for experimental populations from a cross between two inbred lines (Yang et al., 2008).

Web-based QTL analytical tools are also available. Some of the tools developed in other system can potentially serve as a model for plants. WEBQTL (Table 1) provides dense error-checked genetic maps, as well as extensive gene expression data sets (Affymetrix) acquired across more than 35 strains of mice. To map QTLs in out-bred populations, QTL EXPRESS (Seaton et al., 2002) was developed for line crosses, half-sib families, nuclear families and sib-pairs. It provides two options for QTL significance tests: permutation tests to determine empirical significance levels and bootstrapping to estimate empirical confidence intervals of QTL locations.

Association or LD mapping is another mapping tool using unstructured populations of unrelated individuals, germplasm accessions, or randomly selected cultivars. Prior to LD mapping, genotype units are subjected to statistical analysis to remove population structure, which can cause false positive associations due to circumstantial correlations rather than real linkage. To meet the requirement, the STRUCTURE software (Pritchard et al., 2000) can be used. Some software packages have already included the population structure analysis functionality. STRAT, as a companion program to STRUCTURE, uses a structured association method for LD mapping, enabling valid case-control studies even in the presence of population structure (Pritchard et al., 2000). TASSEL can be used for trait analysis by association, evolution and linkage, which performs a variety of genetic analyses including LD mapping, diversity estimation and LD calculation (Zhang, et al., 2006). MIDAS can be used for analysis and visualization of inter-allelic disequilibrium between multi-allelic markers (Gaunt et al., 2006). With PEDGENIE, any size pedigree may be incorporated into this tool, from independent individuals to large pedigrees and independent individuals and families may be analyzed together. GENERECON is another software package for LD mapping using coalescent theory. It is based on a Bayesian Markovchain Monte Carlo method for fine-scale LD mapping using high-density marker maps. Genome-wide association (GWA) studies are used to find the link between genetic variations and common diseases in humans, as well as agronomic traits in plants. A well-powered GWA study will involve the measurement of hundreds of thousands of SNPs in thousands of individuals. Statistical tools developed for GWA studies include GENOMIZER, MAPBUILDER, CATS (Table 1).

3. Application and contributions of DNA markers to cultivar development

3.1 DNA marker utilizations

One of the most successful practical uses of molecular markers to date is gene introgression and pyramiding. Publicly available information on gene-marker association for a number of important agronomic traits can readily be used to introgress and pyramid these genes into elite breeding lines used in cultivar development. Marker-assisted backcrossing (MABC) is a straight forward method to introgress or move target gene(s) from parental donors to parental recipients. It involves successive backcrossing to remove the genetic background of the donor while recovering genetic properties of recurrent parents as much as possible. Statistical methods and schedule of backcrosses to create effective MABC have been reviewed in various papers (Hospital, 2001; Hospital & Charcosset, 1997; Herzog & Frisch, 2011). MABC with marker-based genome scanning has allowed a speedy recovery of most recurrent genome in a few crosses (Frisch et al., 1999; Frisch & Melchinger, 2005). MABC can also be used to develop cleaner near isogenic lines by minimizing carried over donor segments flanking the target locus, providing precise introgression of individual genes for detailed characterization of the QTLs. Marker-assisted gene pyramiding has been successfully utilized to combine multiple genes of male sterility (Nas et al., 2005) or to provide broader-spectrum of resistance against major diseases, such as rice blast and bacterial blight (Yoshimura et al., 1996; Jeung et al., 2006). Individual genes have unique reactions against pathogenic races and some of them have overlapping spectra that make selection based on disease reactions or symptoms more challenging. This problem can easily be overcome using molecular markers linked to individual disease-resistant genes allowing effective selection to be carried out to stack the genes.

No.	Name	Common Use
1	MAPMAKER/EXP	The first and most frequently used mapping software for map construction in the early era of DNA markers developed by the Whitehead Institute (Lander et al., 1987).
2	MAP MANAGER CLASSIC	Provides a graphical presentation and interactive tool to map Mendelian loci for codominant markers, using backcrosses or RILs in plants or animals (Manly, 1993).
3	MAP MANAGER CLASSIC	Provides a graphical presentation and interactive tool to map Mendelian loci for codominant markers, using backcrosses or RILs in plants or animals (Manly, 1993).
4	MAPDISTO	Can be used to address segregation distortion in segregating populations, such as backcross, double haploid (DH) and RIL populations. It computes and draws genetic maps through a graphical interface and analyzes marker data by showing segregation distortion due to differential viability of gametes or zygotes. Maps or data from multiple populations derived from different crosses can be combined into single or consensus maps through joint mapping. (ftp: http://mapdisto.free.fr/)
5	JOINMAP	For construction of genetic linkage maps for several types of mapping populations. (www.kyazma.nl/index. php/ mc.JoinMap /).
6	CMAP	Provides comparative function developed as a web-based tool to allow users to view comparisons of genetic and physical maps.
7	MAPMAKER/ QTL;	A sister software package to MAPMAKER/EXP, developed by Lander et al. (1987) based on maximum likelihood estimation of linkage between marker and phenotype using interval mapping to deal with simple QTLs and several standard populations
8	MAPL	Allows a user to get results on segregation ratio, linkage test, recombination value, group markers, and order of markers by metric multi-dimensional scaling, and to draw a QTL map through interval mapping and analysis of variance (ANOVA). Developed by Ukai et al., 1995),
9	QTL CARTOGRAPHER	Implements several statistical methods using multiple markers simultaneously, including composite interval and multiple composite interval mapping. Interaction between identified QTLs can also be estimated.
10	PLABQTL	Uses composite interval mapping.
11	QGENE	Intended for comparative analyses of QTL mapping data sets, developed in 1991 as a map and population simulation program, to which QTL analyses were added later on (www.qgene.org/)

No.	Name	Common Use
12	MAPQTL	Calculates QTL positions on genetic maps for several types of mapping populations, including BC1s, F₂s, RILs, DHs. It can also be used for QTL interval mapping, composite interval mapping and non-parametric mapping using functions for automatic cofactor selection and permutation test.
13	BQTL	Used for the mapping of genetic traits from line crosses and RILs (Borevitz et al., 2002).
14	BLADE	Used for Bayesian analysis of haplotypes for LD mapping (Liu et al., 2001; Lu, et al., 2003).
15	MULTIMAPPER/ OUTBRED	Has multi uses, including populations derived from out-bred lines.
16	WEBQTL	Used for exploring the genetic modulation of thousands of phenotypes gathered over a 30-year period by hundreds of investigators using reference panels of recombinant inbred strains of mice in web-based applications.
17	TASSEL	Comprehensive LD-based QTL mapping for trait analysis by association, evolution and linkage, which performs a variety of genetic analyses, including LD mapping, diversity estimation and LD calculation (Zhang et al., 2006).
18	MIDAS	For analysis and visualization of inter-allelic disequilibrium between multi-allelic markers (Gaunt et al., 2006)
19	PEDGENIE	Used as a general purpose tool to analyze association and transmission disequilibrium (TDT) between genetic markers and traits in families of arbitrary size and structure (Allen-Brady et al., 2006)
20	GENOMIZER	A platform independent Java program for the analysis of GWA experiments.
21	PLINK	A whole genome LD analysis toolset (Purcell et al., 2007),
22	MAPBUILDER	For chromosome-wide LD mapping (Abad-Grau et al., 2006)
23	CATS	Calculates the power and other useful quantities for two-stage GWA studies (Skol et al., 2006)

Table 1. List of QTL mapping software

3.2 QTL mapping

A long history of breeding suggests that grain yield is controlled by many genes with small effects. For this type of trait, applicability of finding and introgressing QTLs are limited since estimates of QTL effects for minor QTLs are often inconsistence. Even though these minor QTLs could show consistent effects, pyramiding these minor QTLs is increasingly challenging as the number of QTLs pyramided into one line increases (Bernardo, 2008). Inconsistency of estimated QTL effects for complex traits controlled by many minor genes brings the following important consequences. Due to limited transferability of estimated QTL effects across different populations for traits such as grain yield, QTL mapping will have to be repeated for each breeding population. Under this condition, Marker-assisted recurrent selection (MARS) is suitable since genotyping, phenotyping, and construction of selection index are repeated for each population (Koebner, 2003; Campbell et al., 2003). Because GXE interactions have a great influence on complex traits controlled by many QTLs with minor effects, QTL mapping from the same population needs to be conducted in each target set of environments. Finally, because the effects of sampling errors are high, population size of 500 to 1,000 is suggested (Beavis, 1994).

Mapping of multiple trait complexes in multiple environments can be conducted by employing algorithmic models to predict the association of genetic markers with trait of interest based on the effect of variance and covariance of the analyzed. These models allow the designing of new mapping frameworks and simulation tools, and the association to be extrapolated to the progeny of the plant or genetic materials tested in multiple environments. There is no limitation on the number of environments where the traits are scored. Based on simulation studies (Howes et al., 1998; Wang et al., 2007a), combining favorable marker alleles for more than 12 unlinked QTLs appears to be not feasible. The breeder may initially target a large number of QTLs but expects to accept having fewer QTL alleles fixed in a recombinant inbred. Since the improvement can only be targeted in a limited QTL number, breeders need a high level of confidence that the target QTLs do not represent a false positive that implies stringent levels of significance, $P \leq 0.0001$, when identifying the QTLs initially. A stringent significant level, however, can lead to an upward bias in estimating QTL effects (Beavis, 1994; Xu, 2003a) and therefore lead to overly optimistic expectation of response from MAS. Based on empirical and simulation studies, selection responses are increased when less stringent significant levels of $P = 0.20$ to 0.40 were applied in MARS. These relaxed significant levels allow QTLs with smaller effects to be selected and these minor QTLs can exceedingly compensate for the higher frequency of false positive. Less stringent significant levels are acceptable for pointing QTL locations, and when the goal is to predict genotypic performance such as in MARS, more stringent significant levels are required for combining favorable QTLs in recombinant inbred, introgression, and gene discovery. Along this line, QTLs should ideally be tagged by the markers inside the QTLs, or closely linked, or flanking the QTLs. Based on simulation studies in maize, the response of MARS in a population size of 144 plants was highest when about 128 markers are used (Bernardo & Charcosset, 2006), indicating that markers should be placed 10 to 15 cM apart and, therefore, denser markers are not necessary for predicting the performance.

Methods for using genetic markers, such as gene sequence diversity information, to improve plant breeding in developing cultivars by predicting the values of phenotypic traits based on genotypic, phenotypic, and optional family relationship information to identify marker-trait associations in the first population and used to predict the value of the phenotypic trait in the second or target population (Smith et al., 2005). These locally important traits are complex qualitative traits that are affected by many genes, the environment, and interaction between genes and environments. The next wave of QTL mapping should be targeted for locally important QTLs directly associated with cultivar development, including the matrix QTLs. It has been suggested that specific targets need to be clearly defined before embarking into the QTL mapping (Bernardo, 2008). In the context described above, yield potential and its components, quality traits, and local adaptation are among the most important QTL mapping targets. The architecture of genetic matrix of these complex traits could be dissected through QTL mapping to provide critical information on genomic regions and fragment sizes that produce the effects and relative importance of additive and non-additive gene action (Fridman et al., 2004).

Multiple QTL Mapping (MQM Mapping) using haplotyped putative QTL alleles has been used as a simple approach for mapping QTLs in plant breeding population (Jansen & Beavis, 2001). It described a method for mapping a phenotypic trait to correspond to chromosomal location. Statistical methods to correlate pedigree with multiple markers (haplotype) are used to determine identical-by-descent (IBD) data to map the phenotypic traits. The statistical model, HAPLO-IM+, HAPLO-MQM, and HAPLO-MQM+ are used for mapping traits to determine a single gene or QTL. This invention provides an efficient method for mapping phenotypic traits in interrelated plant populations. The basic principle of this method is clustering of the original parental lines into groups on the basis of their haplotypes for multiple genetic markers is the basic principle of this method. The effect of a QTL on the phenotype is modeled per haplotype group instead of per family, allowing an examination of the effects of haplotype-allele across families. Simulations of realistic plant breeding schemes have shown a significant increase in the power of QTL detection. This approach offers new opportunities for mapping and exploitation of QTL in commercial breeding programs. In addition, selection can be performed at any stage of a breeding program, including among genetically distinct breeding populations as a preselection to increase the selection index and to drive up the frequency of favorable haplotypes among the breeding populations, among segregating progeny from breeding population to increase the frequency of favorable haplotypes for the purpose of developing cultivars, among segregating progeny from a breeding population to increase the frequency of the favorable haplotypes prior to QTL mapping within this breeding population, and among parental lines from different heterotic groups in hybrid crops to predict the performance potential of different hybrids.

The index values generated from haplotype window-trait association allows pre-selection, which is widely considered as the next generation of MAS, to further economize breeding by not only removing the need of required phenotypic evaluation but also enabling screening of inbreed lines prior to making crosses. Breeders can initiate their programs by selecting a list of crosses and building a model based on haplotypes carried by each parental line in the cross. Selecting a model from cross to cross and inclusion of target genomic regions in the model will increase the complexity of the models. If it is not controlled, it will compromise the predictive ability and selection gain. For controlling the model's complexity, Automatic Model Picking (AMP) algorithm can be employed. The relative strength of each cross can be predicted using the Best Linear Unbiased Predictions (BLUP) approach, calculated on parental lines using phenotypic data. Once the final model is determined, the full gain of for each trait is calculated and the frequency-adjusted predicted gain can be obtained based on expected allele frequency. An additional optimization step can be included to either decrease or increase the importance of the secondary trait in the model based on frequency-adjusted predicted gain. This method provides haplotype information that allows the breeder to make informed breeding decisions based on genotype rather than phenotype into predictive breeding.

3.3 Channeling molecular information into new cultivar development

Successful marker breeding requires integration of molecular information into cultivar development programs. The development and testing of a QTL mapping population and the development of near-isogenic lines (NIL) can take several years before the results can be utilized (Monforte & Tanksley, 2000; Chaib et al., 2006). Because of that by the time the QTL identified, the recurrent parent used in the population development is probably commercially obsolete. The purified QTLs still require to be reintroduced into a competitive germplasm for commercial use via a time-consuming backcross scheme. Frampton (2008) has proposed a direct integration of genomic technologies into commercial plant breeding by designing specific crossing schemes to allow the development of marker profiles, QTL mapping of major gene loci, and new cultivars to be advanced simultaneously. The method can be applied repeatedly to achieve complete integration. The breeding population is developed through an initial cross, followed by two backcrosses and self-pollination of BCF_1 plants. Molecular marker development consists of QTL identification using the means of BC_2F_2 family, gene fine mapping, and new marker development using bulk-segregant analysis. Therefore, the method provides simultaneous development of a breeding population with molecular marker development and gene mapping, and integration of molecular marker platform with the breeding platform.

4. The bottom line: Prospect and current limitations

Significant progress has been achieved in marker detection methodology in term of speed and cost. Current efforts by various consortiums supported by both public and private entities are underway to push the development of genomic tools to make them more economically and logistically feasible for the breeders. High resolution of marker assay covering all important information across the genomes, such as SNP chip sets being developed, will provide a tremendous asset to mobilize and assemble critical alleles that can improve crop production systems in a significant way. Over 1,200 reports of mapped QTLs are available through various publications in 12 major crop species (Bernardo, 2008; Xu & Crouch, 2008). Each typically reported an average of 3 to 5 QTLs for the trait studied (Bernardo, 2008; Eathington et al., 2007). This large volume of published molecular marker-trait associations will continue to grow as a result from the abundant amount of available markers, high density molecular assays, and development of sophisticated user-friendly computer software, and improved cost and technical efficiency in marker analyses. Despite a significant influx of reported marker-QTL trait associations to date, successful exploitation of available mapped QTLs remains low, indicating a lack of synchronization between the QTLs reported and actual breeding goals in the cultivar development. Successful integration of genomic tools in the cultivar development requires sufficient knowledge of breeding materials from molecular perspectives. This is essential for marker-based accumulation of favorable alleles and the ability to predict the effects of new QTLs assembled during cultivar development. Dissection of individual QTLs will lead to a better understanding of their interaction, discovery of hidden QTLs, and a new way to characterize and classify QTLs to facilitate a speedy assembly of critical genes needed to maximize the end products, such as grain production or other specific quality traits of economic significance. The genes critical for maintaining local adaptation and standard industrial and market quality are also among the most important QTLs.

At present, there is a mounting gap between available QTL mapping information and marker-based QTL applications in cultivar development. This vast mostly unexplored area presents a tremendous opportunity for both public researchers and private industry to tag pivotal genes, including their interactions that play important roles in grain production. Because of the massive nature of the undertaking in both financial and human powers, a number of consortiums have been formed and operated to achieve greater common goals nearly impossible to achieve by any single lab. Major companies that have sufficient resources have shown an increased intensity in their effort in this arena, though only very limited information goes to the public.

Publicly available plant databases provide a large amount of genomic data for a wide range of plant species. This wealth of knowledge, however, has not yet found its way into mainstream plant breeding due to several reasons; first, there is no apparent connection between the primary information generated in plant genomics and real life breeding application. Second, databases storing various bits of supportive information (e.g. pedigree, genotype and phenotype) are usually stored in different places and managed by different groups of scientists. And third, there is a gap between breeders and molecular geneticists in perceiving their focus of interest, i.e. tools and interfaces for bioinformatic data focuses vs. organism level. Integration of the fragmented information, views, and priorities will be one of many challenges to overcome. Bioinformatics data typically consist of cDNA and genomic sequence data, genetic maps of mutants, DNA markers and maps, candidate genes and quantitative trait loci (QTL), physical maps based on chromosome breakpoints, gene expression data and libraries of large inserts of DNA such as bacterial artificial chromosomes and radiation hybrids. Information flows from molecular markers to genetic maps to sequences and to genes. However, the relationship between breeding (i.e. germplasm, pedigree and phenotype) and sequence-based information has not been established. An example of how genetic information can be integrated into plant breeding programs to produce cultivars from molecular variation using bioinformatics and what crop scientists might want from bioinformatics have been previously discussed (Mayes et al., 2005). How to best utilize all relevant genomic information efficiently and comprehensively, and harnessing the power of informatics to support molecular breeding is a challenge to modern plant breeding.

Processing speed and cost of DNA markers have improved substantially in the past several years, resulting in a significant reduction in processing time and cost per data point. This is one of the research areas where most rapid development occurs. However, its current application as a breeding tool has not reached its potential fully. Marker approach involves a separate line of research activities, and almost in all cases, it requires substantial upfront support. In addition to the cost of genotyping and phenotyping, it requires lab facilities and bioinformatics personnel to analyze complex data. This vast mostly unexplored area poses current limitations, but it could also present a tremendous opportunity for both public researchers and private industry to tag pivotal genes, including their interactions that play important roles in grain production, and to subsequently protect their invention.

4.1 Common platform and supporting tools

An appropriate experimental design and data analysis are a critical component for successful application of molecular breeding. Various models of data flowchart and analytical tools to funnel DNA marker data into cultivar development have been proposed. However, they lack of simple-to-use guidelines to allow breeders to confidently select the appropriate design and analysis. Communications between genomics scientists, geneticists, bioinformaticians, and breeders are still limited hampering the development of truly integrated tools for applied molecular breeding design, integrated mapping, and MAS. Decision support tools for marker breeding that can model, simulate, and analyze most of the pre-existing genetic conditions will help breeders design and implement the efficient breeding scheme in term of cost and time using the optimum combination of MAS and phenotypic selection. Similarly important are decision support tools that include sample collection and depositing, retrieving, and tracking data, and also acquiring, collecting, processing, and mining databases.

For information-driven plant breeding, databases and supporting tools that allow an interchangeable flow of information through communicable platforms that required minimum maintenance and updates are critical. The use of universal language within different platforms will strengthen interaction among breeders, database curators, bio-informaticians, molecular biologists and tool developers. Interchangeable format and data content across all plant species are needed to develop a universal database. Current models, such as the one provided by Gene Ontology and Plant Ontology projects, offer a glimpse of future possibility in this area. An automatic ontological analysis has been used to develop biological interpretation of the data (Khatri et al., 2002). Currently, this approach becomes the standard for the secondary analysis of high-throughput experiments. A large number of tools have been developed for this purpose. Khatri and Draghici (2005) provided a review of detailed comparison for the 14 available tools using six different criteria; scope of the analysis, visualization capabilities, statistical model(s) used, correlation for multiple comparisons, reference microarray available, and installation issues and sources of annotation data. These analyses help researchers to select the most appropriate tool for a given type of analysis. Despite a few drawbacks in each tool associated with conceptual limitations of the current state-of-the-art in ontological analysis, this type of analysis has been generally adopted. These limitations are some of the challenges to overcome in order to create the next generation of secondary data analysis tools. Another major challenge is to construct a graphical presentation of systematic biological relationship that integrates gene, protein, metabolite, and phenotype data as suggested by Blanchard (2004). This will include an assembly of large-scale data sets into a more comprehensive presentation by minimizing high false positive rates and validating the existing models using probability and graph theory.

4.2 Added complexity in scope and time management

Crop development is a complex process, and molecular marker information further increases the complexity. To successfully apply molecular marker-assisted breeding, breeders have to structure their specific breeding methodologies to allow for the integration of empirical results from molecular marker analyses. All of molecular activities, including molecular analyses, establishment of genotypic-phenotypic associations, and molecular marker-based decision making, have to be completed in the same limited time frame in conventional breeding. They must be synchronized with seed planting preparation, progeny selection, yield trials, collecting phenotypic data, harvesting, data analyses, and use of off-season nurseries.

While some breeders have the access to computational infrastructure and statistical expertise needed to generate and analyze the gigabytes of genomic data, the majority will depend on the availability of smaller subsets of genomic data that can be analyzed using an MS Excel spreadsheet. In the rice SNP system, for example, one of the current major efforts is to develop low-resolution SNP assay (through Affymetrix's custom-designed SNP genotyping arrays and Illumina's custom-designed SNP oligonucleotide pools assays (OPAs), or other platforms developed by KBiosciences) to address the problem (McCouch et al., 2010). In addition to reduce computation complexity, breeders will eventually be able to request targeted SNP detection assays that can be tailored into their specific breeding purposes or selecting their population base at a fraction the current cost of re-sequencing, particularly when the bioinformatic requirements are taken into account.

Breeders utilize breeding information from many different sources to obtain a description of genetic background and phenotypic traits under specific growing environment. The depth and types of information needed by individual breeders will vary greatly. However, data that critical for individual breeders will include some basic information, such as germplasm information (pedigree, genealogy, genetic stock data, etc.), genotypic information (DNA markers, sequences, and expression information), phenotypic data and environmental information. In addition, historical data preserved timely in the repository system can be used to reanalyze hypotheses and guide new research for molecular marker breeding. To obtain a high quality of mapping, both genotyping and phenotyping have to be conducted effectively. While molecular detection systems are rapidly enhanced, methods of phenotyping have not been improved as fast. Dissection of agronomically important QTLs requires phenotyping under target environments in multiple test sites. Proper techniques to ensure the consistency of phenotyping over multiple growing environments will need to be established.

5. Unraveling genetic potential globally

Providing sufficient food for an increasing world population is a tremendous challenge to overcome. Finding ways to boost the yield potential of major grain crops beyond current productivity levels, therefore, is critically important. One of the keys to solving the problem is to increase the ability to find novel alleles that are not present among cultivated species

Various studies show that wild species have a wider genetic diversity where critical alleles hidden or lost during the early domestication process and along the progression of modern breeding processes can be recovered. Extensive germplasm of various crop plants and their wild relatives are available in various places. In rice, for example, more than 102,547 accessions of Asian cultivated rice O. sativa, 1,651 accessions of African cultivated rice O. glaberrima and 4,508 accessions of wild ancestors are maintained in the International Rice Germplasm Collection (IRGC) at IRRI (McNally et al. 2006) in addition to an extensive rice germplasm collection in Japan, China, Taiwan, India, Korea, the USA, and many other countries. Relatives of rice species, such as Oryza rufipogon appear to have many new putative yield-related QTLs that can potentially be used to improve cultivated rice (Tan et al., 2007). Genome sequencing of wild species and map alignment are current ground breaking projects to provide a basic road to unravel the whole potential of wild species. The Oryza map Alignment Project (OMAP) is set to develop physical maps of 12 wild species to be aligned with the reference genome sequence of Nipponbare (Ammiraju et al., 2006; 2010; Wing et al., 2005; 2007). Sequence data will provide direct evidence of evolutionary path of Oryza genus. However, the most important expected outcomes from this current endeavor are to find new genes and QTLs that can be used to improve grain production, levels of pest and disease tolerance, ability to tolerate stress and other less favorable growing environments. The ideas to unlock wild genetic variation to improve global grain production (McCouch et al., 2010; Fridman et al., 2004; Matsumoto et al., 2005) have, therefore, gained renewal interest from time to time. The precision of DNA markers to unravel the intercalating process of gene expression to determine the productivity of grain crop is needed to rediscover valuable alleles that can be funneled into the pipeline of cultivar development.

6. Practical utilization: Global source, local purpose

To survive in a very competitive market that demands high quality product, breeders have to assemble a series of genes that give rise to high yielding cultivars that have stable grain quality, disease resistance, optimum plant maturity and height, and are very adapted to target growing regions of a typically narrow niche of environments. These quality traits of industrial standards are critical for successful commercial production of crops in a modern era and often are the breeding priorities in current breeding programs. Long breeding selections have resulted in the formation of a specific matrix of complex QTLs that support quality traits required by the market. This matrix provides a skeleton for newer cultivars in grain crop breeding programs. Any efforts to improve current yield potential should, therefore, be built to correspondingly maintain or enhance the trait matrix. To stay competitive, breeders will be required to expand their crop to provide additional traits that are not currently available in their breeding populations. During the introgression of foreign traits into their breeding lines, all necessary matrix traits to produce high quality standards need to be maintained. Breeders have acquired detailed knowledge on the genetics underlying the matrix of these complex traits among individual breeding lines in the pipeline of cultivar development. Should molecular markers be employed in the breeding program, the same in-depth molecular knowledge must be acquired for the QTL matrix, target QTLs, and their individual breeding lines in their programs.

Incorporation of molecular marker-based selections into a conventional breeding program will require breeders to custom their molecular breeding schemes and tailor them directly into their specific breeding objectives. However, understanding molecular properties of the quality matrix requires tremendous investment and undoubtedly represents the current bottleneck as to why successful exploitation of available mapped QTLs into cultivar development remains limited at the present time. With the advancement in molecular techniques, such as high throughput SNP technology, developing a SNP chip to specifically guard the quality matrix will be possible. Once customized chips can be developed for individual breeding programs, any novel traits from the global source (different genetic backgrounds, inter or intra subspecies or wild-related ancestors from global populations) can potentially be incorporated into their breeding programs to add and/or improve specific quality or to boost yield without jeopardizing locally adapted standard qualities. Private companies have developed proprietary methodology that allows their breeders to combine their germplasm knowledge and breeding population objectives with molecular phenotypic trait association in order to develop genetic modeling for multiple marker-assisted selections and obtain rapid increase in the frequency of favorable alleles associated with target traits within the breeding population (Eathington et al., 2007).

7. References

Abad-Grau, M.M.; Montes, R. & Sebastiani, P. (2006). Building chromosome-wide LD maps. *Bioinformatics* 22: 1933–1934, ISSN 1367-48

Aggarwal, R.K.; Hendre, P.S., Varshney, R.K., Bhat, P.R., Krishnakumar, V. & Singh, L. (2007). Identification, characterization and utilization of EST-derived genic microsatellite markers for genome analyses of coffee and related species. *Theoretical and Applied Genetics* 114(2):359-372, ISSN 0040-5752

Allen-Brady, K.; Wong, J. & Camp, N.J. (2006). PedGenie: an analysis approach for genetic association testing in extended pedigrees and genealogies of arbitrary size. *BMC Bioinformatics* 7: 209, ISSN 1471-2105

Alves-Freitas, D.M.T.; Kilian, A. & Grattapaglia, D. (2011). Development of DArT (Diversity Arrays Technology) for high-throughput genotyping of Pinus taeda and closely related species. *BMC Proceedings* 2011 5(Suppl 7):P22. http://www.biomedcentral.com/1753-6561/5/S7/P22

Ammiraju, J.S.S.; Luo, M., Goicoechea, J.L., Wang, W., Kudrna, D., Mueller, C., Talag, J., Kim, H., Sisneros, N.B., Blackmon, B., Fang, E., Tomkins, J.B., Brar, D., MacKill, D., McCouch, S., Kurata, N., Lambert, G., Galbraith, D.W., Arumuganathan, K., Rao, K., Walling, J.G., Gill, N., Yu, Y., SanMiguel, P., Soderlund, C., Jackson, S. & Wing, R.A. (2006). The *Oryza* bacterial artificial chromosome library resource: construction and analysis of deep-coverage large insert BAC libraries that represent the 10 genome types of genus *Oryza. Genome Research* 16(1):140-147, ISSN 1088-9051

Ammiraju, J.S.S.; Luo, M., Sisneros, N., Angelova, A., Kudrna, D., Kim, H., Yu, Y., Goicoechea, J.L., Lorieux, M., Kurata, N., Brar, D., Ware, D., Jackson, S. & Wing, R.A. (2010). The *Oryza* BAC resource: a genus wide and genome scale tool for exploring rice genome evolution and leveraging useful genetic diversity from wild relatives. *Breeding Science* 60: 536–543, ISSN 1344-7610

Andersen, J.R. & Lübberstedt, T. (2003). Functional markers in plants. *Trends in Plant Science* 8: 554–560, ISSN 1360-1385

Baird, N.A.; Etter, P.D., Atwood, T.S., Currey, M.C., Shiver, A.L., Lewis, Z.A., Selker, E.U., Cresko, W.A. & Johnson, E.A. (2008). Rapid SNP discovery and genetic mapping using sequenced RAD markers. *PLoS ONE* 3(10):e3376, ISSN 1932-6203

Beavis, W.D. (1994). The power and deceit of QTL experiment: Lessons from comparative QTL studies. In. *Proc. Corn Sorghum Ind Res Conf,* pp. 250-266, ISBN, Chicago, IL., USA. Dec. 7-8, 1994

Bernardo, R. (2008). Molecular markers and selection for complex traits in plants: Learning from the last 20 years. *Crop Science* 48:1649-1664, ISSN 0011-183X

Bernardo, R. & Charcosset, A. (2006). Usefulness of gene information in marker-assisted recurrent selection: A simulation appraisal. *Crop Science* 46:614-621, ISSN 0011-183X

Blanchard, J.L. (2004). Bioinformatics and systems biology, rapidly evolving tools for interpreting plant response to global change. *Field Crops Research* 90:117–131, ISSN 0378-4290

Bordes, J.; Charmet, G., Dumas de Vaulx, R., Pollacsek, M., Beckert, M. & Gallais, A. (2006). Doubled haploid versus S1 family recurrent selection for testcross performance in a maize population, *Theoretical and Applied Genetics* 112: 1063–1072, ISSN 0040-5752

Borevitz, J.O.; Maloof, J.N., Lutes, J., Dabi, T., Redfern, J.L., Trainer, G.T., Werner, J.D., Asami, T., Berry, C.C., Weigel, D. & Chory, J. (2002). Quantitative trait loci controlling light and hormone response in two accessions of *Arabidopsis thaliana*. *Genetics* 160:683–696, ISSN 0016-6731

Campbell, B.T.; Baenziger, P.S., Gill, K.S., Eskridge, K.M., Budak, H., Erayman, M., Dweikat, I. & Yen, Y. (2003). Identification of QTLs and environmental interactions associated with agronomic traits on chromosome 3A of wheat. *Crop Science* 43:1493–1505, ISSN 0011-183X

Chaib, J.; Lecomte, L., Buret, M. & Causse, M. (2006). Stability over genetic backgrounds, generations and years of quantitative trait locus (QTLs) for organoleptic quality in tomato. *Theoretical and Applied Genetics* 112:934-944, ISSN 0040-5752

Choo, T.M.; Reinbergs, E. & Kasha, K.J. (1985). Use of haploids in breeding barley. *Plant Breeding Reviews* 3:219–252, ISSN 0730-2207

CoGePedia. (2011). Sequenced plant genomes. (Verified: 11/21/2011) http://genomevolution.org/wiki/index.php/Sequenced_plant_genomes

Davey, J.W.; Hohenhole, P.A., Etter, P.D., Boone, J.O., Catchen, J.M., & Blaxter, M.L. (2011). Genome-wide genetic marker discovery and genotyping using next-generation sequencing. *Nature Reviews Genetics* 12:499-510, ISSN 1471-0056

Delseny, M.; Bin Han, B. & Hsing, Y.I. (2010). High throughput DNA sequencing: The new sequencing revolution. *Plant Science* 179:407–422, ISSN 0168-9452

Eathington, S.R.; Crosbie, T.M., Edwars, M.D., Reiter, R. & Bull, J.K. (2007). Molecular markers in commercial breeding program. *Crop Science* 47(S3):S154-S163, ISSN 0011-183X

Edwards, D.; Forster, J.W., Chagné, D. & Batley, J. (2007). What is SNPs? In: *Association Mapping in Plants*, Oraguzie, N.C., Rikkerink, E.H.A., Gardiner, S.E. and De Silva, H.N. (eds), pp. 41–52, Springer, ISBN 978-0-387-35844-4, Berlin, Germany

Edwards, D. & Batley, J. (2010). Plant genome sequencing: applications for crop improvement. *Plant Biotechnol. J.* 7:2–9, ISSN 1467-7644

Eujayl, I.; Sorrels, M.E., Baum, M., Wolters, P. & Powell, W. (2002). Isolation of EST-derived microsatellite markers for genotyping the A and B genomes of wheat. *Theoretical and Applied Genetics* 104:399–407, ISSN 0040-5752

Feuillet, C.; Leach, J.E., Rogers, J., Schnable, P.S. & Eversole, K. 2011. Crop genome sequencing: lessons and rationales. *Trends in Plant Science* 16(2):77-88, ISSN 1360-1385

Forster, B.P. & Thomas, W.T.B. (2004). Doubled haploids in genetics and plant breeding. *Plant Breeding Reviews* 25:57–88, ISSN 0730-2207

Frampton, A. (2008). Integration of commercial plant breeding and genomic technologies. United States Patent Application 20080034450

Fridman, E.; Carrari, F., Liu, Y-S, Fernie, A.R. & Zamir, D. (2004). Zooming in on quantitative trait for the tomato yield using interspecific introgression. *Science* 305:1786-1789, ISSN 0036-8075

Frisch, M.; Bohn, M. & Melchinger, A.E. (1999). Comparison of selection strategies for marker-assisted backcrossing of gene. *Crop Science* 39(5):1295-1301, ISSN 0011-183X

Frisch, M. & Melchinger, A.E. (2005). Selection theory for marker-assisted backcrossing. *Genetics* 170: 909–917, ISSN 0016-6731

Fukuoka, S.; Ebana, K., Yamamoto, T. & Yano, M. (2010). Integration of Genomics into Rice Breeding. *Rice* 3:131-137, ISSN 1939-8425

Garcia, A.A.; Kido, E.A., Meza, A.N., Souza, H.M., Pinto, L.R., Pastina, M.M., Leite, C.S., Silva, J.A., Ulian, E.C., Figueira, A. & Souza, A.P. (2006). Development of an integrated genetic map of a sugarcane (*Saccharum* spp.) commercial cross, based on a maximum-likelihood approach for estimation of linkage and linkage phases. *Theoretical and Applied Genetics* 112:298–314, ISSN 0040-5752

Gaunt, T.R.; Rodriguez, S., Zapata, C. & Day, I.N.M. (2006). MIDAS: software for analysis and visualization of interallelic disequilibrium between multiallelic markers. BMC *Bioinformatics* 7:227, ISSN 1471-2105

Harlan, J.R. (1992). Crops and Man. American Society of Agronomy and Crop Science Society of America, Madison, WI, ISBN 0-89118-107-5

Haupt, W.; Fischer T.C., Winderl, S., Fransz, P. & Torres-Ruiz, R.A. (2001). The centromere1 (CEN1) region of *Arabidopsis thaliana*: Architecture and functional impact of chromatin. *Plant Journal* 27:285–296, ISSN 0960-7412

Herzog, E. & Frisch, M. (2011). Selection strategies for marker-assisted backcrossing with high-thoughput marker systems. *Theoretical and Applied Genetics* 123:251-260, ISSN 0040-5752, ISSN 1344-7610

Hohenlohe, P.A.; Bassham, S., Etter, P.D., Stiffler, N., Johnson, E.A. & Cresko, W.A. (2010). Population genomics of parallel adaptation in threespine stickleback using sequenced RAD tags. *PLoS Genetics* 6(2):e1000862, ISSN 1553-7390

Hospital, F. (2001). Size of donor chromosome segments around introgressed loci and reduction of linkage drag in marker-assisted backcross programs. *Genetics* 2001; 158(3):1363-1379, ISSN 0016-6731

Hospital, E. & Charcosset, A. (1997). Marker-assisted introgression of qualitative trait loci. *Genetics* 1997; 147(3):1469-1485, ISSN 0016-6731

Howes, N.K.; Woods, S.M. & Townley-Smith, T.F. (1998). Simulations and practical problems of applying multiple marker-assisted selection and doubled haploid to wheat breeding programs. *Euphytica* 100:225-230, ISSN 0014-2336

Hyten, D.L.; Cannon, S.B., Song, Q., Weeks, N., Fickus, E.W., Shoemaker, R.C., Specht, J.E., Farmer, A.D., May, G.D. & Cregan, P.B. (2010). High-throughput SNP discovery through deep resequencing of a reduced representation library to anchor and orient scaffolds in the soybean whole genome sequence. *BMC Genomics* 11:38 (http://www.biomedcentral.com/1471-2164/11/38)

Jaccoud,D.; Peng, K., Feinstein, D. & Kilian, A. (2001). Diversity arrays: A solid state technology for sequence information independent genotyping. *Nucleic Acids Research* 29: e25, ISSN 1362-4962

Jansen, R.C. & Beavis, W. (2001). *MQM Mapping using haplotyped putative QTL-alleles; a simple approach for mapping QTL's in plant breeding populations.* Patent EP 1265476

Jenkins, S. & Gibson, N. (2002). High-throughput SNP genotyping. *Comparative and Functional Genomics* 3, 57–66, ISSN 1532-6268

Jeung, J.U.; Heu, S.G., Shin, M.S., Vera Cruz, C.M. & Jena, K.K. (2006). Dynamics of Xanthomonas oryzar pv populations in Korea and their relationship to known bacterial blight resistant genes. *Phytopathology* 96(8):867-875, ISSN 0031-949X

Jourjon, M.F.; Jasson, S., Marcel, J., Ngom, B. & Mangin, B. (2005). MCQTL: multi-allelic QTL mapping in multi-cross design. *Bioinformatics* 21:128–130, ISSN 1367-48

Khatri, P. & Draghici, S. (2005). Ontological analysis of gene expression data: current tools, limitations and open problems. *Bioinformatics* 21:3587–3595, ISSN 1367-48

Khatri, P.; Draghici, S., Ostermeier, G.C. & Krawetz, S.A. (2002) Profiling gene expression using OntoExpress. *Genomics* 79:266–270, ISSN 0888-7543

Koebner, R. (2003). MAS in cereals: Green for maize, amber for rice, still red for wheat and barley. In: *Marker assisted selection: A fast tract to genetic gain in plant and animal breeding?* 17-18 Oct. 2003; pp. 12-17, Turin, Italy. FAO, Rome. (www.fao.org/biotech/docs/Koebner.pdf)

Lander, E.S.; Green, P., Abrahamson, J., Barlow, A., Daly, M.J., Lincoln, S.E. & Newburg, L. (1987). MAPMAKER: an interactive computer package for constructing primary genetic linkage maps of experimental and natural populations. *Genomics* 1:174–181, ISSN 0888-7543

Levinson, G. & Gutman, G.A. (1987). Slipped-strand mispairing: a major mechanism for DNA sequence evolution. *Molecular Biology and Evolution* 4:203–221, ISSN 0737-4038

Lewis, Z.A.; Shiver, A.L., Stiffler, N., Miller, M.R., Johnson, E.A. & Selker, E.U. (2007). High density detection of restriction site associated DNA (RAD) markers for rapid mapping of mutated loci in Neurospora. *Genetics* 177(2):1163-1171, ISSN 0016-6731

Li, F.; Kitashiba, H., Inaba, K. & Nishio, T. (2009). A *Brassica rapa* linkage map of EST-based SNP markers for identification of candidate genes controlling flowering time and leaf morphological traits, *DNA Research* 16:311–23, ISSN 1756-1663

Liu, J.S.; Sabatti, C., Teng, J., Keats, B.J.B. & Risch, K. (2001). Bayesian analysis of haplotypes for linkage disequilibrium mapping. *Genome Research* 11:1716–1724, ISSN 1088-9051

Lu, X.; Niu, T. & Liu, J.S. (2003). Haplotype information and linkage disequilibrium mapping for single nucleotide polymorphisms. *Genome Research* 13:2112–2117, ISSN 1088-9051

Lübberstedt, T.; Melchenger, A.E., Fähr, S., Klein, D., Dally, A. & Westhoff, P. (1998). QTL mapping in test crosses of flint lines of maize: III. Comparison across populations for forage traits. *Crop Science* 38:1278–1289

Manly, K.F. (1993). A Macintosh program for storage and analysis of experimental genetic mapping data. *Mammalian Genome* 4:303–313, ISSN 0938-8990

Martinez, V.; Thorgaard, G., Robison, B. & Sillanpää, M.J. (2005) An application of Bayesian QTL mapping to early development in double haploid lines of rainbow trout including environmental effects. *Genetical Research* 86:209–221, ISSN 0016-6723

Matsumoto, T.; Wu J.Z., Kanamori, H., et al. (2005). The map-based sequence of the rice genome. *Nature* 436:793–800, ISSN 0028-0836

Mayes, S.; Parsley, K., Sylvester-Bradley, R., May, S. & Foulkes, J. (2005). Integrating genetic information into plant breeding programs: how will we produce varieties from molecular variation, using bioinformatics? *Annals of Applied Biology* 146:223–237, ISSN 0003-4746

McCouch S.R.; Teytelman, L., Xu, Y., Lobos, K.B, Clare, K., Walton, M., Fu, B., Maghirang, R., Li, Z., Xing, Y., Zhang, Q., Kono, I., Yano, M., Fjellstrom, R., DeClerck, G.G., Schneider, D., Cartinhour, S., Ware, D. & Stein, L. (2002). Development and mapping of 2240 new SSR markers for rice (*Oryza sativa* L.), *DNA Research* 9(6):199-207, ISSN 1340-2838

McCouch, S.R.; Zhao, K., Wright, M., Tung, C.W, Ebana, K., Thomson, M., Reynolds, A., Wang, D., DeClerck, G., Ali,M.L., McClung, A., Eizenga, G. & Bustamante, C.

(2010). Development of genome-wide SNP assays for rice, *Breeding Science* 60: 524–53, ISSN 1344-7610

McNally, K.L.; Bruskiewich, R., Mackill, D., Buell, C.R., Leach, J.E. & Leung, H. (2006). Sequencing multiple and diverse rice varieties. Connecting whole genome variation with phenotypes, Plant Physiol. 141:26–33, ISSN 0032-0889

McNally, K.L.; Childs, K.L., Bohnert, R., Davidson, R.M., Zhao, K., Ulat, V.J., Zeller, G., Clark, R.M.,Hoen, D.R., Bureau, T.E., Stokowski, R., Ballinger, D.G., Frazer, K.A., Cox, D.R., Padhukasahasram, B., Bustamante, C.D., Weigel, D., Mackill, D.J., Bruskiewich, R.M., Rätsch, G., Buell, C.R., Leung, H. & Leach. J.L. (2009) Genome-wide SNP variation reveals relationships among landraces and modern varieties of rice. *Proc. Natl. Acad. Sci. USA* 106:12273–12278, ISSN-0027-8424

Miller, M.R.; Atwood, T.S., Eames, B.F., Eberhart, J.K., Yan, Y.L., Postlethwait, J.H. & Johnson, E.A. (2007a). RAD marker microarrays enable rapid mapping of zebrafish mutations. *Genome Biololgy.* 8(6):R105, ISSN 1465-6906

Miller, M.R.; Dunham, J.P., Amores, A., Cresko, W.A. & Johnson, E.A. (2007b). Rapid and cost-effective polymorphism identification and genotyping using restriction site associated DNA (RAD) markers. *Genome Research* 17(2):240-248, ISSN 1088-9051

Mochida, K. & Shinozak, K. (2010). Genomics and Bioinformatics Resources for Crop Improvement. *Plant Cell Physiology* 51(4):497-523, ISSN 0032-0781

Monforte, A.J. &Tanksley, S.D. (2000). Fine mapping of a quantitative trait locus (QTL) from Lycopersicon hirsutum chromosome 1 affecting fruit characteristics and agronomic traits: Breaking linkage among QTLs affecting different traits and dissection of heterosis for yield. *Theoretical and Applied Genetics* 100:471–479, ISSN 0040-5752

Nas, T.M.S.; Sanchez, D.L., Diaz, G.Q., Mendioro, M.S. & Virmani, S.S. (2005). Pyramiding of thermosensitive genetic male sterility (TGMS) genes and identification of candidate tms5 gene in rice. *Euphytica* 145(1-2):67-75, ISSN 0014-2336

Oliveira, K.M.; Pinto, L.R., Marconi, T.G., Mollinari, M., Ulian, E.C., Chabregas, S.M., Falco, M.C., Burnquist, W., Garcia, A.A.F. & Souza, A.P. (2009). Characterization of ne polymorphic functional markers for sugarcane. *Genome* 52:191-209, ISSN 0831-2796

Paux, E.; Sourdille, P., Salse, J., Saintenac, C., Choulet, F., Leroy, P., Korol, A, Michalak, M., Kianian, S., Spielmeyer, W., Lagudah, E., Somers, D., Kilian, A., Alaux, M., Vautrin, S., Berges, H., Eversole, K., Appels, R., Safar, J., Simkova, H., Dolezel, J., Bernard, M. & Feuillet, C. (2008). A physical map of the 1-gigabase bread wheat chromosome 3B. *Science* 322:101-104, ISSN 0036-8075

Primmer, C.R.; Ellengren, H., Saino, N. & Moller, A.P. (1996). Directional evolution in germline microsatellite mutations. *Nature Genetics* 13:391–393, ISSN 1061-4036

Pritchard, J.K.; Stephens, M., Rosenberg, N.A. & Donnelly, P. (2000). Association mapping in structured populations. *American Journal of Human Genetics* 67:170–181, ISSN 0002-9297

Purcell, S.; Neale, B., Todd-Brown, K., Thomas, L., Ferreira, M.A.R., Bender, D., Maller, J., de Bakker, P.I.W.; Daly, M.J. & Sham, P.C. (2007). PLINK: a toolset for whole-genome association and population-based linkage analysis. *American Journal of Human Genetics* 81:559-575, ISSN 0002-9297

Ramchiary, N.; Nguyen, V.D., Li, X., Hong, C.P., Dhandapani, V., Choi, S.R., Yu, G., Piao, Z.Y, & Lim, Y.P. (2011). Genic microsatellite markers in Brassica rapa: Development, characterization, mapping, and their utility in other cultivated and wild Brassica relatives, *DNA Research* pp. 1–16, doi:10.1093/dnares/dsr017, ISSN 1756-1663

Seaton, G.; Haley, C.S., Knott, S.A., Kearsey, M. & Visscher, P.M. (2002) QTL Express: mapping quantitative trait loci simple and complex pedigrees. *Bioinformatics* 18:339–340, ISSN 1367-4803

Semagn, K., Bjørnstad, A. & Xu Y. (2010). The genetic dissection of quantitative traits in crops. *Electronic Journal of Biotechnology* 13:5, http://dx.doi.org/10.2225/vol13-issue5-fulltext-14, ISSN 0717-3458

Shen, Y.-J.; Jiang, H., Jin, J.-P., Zhang, Z.-B., Xi, B., He, Y.-Y., Wang, G., Wang, C., Qian, L., Li, X., Yu, Q.-B., Liu, H.-J., Chen, D.-H., Gao, J.-H., Huang, H., Shi, T.-L. & Yang, Z.-N. (2004). Development of genome-wide DNA polymorphism database for map-based cloning of rice genes. *Plant Physiology* 135:1198–1205, ISSN 0032-0889

Shirasawa, K.; Oyama, M., Hirakawa, H., Sato, S., Tabata, S., Fujioka, T., Kimizuka-Takagi, C., Sasamoto, S., Watanabe, A, Kato, M., Kishida, Y. Kohara, M., Takahashi, C., Tsuruoka, H., Wada, T., Sakai, T. & Isobe, S. (2011). An EST-SSR linkage map of Raphanus sativus and comparative genomics of the Brassicaceae. *DNA Research* 18:221–232, ISSN 1756-1663

Singer, T.; Fan, Y., Chang, H.S., Zhu, T., Hazen, S.P., & Briggs, S.P. (2006). A high-resolution map of *Arabidopsis* recombinant inbred lines by whole-genome exon array hybridization. *PLoS Genet* 2(9): e144. DOI: 10.1371/journal.pgen.0020144, ISSN 1932-6203

Skol, A.D.; Scott, L.J., Abecasis, G.R. & Boehnke, M. (2006). Joint analysis is more efficient than replication-based analysis for two-stage genome-wide association studies. *Nature Genetics* 38:209–213, ISSN 1061-4036

Smith, O.S.; Cooper, M., Tingey, S.V., Rafalski, A.J., Luedtke, R. & Niebur, W.S. (2005). Plant breeding method, WIPO Patent Application WO05000006

Sonah, H.; Deshmukh, R.K., Sharma, A., Singh, V.P., Gupta, D.K., Gacche, R.N., Rana, J.C., Singh, N.K. & Sharma, T.R. (2011). Genome-wide distribution and organization of microsatellites in plants: an insight into marker development in *Brachypodium*. *PLoS ONE* 6(6):e21298. doi:10.1371/journal.pone.0021298, ISSN 1932-6203

Tan, L.; Liu, F., Xue, W., Wang, G., Ye, S., Zhu, Z., Fu, Y., Wang, X. & Sun, C. (2007), Development of *Oryza rufipogon* and *O. sativa* introgression lines and assessment for yield related quantitative trait loci. Journal of Integrative Plant Biology, 49: 871–884

Tautz, D. & Renz, M. (1984) Simple sequences are ubiquitous repetitive components of eukaryotic genomes. *Nucleic Acids Research* 12:4127–4138

Thiel, T.; Michalek, W., Varshney, R.K. & Graner, A. (2003). Exploiting EST data bases for the development and characterization of gene-derived SSR-markers in barley (*Hordeum vulgare* L.). *Theoretical and Applied Genetics* 106:411–422, ISSN 0040-5752

Ukai, Y.; Osawa, R., Saito, A. & Hayashi, T. (1995). MAPL: a package of computer programs for construction of DNA polymorphism linkage maps and analysis of QTL (in Japanese). *Breeding Science* 45:139–142, ISSN 1344-7610

Utz, H.F. & Melchinger, A.E. (1996). PLABQTL: a program for composite interval mapping of QTL. *Journal of Agricultural Genomics* 2(1), Available at:www.cabi-publishing.org/jag/papers96/paper196/indexp196.html

Van Ooijen, J.W. & Voorrips, R.E. (2001). JoinMap® version 3.0: software for the calculation of genetic linkage maps. Wageningen: Plant Research International

Varshney, R.K.; Graner, A. & Sorrells, M.E. (2005). Genic microsatellite markers in plants: features and applications. *Trends in Biotechnology* 23:48–55, ISSN 0167-7799

Varshney, R.K.; Nayak, S.N., May, G.D. & Jackson, S.A. (2009). Next-generation sequencing technologies and their implications for crop genetics and breeding. *Trends in Biotechnology* 27: 522 – 530, ISSN 0167-7799

Vigouroux, Y.; Mitchell, S., Matsuoka, Y., Hamblin, M., Kresovich, S., Smith, J.S.C., Jaqueth, J., Smith, O.S. & Doebley, J. (2005). An analysis of genetic diversity across the maize genome using microsatellites. *Genetics* 169:1617–1630, ISSN 0016-6731

Vision, T.J.; Brown, D.G., Shmoys, D.B., Durrett, R.T. & Tanksley, S.D. (2000). Selective mapping: a strategy for optimizing the construction of high-density linkage maps. *Genetics* 155:407–420, ISSN 0016-6731

Wang, J.; Chapman, S.C., Bonnet, D.G., Rebetzke, G.J. & Crouch, J. (2007a). Application of population genetic theory and simulation models to efficiently pyramid multiple genes via marker-assisted selection. *Crop Science* 47:582-588, ISSN 0011-183X

Wang, Z.; Jia, Y., Rutger, J.N. & Xia, Y. (2007b). Rapid survey for presence of a blast resistance gene Pi-ta in rice cultivars using the dominant DNA markers derived from portions of the Pi-ta gene. *Plant Breeding* 126(1): 36-42, ISSN 0179-9541

Wang, Z.; Li, J., Luo, Z., Huang, L., Chen, X., Fang, B., Li, Y., Chen, J. & Zhang, X. (2011). Characterization and development of EST-derived SSR markers in cultivated sweet potato (*Ipomoea batatas*). *BMC Plant Biology* 2011, 11:139, ISSN 1471-2229

Ware, D.H.; Jaiswal, P., Ni, J., Yap, I.V., Pan, X., Clark, K.Y., Teytelman, L., Schmidt, S.C., Zhao, W., Chang, K., Cartinhour, S., Stein, L.D. & McCouch, S.R. (2002) Gramene, a tool for grass genomics. *Plant Physiology* 130:1606–1613, ISSN: 0032-0889

Wei, X.M.; Jackson, P.A., Hermann, S., Kilian, A., Heller-Uszynska, K., & Deomano, E. (2010). Simultaneously accounting for population structure, genotype by environment interaction, and spatial variation in marker-trait associations in sugarcane. Genome 2010, 53(11):973-981, ISSN 0831-2796

Wei, B.; Jing, R., Wang, C., Chen, J., Mao, X., Chang , X. & Jia, J. (2009). Dreb1 genes in wheat (*Triticum aestivum* L.): development of functional markers and gene mapping based on SNPs. *Molecular Breeding* 23: 13-22, ISSN 1380-3743

Wenzl, P.; Carling, J., Kudrna, D., Jaccoud, D., Huttner, E., Kleinhofs, A. & Kilian, A. (2004). Diversity Arrays Technology (DArT) for whole-genome profiling of barley, *PNAS* 101 (26):9915–9920, ISSN-0027-8424

Wing, R.A.; Ammiraju, J.S.S., Luo, M., Kim, H., Yu, Y., Kudrna, D., Goicoechea, J.L., Wang, W., Nelson, W., Rao, K., Brar, D., Mackill, D.J., Han, B., Soderlund, C., Stein, L., SanMiguel, P. & Jackson, S. (2005). The *Oryza* map alignment project: The golden path to unlocking the genetic potential of wild rice species. *Plant Molecular Biology* 59(1):53-62, ISSN 0167-4412

Wing, R.; Kim, H., Foicoechea, J., Yu, Y., Kudrna, D., Zuccolo, A., Ammiraju, J., Luo, M., Nelson, W. & Ma, J. (2007). The Oryza map alignment project (OMAP): a new resource for comparative genome studies within *Oryza*. In: Upadhyaya, N.M. (ed.) Rice Functional Genomics, Springer, New York, pp. 395–409. ISBN 978-0-387-48903-2

Xia, L.; Peng, K., Yang, S., Wenzl, P., de Vicente, C., Fregene, M., Kilian, A. (2005). DArT for high-throughput genotyping of cassava (*Manihot esculenta*) and its wild relatives. *Theoretical and Applied Genetics* 110:1092–1098, 411–422, ISSN 0040-5752

Xu, S. (2003a). Theoretical basis of the Beavis effect. *Genetics* 165:2259-2268, ISSN 0016-6731

Xu, Y. (2003b). Developing marker-assisted selection strategies for breeding hybrid rice. *Plant Breeding Reviews* 23:73–174, ISSN 0730-2207

Xu, Y., McCouch, S.R. & Zhang, Q. (2005). How can we use genomics to improve cereals with rice as a reference genome? *Plant Molecular Biology* 59:7–26, ISSN 0167-4412

Xu, Y. & Crouch, J.H. (2008). Marker-assisted selection in plant breeding: From publication to practice. *Crop Science* 48:391-407, ISSN 0011-183X

Yan, J., Yang, X., Shah, T., Sánchez-Villeda, H., Li, J., Warburton, M., Zhou, Y., Crouch, J.H. & Xu, Y. (2010). High-throughput SNP genotyping with the GoldenGate assay in maize. *Molecular Breeding* 25:441-451, ISSN 1380-3743

Yang, J., Hu, C., Hu, H., Yu, R., Xia, Z., Ye, X. & Zhu, J. (2008). QTLNetwork: mapping and visualizing genetic architecture of complex traits in experimental populations. *Bioinformatics* 24:721–723, ISSN 1367-48

Yang, S.Y.; Saxena, R.K., Kulwal, P.L., Ash, G.J., Dubey, A., Harpe, J.D.I., Upadhyaya, H.D., Gothalwal, R., Kilian, A. & Varshney, R.K. (2011). The first genetic map of pigeonpea based on Diversity Arrays Technology (DArT) markers. *Journal of Genetics* 90(1):103-109, ISSN 0022-1333

Yoshimura, A.; Lei, J.X., Matsumoto, T., Tsunematsu, H., Yoshimura, S., Iwata, N., Baraoidan, M.R., Mew, T.W. &. Nelson, R.J. (1996). Analysis and pyramiding of bacterial blight resistance genes in rice by using DNA markers. 577–581. In: *Rice genetics III. Proceedings of the third international rice genetics symposium. International Rice Research Institute*, G.S. Khush (ed.), pp. 577-581, ISBN 971-22-0087-6, Manila, Philippines

Yu, H.; Xie, W., Wang, J., Xing, Y., Xu, C., Li, X., Xiao, J. & Zhang, Q. (2011). Gains in QTL detection using an ultra-high density SNP map based on population sequencing relative to traditional RFLP/SSR markers. *PLoS ONE* 6(3):e17595. doi:10.1371/journal.pone.0017595, ISSN1932-6203

Yu, J.K.; La Rota, M., Kantety, R.V. & Sorrells, M.E. (2004). EST-derived SSR markers for comparative mapping in wheat and rice. *Molecular Genetics and Genomics* 271:742–751, ISSN 1617-461

Yu, J.; Wang, J., Lin, W., Li, S ., Li, H., et al. (2005). The genomes of Oryza sativa: A history of duplications. *PLoS Biol* 3(2): e38. DOI: 10.1371/journal.pbio.0030038, ISSN 1932-6203

Zhang, Z.; Bradbury, P.J., Kroon, D.E., Casstevens, T.M. & Buckler, E.S. (2006). *TASSEL 2.0: a software package for association and diversity analyses in plants and animals.* Poster presented at Plant and Animal Genomes XIV Conference, 14–18 January 2006, San Diego, California

2

Impacts of Ozone (O₃) and Carbon Dioxide (CO₂) Environmental Pollutants on Crops: A Transcriptomics Update

Abhijit Sarkar[1], Ganesh Kumar Agrawal[2],
Kyoungwon Cho[3], Junko Shibato[4] and Randeep Rakwal[2,4,5*]
[1]CSIR-SRF, Laboratory of Air Pollution and Global Climate Change, Ecology Research
Circle, Department of Botany, Banaras Hindu University, Varanasi, Uttar Pradesh,
[2]Research Laboratory for Biotechnology and Biochemistry (RLABB), Kathmandu,
[3]KRFC Research Fellow, Seoul Center, Korea Basic Science Institute, Seoul,
[4]Department of Anatomy I, School of Medicine, Showa University, Tokyo,
[5]Graduate School of Life and Environmental Sciences, University of Tsukuba, Ibaraki,
[1]India
[2]Nepal
[3]South Korea
[4,5]Japan

1. Introduction

The human race has evolved through centuries and civilized through many ways on their home planet Earth. Although travelling through the 'ages' – 'man' had learned the use of fire, utilized the nature and natural resources, gathered knowledge, practiced agriculture, developed industries, and gradually moved towards a superior modernized life. While climbing the steps of this 'modernized civilization', 'man' introduced a new term – 'pollution' to the world's vocabulary. By definition, pollution is the "undesirable state of the natural environment being contaminated with harmful substances as a consequence of human activities" (source - http://wordnet.princeton.edu). But, at present, this 'undesirable state of natural environment' has turned into a major concern for the survival of life on Earth. Air, water, and soil – the three major natural resources, and fundamental backbone of Earth's environment, have been found to be heavily 'contaminated with harmful substances' throughout the world. This does not imply all is contaminated and lost, but just to highlight how precarious the situation is for us humans.

Though, there is no historical account on the pollution on Earth, it was found while reviewing the available literatures that the incident of 'air pollution' is not a new event to our society. In an article, published in 'Science', John D. Spengler and Ken Saxton commented that – "….soot found on ceilings of prehistoric caves provides ample evidence of the high levels of pollution that was associated with inadequate ventilation of open fires…." (Spengler and Sexton, 1983). During the past couple of decades, rapid urbanization, unplanned industrialization, fast growth in vehicle use, uncontrolled fossil fuel burning, and injudicious management of natural resources have transformed this indoor problem of 'prehistoric' origin into a serious environmental hazard of the present century. The 'clean air' of today (and tomorrow) is no more as like was a few decades ago. Looking back, this may have been unavoidable as the human race grew and progressed, paying a price for modernization. In a report on the health effects of air pollution, the World Health Organization (WHO) stated that about two million people die every year because of air pollution throughout the world, while many more suffer from breathing ailments, heart diseases, lung infections, and even cancer (WHO, 2008).

In general, air pollutants can be largely divided in two major categories depending on their formation. The first category is of the primary pollutants, which are emitted directly into the atmosphere, and are mostly present at higher concentrations in urban areas and close to large point emission sources, like carbon dioxide (CO_2). The second category is of the secondary pollutants, which are created by the reactions of primary pollutants under favorable environmental conditions, like tropospheric ozone (O_3) formed due to a series of photochemical reactions involving nitrogen oxides (NOx) and volatile organic compounds (VOCs) under bright sunlight. However, over past decades, it has been quite clearly understood that both these above mentioned air pollutants are also two major components of much discussed 'global warming', and hence 'global climate change' phenomenon. Tropospheric O_3, along with methane and black carbon, are key contributors to global warming, augmenting the radiative forcing of CO_2 by 65% (Penner et al., 2010). Global agri-production, hence food security, is under severe crisis due to the direct and indirect effects of both the pollutants. Studies have shown that O_3 and CO_2 generally enter plants through stomata, and subsequently affect the inter- and intra-cellular system by modifying the gene expression behavior (Fiscus et al., 2005; Bokhari et al., 2007; Cho et al., 2008; Sarkar et al., 2010). High throughput '-omics' is a combination of potential and central techniques, which can answer many key biological questions in both plants as well as animals. In the present chapter, we mainly focused our ideas towards describing the in depth and up-to-date '*transcriptomics*' analyses of major agricultural crops under the influence of O_3 and CO_2 rise. We hope that the overall coordinated picture of O_3 and CO_2 – responsive crop transcriptome might help to construct a future roadmap towards the development of next generation crops with optimized yield and other functions for future high-O_3 and CO_2-world.

1.1 Defining 'global climate change': Is it really changing?

According to the IPCC report, climate change refers to a statistically significant variation in either the mean/stable state of the climate and in its components/variability, persisting for an extended period (typically decades or longer). It might happen due to the natural internal processes of atmospheric components and/or external forcing and/or to persistent anthropogenic changes in the composition of the atmosphere and/or in the use and management of natural resources (source: http://www.ipcc.ch/ipccreports/tar/ wg1/518.htm). In 2007, IPCC for the first time reported to the United Nations (UN) that the earth's climate system has undoubtedly and significantly got warmer in the past years and will continue to be. According to the action groups of IPCC, the average annual temperature in the Pacific Northwest rose by 1.5° F in the 20th century and is expected to rise 0.5° F per decade in the first half of the 21st century. Now, this climate warming, hence climate change at global level, has become an important issue and a hot debatable topic throughout the world, in both the developed and developing countries. Initially, it was only an 'issue' to be considered, then as a major 'scientific issue' to be studied, and after that as an principle 'environmental policy issue' to be endlessly debated. But in the past few years, it has been metamorphosed into a significant risk factor to be addressed by the global community, especially for those in the energy sector (Sioshansi and Oren, 2007). Although there are thousands of events that can be directly or indirectly correlated to the global warming and climate change, the prime incidents are as follows -

a. Retreating mountain glaciers on all continents
b. Thinning ice caps in the Arctic and Antarctic
c. Rising sea level – about 6-7 inches in the 20th century
d. More frequent heavy precipitation events (rainstorms, floods or snowstorms) in many areas, which affects the survival of both animal and plant life
e. More intense and longer droughts over wider areas, especially in the tropics and subtropics
f. Reduction in the yield production of major agricultural crops around the world, hence affecting food security and safety

1.2 Past and present trends in the concentrations of ambient O_3 and CO_2: Where we stop?

Interestingly, both the above mentioned gases are present in the earth's atmosphere from ancient periods. However, according to the IPCC (2007) report, mean daily O_3-concentration is estimated to have increased from around 10 ppb, prior to the industrial revolution, to a current level of approximately 60 ppb during the summer months. Projections show that the level will rise 20 to 40% more by the year 2050 in the industrializing countries of the Northern hemisphere. As per the reports and projections, it is quite apparent that this secondary air pollutant will be a far more critical crisis in the coming future than the present time. On the other hand, CO_2 also increased in the atmosphere primarily since the industrial revolution, through the burning of fossil fuels in energy industries, transportation, households, and others. Both these O_3 and CO_2 are the principle greenhouse gases (GHGs) too. In the following sections, we will mainly discuss on the global trends in temporal and spatial distributions of both these pollutants.

1.2.1 Tropospheric O_3 trend

In the ambient air, O_3 precursors play an important role during long range transport downwind from the sources. Polluted air masses from urban and industrial areas can affect suburban and rural areas, even reaching to remote rural areas traveling for considerable distances. High O_3 levels from one particular urban area can extend as far as 48 to 80 km (Krupa and Manning, 1988). Ozone formation also depends largely upon prevailing meteorological conditions of the area. Tiwari et al. (2008) reported positive correlation between mean maximum temperature/sunlight with O_3 concentration.

Background O_3 concentrations have more than doubled in the last century (Meehl et al., 2007). There is an increase in annual mean values of O_3 ranging from 0.1 to 1 ppb per year^{-1} (Coyle et al., 2003). In the Northern hemisphere, O_3 is also influenced by the influx from the stratosphere (Grewe, 2007). However, O_3 varies strongly with episodic peak concentrations during the warmest months in summer in the most polluted regions and maximum concentrations during spring prevailing at background sites (Vingarzan, 2004). In regions such as East Asia exposed to the summer monsoon which transports oceanic air with less O_3, the seasonal patterns show a peak during pre- and post-monsoon periods (He et al., 2008). During the day, O_3 concentration pattern depends on elevation and shows strong diurnal variations at lowland sites where its destruction dominates during the night and vertical mixing together with photochemical activity causes highest levels in the afternoon.

In rural agricultural areas of the USA, mean O_3 concentrations reach between 50 and 60 ppb (90th percentile) (USEPA, 2006). Concentrations over the mid and high-latitude of the Eurasian and North American continents were 15 - 25 ppb in 1860, but increased to 40 - 50 ppb even in remote areas and from 10 - 15 ppb to 20 - 30 ppb over the mid- and high-latitude Pacific Ocean, respectively (Lelieveld and Dentener, 2000). Measures taken to reduce O_3 precursor emissions, led to changes in O_3 levels in many rural and urban areas of Europe, North America, and Japan; the frequency of the highest values shows a declining trend, while lowest values are increasing (Jenkin, 2008). The US EPA has reported that emission reduction in O_3 precursors has been substantial over the past 29 years (US EPA, 2009). The percent change in emissions of NOx and VOCs were 40 and 47%, respectively, for the period 1980 - 2008.

In India, despite of the favorable climatic conditions for O_3 formation, very limited data from systematic monitoring of O_3 are available. O_3 concentrations are continuously increasing from 1992 to 2008 with higher peaks in rural areas. In a field transect study at urban sites of Varanasi, O_3 concentrations varied from 6 to 10.2 ppb during 1989-1991 (Pandey and Agrawal, 1992). During the same period, daytime O_3 concentrations (9 hr mean) were reported to vary from 9.4 to 128.3 ppb at an urban site in Delhi (Varshney and Agrawal, 1992). It was observed that 10 h ground level mean O_3 concentrations in Delhi varied between 34 to 126 ppb during the winter of 1993 (Singh et al., 1997). At Pune, an annual average daytime O_3 concentration of 27 ppb and hourly concentration between 2 and 69 ppb were reported during August 1991 to July 1992 (Khemani et al., 1995). Lal et al. (2000) studied the pattern of O_3 concentrations from 1991 to 1995 at an urban site of Ahmedabad (India), and reported that daytime mean O_3 concentrations exceeding 80 ppb were rarely observed. The monthly average O_3 concentrations ranged between 62 and 95 ppb in summer (April - June) and 50 and 82 ppb in autumn (October - November) at New Delhi (Jain et al., 2005). Emberson et al. (2009) reported that large parts of South Asia experience up to 50 - 90 ppb mean 7 h (M 7) O_3 concentration. Mittal et al. (2007) using the HANK model reported O_3 concentration varying from 25 to 100 ppb over the entire Indian region.

1.2.2 Tropospheric CO_2 trend

Atmospheric CO_2 is accelerating upward from decade to decade. For the past ten years, the average annual rate of increase is 2.04 ppm. This rate of increase is more than double the increase in the 1960s. However, other than being a potent GHG and the main basis of global warming, CO_2 is also a key substrate for plant growth. Interestingly, scientists observed nearly similar trends in the overall concentrations of ambient CO_2 throughout the globe, which means that everybody is under similar crisis. Uncontrolled anthropogenic activities throughout the world have caused the atmospheric CO_2 to increase continuously from about 280 ppm at the beginning of the 19th century to 369 ppm at the beginning of the 21st century (Figure 1). Projections also range between about 450 and 600 ppm by the year 2050, but strongly depend on future scenarios of anthropogenic emissions (Woodward, 2002).

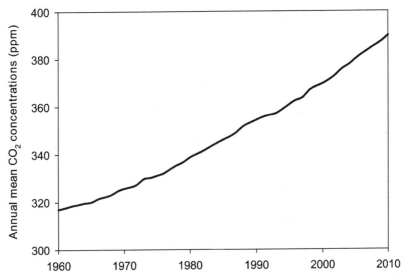

Fig. 1. Increasing trends in atmospheric CO_2 concentrations at global level. [data adopted from http://www.esrl.noaa.gov/gmd/ccgg/trends/]

2. Evaluation of O₃- and CO₂-effect on agricultural crops through modern OMICS approaches: A face-to-face mêlée

The effect of both O_3 and CO_2 on the various levels of crop's responses till the yield has been very well studied so far (for detail, see review Cho et al., 2010). However, their specific effects on the genome, hence – transcriptome and proteome, has not been evaluated to a large extent. In the next section, we have made an attempt to portray the impact of both these pollutants on the agricultural crops through modern OMICS approaches depending on the available reports.

2.1 Multi-parallel OMICS approaches in modern biology

We are running through the golden era of genomics (study of whole 'genome' is loosely called 'genomics'). The genomics era is also in position to use multiple parallel approaches for the functional analysis of genomes in a high-throughput manner. These parallel approaches surely results in an exceptionally swift and effective system for the analyses and deductions of gene(s) function in a wide range of plants, at the level of transcript (transcriptomics), protein (proteomics), and metabolite (metabolomics) (Figure 2). All together, these four approaches are commonly referred as the multi-parallel OMICS approaches in modern biology.

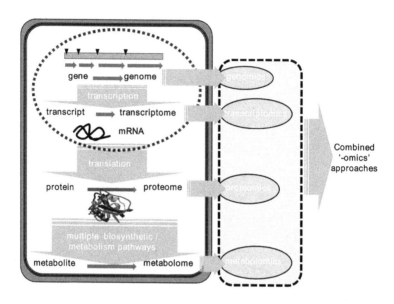

Fig. 2. Multi-parallel OMICS approaches in modern plant/crop biology.

2.2 Case studies

2.2.1 Effect of ozone (O$_3$) on crops

2.2.1.1 Rice (*Oryza sativa* L.)

Among all the major crops, rice (*Oryza sativa* L.) has been studied most for its response to O$_3$-stress (Agrawal et al., 2002; Cho et al., 2008, Feng et al., 2008; Frei et al., 2010). Agrawal et al. (2002) first reported a detailed combined trancriptomics and proteomics response of rice plants under elevated O$_3$-exposure. Two-weeks-old rice (*cv.* Nipponbare) seedlings were exposed to 200 ppb of O$_3$ for three days in a controlled fumigation chamber. A drastic visible necrotic damage in O$_3$-exposed leaves and consequent increase in ascorbate peroxidase (APX) protein(s) accompanied by rapid changes in the immunoblotting analysis and two-dimensional gel electrophoresis (2-DGE) protein profiles were observed. They also reported nearly 52 differentially expressed proteins. Among which were the O$_3$-caused drastic reductions in the major leaf photosynthetic proteins, including the abundantly present ribulose-1, 5-bisphosphate carboxylase/oxygenase (RuBisCO) and induction of various defense/stress-related proteins. Most prominent change in the rice leaves, within 24 h post-treatment with O$_3$, was the induced accumulation of a pathogenesis related (PR) class 5 protein, three PR 10 class proteins, APX(s), superoxide dismutase (SOD), calcium-binding protein, calreticulin, a novel ATP-dependent CLP protease, and an unknown protein. Feng et al. (2008) also followed the established two-week-old rice seedlings experimental model of Agrawal and co-workers (2002) and exposed plants to 0, 40, 80, and 120 ppb O$_3$ for nine days. A drastic damage in the photosynthetic proteins (mainly large and small subunits of RuBisCO) and primary metabolism related proteins was found, whereas an induced expression of some major antioxidant (i.e., glutathione-S-transferase and MnSOD) and defense/stress-related proteins (i.e., PR5 and two PR10 proteins OsPR10/PBZ1 and RSOsPR10) were reported. In another independent study, Cho et al. (2008) also checked the expression profiles of genes in leaves of two-week-old rice seedlings exposed to 200 ppb O$_3$ for 1, 12, and 24 h using a 22K rice DNA microarray chip. A total of 1,535 genes were differentially expressed more than five-fold over the control. Their functional categories revealed genes involved in transcription, pentose phosphate pathway, and signal transduction at 1 h. Genes related to antioxidant enzymes, ribosomal protein, post-translational modification (PTM), signal transduction, jasmonate, ethylene, and secondary metabolism at 12 and 24 h play a crucial role in O$_3$-response (Cho et al., 2008). Recently, Frei et al. (2010) have tried to identify the possible mechanism of O$_3$-response in rice seedlings by characterizing two important quantitative trait loci (QTL), in two different chromosome segment substitution lines (SL15 and SL41). Their findings suggest that the activity of some major antioxidant genes might contribute significantly in the response strategy of rice plant under higher O$_3$-stress.

In contrast with the above laboratory-based experimental models, Sarkar and Agrawal (2010) had applied 'field-based integrated OMICS' approach to understand the background of O$_3$ response in two high-yielding cultivars (*Malviya dhan 36* and *Shivani*) of mature rice plants under natural conditions. They found dependable phenotypical response, in the form of foliar injury, followed by definite changes in leaf proteome. Major damage was in the photosynthetic proteins (large and small subunits of RuBisCO) and primary metabolism-related proteins. Moreover, an induced expression of some antioxidant and defense/stress-related proteins were reported in the rice leaf proteome.

2.2.1.2 Wheat (*Triticum aestivum* L.)

Wheat (*Triticum aestivum* L.) is the third most important crop around the globe, and nearly two thirds of the world population depends on this crop for their primary nutrition supplement. Sarkar et al. (2010) employed 'field-based integrated OMICS' approach to understand the background of O$_3$ response in two wheat cultivars (cvs Sonalika and HUW 510) against elevated O$_3$ concentrations (ambient + 10 and 20 ppb) under near natural conditions using open top chambers (OTCs). Results of their study showed drastic reductions in the abundantly present RuBisCO large and small subunits. Western blot analysis confirmed induced accumulation of antioxidative enzymes like SOD and APX protein(s) and common defense/stress-related thaumatin-like protein(s). 2-DGE analysis revealed a total of 38 differentially expressed protein spots, common in both the wheat cultivars. Among those, some major leaf photosynthetic proteins (including Rubisco and Rubisco activase) and important energy metabolism proteins (including ATP synthase, aldolase, and phosphoglycerate kinase) were drastically reduced, whereas some stress/defense-related proteins (such as harpin-binding protein and germin-like protein) were induced.

2.2.1.3 Maize (*Zea mays* L.)

Maize (*Zea mays* L.) is another important crop at global context. Being a C$_4$ crop, its response to climate change has been always been a bit different from the other plants/crops. Torres et al. (2007) have performed a detailed investigation of O$_3$ response in maize (cv. Guarare 8128) plants through gel-based proteomics approach. In that experiment, 16-day-old maize plants (grown in controlled environment at green house) were exposed to 200 ppb O$_3$ for 72 h, and then the response was compared with a controlled plant (grown under filtered pollutant-free air). Results showed that nearly 12 protein spots were differentially expressed under O_3 exposure, and can be exploited as marker proteins. Expression levels of catalase (increased), SOD (decreased), and APX (increased) were drastically changed by O_3 depending on the leaf stage, whereas cross-reacting heat-shock proteins (HSPs; 24 and 30 kDa) and naringenin-7-O-methyltransferase (NOMT; 41 kDa) proteins were strongly increased in O_3-stressed younger leaves. The study also enumerated leaf injury as biomarker under O_3 stress in maize leaves.

2.2.1.4 Bean (*Phaseolus vulgaris* L.)

Torres et al. (2007) also conducted a study on response of cultivated bean (*Phaseolus vulgaris* L. cv. IDIAP R-3) against O_3 stress using the same experimental protocol as for maize (see above) and the effects were evaluated using gel-based proteomics followed by MS and immunoblotting. Results showed that in bean leaves two SOD proteins (19 and 20 kDa) were dramatically decreased, while APX (25 kDa), small HSP (33 kDa), and a NOMT (41 kDa) were increased after O_3 fumigation.

2.2.1.5 Pepper (*Capsicum annuum* L.)

Lee and Yun (2006) applied cDNA microarrays to monitor the transcriptome of O_3 stress-regulated genes in two pepper cultivars [*Capsicum annuum* cv. Dabotop (O_3-sensitive) and cv. Buchon (O_3-tolerant)]. Ozone stress up-/down-regulated 180 genes more than three-fold with respect to their controls. Transcripts of 84 genes increased, transcripts of 88 others diminished, and those of eight either accumulated or diminished at different time points in the two cultivars or changed in only one of the cultivars. A total of 67% (120 genes) were regulated differently in O_3-sensitive and O_3-tolerant pepper cultivars, most being specifically up-regulated in the O_3-sensitive cultivar.

2.2.1.6 Linseed (*Linum usitassimum* L.)

Tripathi et al. (2011) analyzed the response of linseed plants under elevated O_3-stress through combined genomics and proteomics approaches. The results showed that 10 ppb elevation over ambient O_3 concentration can cause 50% damage in the genome stability of linseed plants. In line to the genome response, leaf proteome was also severely affected under O_3 stress, and the damages were mainly observed on the photosynthetic and primary metabolism-related proteins.

2.2.2 Effect of carbon dioxide (CO_2) on crops

2.2.2.1 Rice

Bokhari et al. (2007) had exposed 10-day-old rice (*O. sativa* L. spp *Indica* cv. 93-11) seedlings to 760, 1140, and 1520 ppm of CO_2 for 24 h, and assessed the response of test plants through 2-D gel-based proteomics followed by protein identification. Comparative analysis of leaf proteome revealed 57 differentially expressed proteins under elevated CO_2 in the rice leaf proteome. Majority of these differentially expressed proteins belonged to photosynthesis (34%), carbon metabolism (17%), protein processing (13%), energy pathway (11%), and antioxidants (4%). Several molecular chaperones and APX were found to be up-regulated under higher CO_2, whereas major photosynthetic proteins like RuBisCO and RuBisCO activase, and different proteins of Calvin cycle were down-regulated.

2.2.2.2 Wheat

Hogy et al. (2009) have studied the effect of elevated CO$_2$ on the grain proteome of wheat (*T. aestivum* L. cv. Triso) in a completely free-air CO$_2$ enrichment (FACE) setup. Results of this experiment revealed a total of 32 proteins were affected. Out of them, 16 proteins were up-regulated and 16 proteins were down-regulated. Among the up-regulated proteins, triticin precursor, putative avenin-like beta precursor, serpin, peroxidase 1, alpha-chain family 11 xylanase, starch synthase I, and cytosolic glyceraldehyde-3-phosphate dehydrogenase (GAPDH) were the major proteins, whereas among down-regulated proteins, globulin (Glb 1) storage protein, low-molecular weight glutenin, ATP synthase β subunit, and alpha-tubulin were the major changed proteins.

3. Concluding remarks

The projected levels of O$_3$ and CO$_2$ are critically alarming, and have become a major issue of concern for food security worldwide. Scientific evidences indicate that crop plants are in general sensitive to both these air pollutants, but in different ways.

Plant resistance to O$_3$ involves a wide array of response ranging from the molecular and cellular level to the whole plant level. Significant effects of O$_3$ are early leaf senescence, decreased photosynthetic assimilation, altered stomatal behavior, decreased growth and productivity, and reduced carbon allocation to roots and changes in metabolic pathways. Genotype differences in response to O$_3$ are related to stomatal behavior, anti-oxidative potential hormonal regulation, and carbon allocation during reproduction affecting the yield responses. Detailed understanding of genotypic response is crucial in predicting the long-term impacts of O$_3$ on agriculture in global context, including the breeding of resistant cultivars. Several potential O$_3$ biomarkers have been identified, which could be exploited to screen and develop O$_3$-tolerant varieties in future (Figure 3). However, in case of CO$_2$, it is an integrated compound for plant's survival. So, at the initial stages, any increment in the ambient CO$_2$ levels showed a positive response towards plants yield, but also raised many questions. The behavior of RuBisCO, key enzyme of photosynthesis, is still under debate at higher CO$_2$ levels (Figure 3).

While reviewing the available reports on O$_3$ and CO$_2$ air pollutants, we found that both the stresses leave some kind of specific 'signature' on the crops response and that the 'signature' is not crop dependent. However, it must be emphasized that there are only limited OMICS studies available on crop responses to O$_3$ and CO$_2$. Future work in our laboratories and those around the world will help provide new and much needed insight into the nature of the plant response to air pollutants and ways and means to help circumvent their deleterious effects. It is quite clear that we will need proper engineering of crops to combat the emerging problem, and researches, analysis, and reviews on initial crop-pollutants interaction have pointed toward some important functional traits required while considering the next-generation crops:

i. Crops should have efficient and effective stomatal behavior to properly maintain the balance of external gas influx. As per the present research outcomes, we can see that crops prefer avoidance more than developing resistance towards any stress.
ii. Crops should possess efficient photosynthetic system with higher catalytic ability to generate energy for combating the prevailing unfavorable atmosphere.
iii. Crops should have improved detoxification system and superior stress tolerant molecular network within the cell.

Finally, as both O_3 and CO_2 stress leave a specific signature on the crop plant transcriptome, inputs from genome-wide analysis could be effectively exploited for further crop improvement *in vivo*, and the objective of our on-going studies.

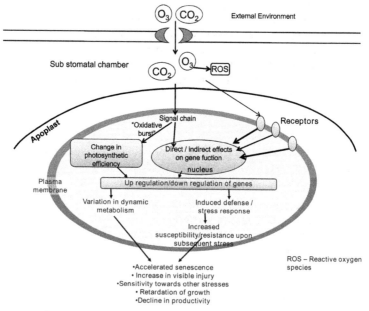

Fig. 3. Diagrammatic representation of O_3- and/CO_2-effects on plants, at cellular levels.

4. Acknowledgements

AS also acknowledge the support of Prof. S. B, Agrawal for promoting collaborative work, and financial help from CSIR, New Delhi, India in the form of Senior Research Fellowship. KC appreciates the support of Dr. Akihiro Kubo for his stay and research at the National Institute for Environmental Studies (NIES) in Tsukuba, Japan, where he worked extensively on ozone responses in rice as an Eco-Frontier Fellow (09-Ba086-02). RR acknowledges the great support of Professor Seiji Shioda and Dr. Tetsuo Ogawa (Department of Anatomy I, Showa University School of Medicine), and the Provost, Professor Yoshihiro Shiraiwa of the Graduate School of Life and Environmental Sciences (University of Tsukuba) in promoting interdisciplinary research and unselfish encouragement.

5. References

Agrawal, G.K., R. Rakwal, M. Yonekura, A. Kubo, and H. Saji. 2002. Proteome analysis of differentially displayed proteins as a tool for investigating ozone stress in rice (Oryza sativa L.) seedlings. Proteomics 2:947-959.

Beig, G., Ali, K. 2006. Behaviour of boundary layer ozone and its precursors over a great alluvial plain of the world: Indo-gangetic plains. Geophysical Research Letters 33: L 24813.

Bokhari, S.A., Wan, X., Yang, Y., Zhou, L., Tang, W., Liu, J. 2007. Proteomic response of rice seedling leaves to elevated CO_2 levels. Journal of Proteome Research 6: 4624-4633.

Cho, K., Shibato, J., Agrawal, G.K., Jung, Y., Kubo, A., Jwa, N., Tamogami, S., Satoh, S., Higashi., Kimura, S., Saji, H., Tanaka, Y., Iwahashi, H., Masuo, Y., Rakwal, R., 2008. Integrated transcriptomics, proteomics, and metabolomics analyses to survey ozone responses in the leaves of rice seedling. Journal of Proteome Research 7: 2980-2998.

Cho, K., Tiwari, S., Agrawal, S.B., Torres, N.L., Agrawal, M., Sarkar, A., Shibato, J., Agrawal, G. K., Kubo, A., Rakwal, R., 2011. Tropospheric ozone and plants: absorption, responses, and consequences. Reviews of Environmental Contamination and Toxicology 212, 61-111.

Coyle, M., Flower, D., Ashmore, M.R. 2003. New directions: implications of increasing tropospheric background ozone concentrations for vegetation. Atmospheric Environment 37: 153-154.

Derwent, R.G., Simmonds, P.G., O'Doherty, S., et al. 2005. External influences on Europe's air quality: methane, carbon monoxide and ozone from 1990 to 2030 at Mace Ireland. Atmospheric Environment 40: 844-855.

Emberson, L.D., Buker, P., Ashmore, M.R., et al. 2009. A comparison of North- America and Asian exposure- response data for ozone effects on crop yields. Atmospheric Environment 43: 1945-1953.

EMEP (2004) EMEP Assessment Part I-European perspective. In: Lövblad G, Tarrasón L, Tørseth K, Dutchak S (eds). http://www.emep.int, last accessed Oct 2008.

Feng, Y.W., Komatsu, S., Furukawa, T., Koshiba, T., Kohno, Y. 2008. Proteome analysis of proteins responsive to ambient and elevated ozone in rice seedlings. Agriculture Ecosystems & Environment 125: 255-265.

Fiscus, E. L., Booker, F. L., Burkey, K. O. 2005. Crop responses to ozone: uptake, modes of action, carbon assimilation and partitioning. Plant, Cell and Environment 28: 997-1011.

Frei, M., Tanaka, J. P., Chen, C. P., Wissuwa, M. 2010. Mechanisms of ozone tolerance in rice: characterization of two QTL affecting leaf damage by gene expression profiling and biochemical analyses. Journal of Experimental Botany 61: 1405-1417.

He, Y.J, Uno, I., Wang, Z.F., et al. 2008. Significant impact of the East Asia monsoon on ozone seasonal behavior in the boundary layer of Eastern China and the west Pacific region. Atmospheric Chemistry Physics Discussion 8:14927-14955.

Hogy P., Zorb, C., Langenkamper, G., et al. 2009. Atmospheric CO₂ enrichment changes the wheat grain proteome. Journal of Cereal Science 50: 248-254.

http://wordnet.princeton.edu

ICP Vegetation International Cooperative Programme on Effects of Air Pollution on natural vegetation and crops, Annual Report 2007 (http://icpvegetation.ceh.ac.uk/, last accessed Oct 2008.

IPCC climate change 2007: The physical science basis. In Solomon, S., Qin, D., Manning, M., Chen, Z., Marquis, M., Averyt, K.B., Tignor, M., Miller, H.L. (eds.). *Contribution of Working Group I to the Fourth Annual Assessment Report of the Intergovernmental Panel on Climate Change*. Cambridge University Press, Cambridge, UK, pp 996.

Jenkin, M.E. 2008. Trends in ozone concentration distributions in the UK since 1990: local, regional and global influences. Atmospheric Environment 42: 5434-5445.

Krupa, S.V., Manning, W.J. 1988. Atmospheric ozone: formation and effects on vegetation. Environment Pollution 50: 101-137.

Lelieveld, J., Dentener, F.J. 2000. What controls tropospheric ozone. Journal Geophysical Research 105(D3): 3531-3551.

Meehl, G.A., Stocker, T.F., Collins, W.D., et al. 2007. Global Climate Projections. In: Solomon S, Qin D, Manning M, Chen Z, Marquis M, Averyt KB, Tignor, M, Miller, HL (Eds.), Climate Change 2007: The physical basis. Contribution of working group I to fourth assessment report of IPCC on climate change. Cambridge University Press, Cambridge, UK, NY, USA.

Mills, G., Buse, A., Gimeno, B., et al. 2007. A synthesis of AOT40-based response functions and critical levels of ozone for agricultural and horticultural crops. Atmospheric Environment 41: 2630-2643.

Pandey, J., Agrawal, M. 1992. Ozone: concentration variabilities in a seasonally dry tropical climate. Environment International 18: 515-520.

Pang, J., Kobayashi, K., Zhu, J. 2009. Yield and photosynthetic characteristics of flag leaves in chinese rice (Oryza sativa L.) varieties subjected to free- air release of ozone. Agriculture, Ecosystems and Environment (doi. 10.1016/j.agee.2009.03. 012).

Pasqualini, S., Antonielli, M., Ederli, L., et al. 2002. Ozone uptake and its effect on photosynthetic parameters of two tobacco cultivars with contrasting ozone sensitivity. Plant Physiology and Biochemistry 40: 599-603.

Rai, R., Agrawal, M., Agrawal, S.B. 2007. Assessment of yield losses in tropical wheat using open top chambers. Atmospheric Environment 41: 9543-9554.

Rai, R., Agrawal, M. 2008. Evaluation of physiological and biochemical responses of two rice (Oryza sativa L.) cultivars to ambient air pollution using open top chambers at a rural site in India. Science of the Total Environment 407: 679-691.

Sarkar, A., Agrawal, S.B. 2010. Identification of ozone stress in Indian rice through foliar injury and differential protein profile. Environmental Monitoring Assessment 161: 283-302.

Sarkar, A., Rakwal, R., Agrawal, S.B, et al. 2010. Investigating the impact of elevated levels of O_3 on tropical wheat using integrated phenotypical, physiological, biochemical and proteomics approaches. Journal of Proteome Research 9: 4565-4584.

Singh, A., Sarin, S.M, Shanmugam, P., et al. 1997. Ozone distribution in the urban environment of Delhi during winter months. Atmospheric Environment 31: 3421-3427.

Sioshansi, R., Oren, S. 2007. How good are supply function equilibrium models: An empirical analysis of the ERCOT balancing market. Journal of Regulatory Economics 31: 1-35.

Tiwari, S., Rai, R. Agrawal, M., 2008. Annual and seasonal variations in tropospheric ozone concentrations around Varanasi. International Journal of Remote Sensing 29, 4499- 4514.

Torres, N.L, Cho, K., Shibato, J., Kubo, A., Iwahashi, H., Jwa, N.S., Agrawal, G.K., Rakwal, R. 2007. Gel based proteomics reveals potential novel protein markers of ozone stress in leaves of cultivated bean and maize species of Panama. Electrophoresis 28: 4369-4381.

Tripathi, R., Sarkar, A., Pandey Rai, S., Agrawal, S.B., 2011. Supplemental ultraviolet-B and ozone: Impact on antioxidants, proteome and genome of linseed (Linum usitatissimum L. cv Padmini). Plant Biology 13: 93-105.

US Environmental Protection Agency U.S. EPA. 2006. Air quality criteria for ozone and related photochemical oxidants (Final). U.S. Environmental Protection Agency, Washington, DC, EPA/600/R-05/004aF-cF.

Varshney, C.K, Aggarwal, M. 1992. Ozone pollution in the urban atmosphere of Delhi. Atmospheric Environment 26: 291-294.

Vingarzan, R. 2004. A review of surface ozone background levels and trends. Atmospheric Environment 38: 3431-3442.

Woodward, F.I. 2002. Potential impacts of global elevated CO_2 concentrations on plants. Current Opinion in Plant Biol. 5: 207-211.

Xu, X., Lin, W., Wang, T., et al. 2008. Long term trend of surface ozone at a regional background station in eastern China 1991–2006: enhanced variability. Atmospheric Chemistry Physics Discussion 8: 215-243

Phenomenal RNA Interference: From Mechanism to Application

Pallavi Mittal[1,*], Rashmi Yadav[2], Ruma Devi[3],
Shubhangini Sharma[4] and Aakash Goyal[5]
[1]ITS Paramedical College, Ghaziabad,
[2]All India Institute of Medical Science, Delhi,
[3]PAU Regional Station, Gurdaspur,
[4]Aptara (Techbook International), Delhi
[5]Bayer Crop Science Saskatoon,
[1,2,3,4]India
[5]Canada

1. Introduction

The phenomenon of dsRNA-mediated interference (RNAi), was first demonstrated in nematodes in 1998 by *Professor* Andrew Z. Fire *at Stanford University, California, USA* and *Professor* Craig C. Mello at University of Massachusetts *Medical School in Worcester, USA*. It is thought to have evolved as a type of "genetic immune system" to protect organisms from the presence of foreign or unwanted genetic material. To be more specific, RNAi probably evolved as a mechanism to block the replication of viruses and/or to suppress the movements of transposons within the genome, because both of these potentially dangerous processes typically involve the formation of dsRNA intermediates. Cells can recognize dsRNAs as "undesirable" because such molecules are not produced by the cell's normal genetic activities.

The RNAi was introduced to the public in mid-2001 and in just about few years it became one of the most widely used technologies in both academic and industrial research environments. In recognition of overwhelming importance of RNAi is an biological process and universally applicable tool, the leading Journal Science proclaimed it "The breakthrough of the year, 2002". During the last decade, our knowledge repertoire of RNA-mediated functions has hugely increased, with the discovery of small non-coding RNAs which play a central part in a process called RNA silencing. Ironically, the very important phenomenon of co-suppression has recently been recognized as a manifestation of RNA interference (RNAi), an endogenous pathway for negative post-transcriptional regulation. RNAi has revolutionized the possibilities for creating custom "knock-downs" of gene activity. RNAi operates in both plants and animals, and uses double stranded (dsRNA) as a trigger that targets homologous mRNAs for degradation or inhibiting its transcription or

---

* Corresponding author

translation (Almedia and Allshire 2005; Cotta-Ramusion et al., 2011) whereby susceptible genes can be silenced. Hence, RNA interference is the newest kid on the genetic block, allows the scientists to selectively turn off genes and finally promises to set the scientific world alight with its therapeutic potential and it has provided new platforms for developing eco-friendly molecular tools for crop improvement by suppressing the genes responsible for various stresses and improving novel traits in plants including disease resistance.

1.1 Landmarks in RNAi discovery

The discovery of RNAi was preceded first by observations of transcriptional inhibition by antisense RNA expressed in transgenic plants and more directly by reports of unexpected outcomes in experiments performed *in 1990s*. In an attempt to produce more intense purple coloured *Petunias*, researchers introduced additional copies of a transgene encoding chalcone synthase (a key enzyme for flower pigmentation). However, they were surprised at the result that instead of a darker flower, the *Petunias* were either variegated or completely white (Figure 1). They called this phenomenon co-suppression of gene expression (Napoli et al., 1990), since both the expression of the existing gene (the initial purple colour) and the introduced gene / transgene (to deepen the purple) were suppressed. It was subsequently reported by Christine, 2008 that suppression of gene activity could take place at the transcriptional level (transcriptional gene silencing, TGS) or at the posttranscriptional level (posttranscriptional gene silencing, PTGS).

Fig. 1. Upon injection of the transgene responsible for purple colouring in *Petunias*, the flowers became variegated or white rather than deeper purple as was expected.

Similar PTGS like effect "Quelling" (Cogoni and Macino, 1999) was also seen in fungi *Neurospora crassa*. Another RNAi related phenomenon, coat protein mediated protection (CPMP) in plants gave insight into the mechanism of PTGS. In 1995 Guo and Kempheus first studied RNA silencing in animals. They used antisense RNA technique to silence par1 mRNA expression in *C. elegans* but found that par1 mRNA itself repressed *par1* gene and concluded that both sense and antisense RNA could cause silencing. Their observation inspired the experiment of Fire, Mello and colleagues. Thus began the journey of the newly dubbed technology RNA Interference. The mystery of molecular mechanism responsible for gene silencing now known as RNA interference (RNAi) exploded in 1998. It was discovered that PTGS was triggered by double-stranded RNA (dsRNA), provided most unexpected explanation with many profound consequences.

Andrew Fire and Craig Mello published their break-through study on the mechanism of RNA interference in 1998, *Nature*. It was earlier known that antisense RNA, but remarkably also sense RNA could silence genes, but the results were inconsistent and the effects usually modest. However, due to the fact that both sense and antisense RNA could cause silencing Mello argued that the mechanism could not just be a pairing of antisense RNA to mRNA, and he coined the term RNA interference for the unknown mechanism. In their *Nature* paper, Fire and Mello tested the phenotypic effect of RNA injected into the worm *C. elegans*. They established that annealed sense/antisense RNA, but neither antisense nor sense RNA alone, caused the predicted phenotype. Furthermore, only injection of double-stranded RNA (dsRNA) led to an efficient loss of the target mRNA. Fire and Mello made the remark that RNAi could provide an explanation for a phenomenon studied in plants for several years: posttranscriptional gene silencing (PTGS). Finally, they ended their paper by speculating about the possibility that "dsRNA could be used by the organism for physiological gene silencing". This discovery later won Fire and Mello the 2006 Nobel Prize in Physiology or Medicine. Thus it was clear that co-suppression in plants, quelling in fungi and RNAi in nematodes all shared a common mechanism. Further work showed that this effect was even more widespread, occurring in fruitflies and mammals too.

The biochemistry of RNAi was further elucidated in an *in vitro* system, built on *Drosophila* cultured cells, Elbashir et al., 2011 demonstrated that Double-stranded RNA is cut into short interfering RNA (siRNA) by the endonuclease Dicer (Lee et al., 2002). The antisense strand is loaded into the RISC (RNA-induced silencing complex) and links the complex to the endogenous mRNA by base-pairing (Martinez et al., 2002). The RISC complex cuts the mRNA strand, and the mRNA is subsequently degraded. In certain systems, in particular plants, worms and fungi, an RNA dependent RNA polymerase (RdRP) plays an important role in generating and/or amplifying siRNA. Thus, within few years a vast amount of information accumulated on the specific proteins and protein complexes involved in RNAi and molecular machinery involved in RNAi was subsequently revealed (Thakur, 2003).

2. Proteins involved in RNAi/ PTGS/Quelling

To understand the basis of RNA silencing both genetic and biochemical approaches have been undertaken. Genetic screens were carried out to search for mutants defective in quelling, RNAi or PTGS and a large number of genes whose products are somehow implicated in RNA silencing have been identified in *C. elegans*, *D. melanogaster*, *Homo sapiens*,

Dictyostelium discoideum, N. crassa, Chlamydomonas reinhardtii and *A. thaliana*. The identified genes encode various components some of which identified as initiators while others serve as effectors, amplifiers and transmitters of the gene silencing process. In the years to come, many other components as well as their interrelations will be revealed. Here, we outline what is known so far.

2.1 Dicer

The endonuclease enzyme called Dicer was first discovered in *Drosophila* by Bernstein et al., 2001. It belongs to the RNase III-family that shows specificity for dsRNAs and cleaves them with 3' overhangs of 2 to 3 nucleotides and 5'-phosphate and 3'-hydroxyl termini (Elbashier et al., 2001; Nicholson, 1999). Dicer is involved in the first step of RNA silencing -the production of siRNAs. Owing to its ability to digest dsRNA into uniformly sized small interfering RNAs (siRNA), this enzyme was named Dicer (DCR). Dicer is ATP-dependent and contains four distinct domains an N-terminal helicase domain, a PAZ domain, a 110-amino-acid domain conservative throughout evolution found in Piwi/Argonaute/Zwille proteins in *Drosophila, Arabidopsis* and involved in developmental control (Catalanotto et al., 2000; Tabara et al., 1999) dual RNase III domains and a double stranded RNA-binding domain, ruler helix. Evolutionarily conserved Dicer homologues from many different sources were also identified and tested in plants, fungi and mammals (Bernstein et al., 2001; Ketting et al., 2000). Furthermore, some recombinant Dicers have also been examined *in vitro*, and phylogenetic analysis of the known Dicer-like proteins indicates a common ancestry of these proteins (Golden et al., 2002).

2.2 RNA-induced silencing complex (RISC)

During studies on the biochemistry of RNAi several proteins engaged in RISC formation were characterised. After partial purification of crude extracts from *Drosophila* embryolysate and human HeLa cells through differential centrifugation and anion exchange chromatography, the nuclease cofractionated with a discrete ≈25-nucleotide RNA species (siRNAs) are part of an effector nuclease which targets homologous RNAs for degradation (Hammond et al., 2000). This complex is referred to as the RNA-induced silencing complex (RISC). It is made up of a group of proteins which use the siRNA as a guide, presumably identifying the substrate through Watson and Crick base-pairing. The proteins in this complex are members of the Argonaute protein family, which are defined as having a PAZ and PIWI domains. The Argonaute PAZ domain most likely holds the 3' end of siRNA, providing the proper orientation for recognition and cleavage of mRNA. PIWI contains the active site for cleaving the mRNA, shown by the scissors in the schematic (Nykanen et al., 2001). The Argonaute family members have been linked both to the gene-silencing phenomenon and to the control of development in diverse species. The first link between Argonaute protein and RNAi was shown by isolation of *rde1* mutants of *C. elegans* in a screen for RNAi-deficient mutants. Argonaute family members have been shown to be involved in RNAi in *Neurospora crassa* (QDE3) as well as in *A. thaliana* AGO1 (Fagard et al., 2000).

2.3 RNA-dependent RNA polymerase

As a result of screening for genes involved in RNAi a family of proteins that exhibit the activity of RNA-dependent RNA polymerase (RdRP) was also identified (Birchler, 2009).

The identification of the quelling-defective gene *qde-1* in *Neurospora* was the first experimental evidence of the involvement of an RdRP in PTGS. The *C. elegans* nuclear genome also contains four members of this gene family: *ego-1*, *rrf-1*, *rrf-2* and *rrf-3*. Among all three of these *rrf-1* was found as the gene coding for RdRp involved in RNAi (Correa et al., 2010). Therefore, RdRP may also be responsible for the amplification and maintenance of the silencing signal by synthesis of secondary dsRNA trigger molecules, which in turn would be processed into secondary siRNAs. However, no RdRp has been identified by homology in the genomes of either flies or humans.

3. Mechanism of action

RNA interference is a classical mechanism of gene regulation found in eukaryotes as diverse as in yeast and mammals and, probably plays a central role in controlling gene expression, by inhibiting gene expression at the stage of translation or by hindering the transcription of specific genes. The RNAi pathway is initiated by the enzyme dicer, which trims long double stranded RNA, to form small interfering RNA (si RNA) or microRNA (miRNA). These processed RNAs are incorporated into the RNA-induced silencing complex (RISC), which targets messenger RNA to prevent translation. Using a recently developed *Drosophila in vitro* system, molecular mechanism underlying RNAi was examined. It was found that RNAi is ATP dependent yet uncoupled from mRNA translation. RNAi pathway can be divided into three major steps:

3.1 Initiator step: dsRNA cleavage

This is the first step in which, dsRNA is converted into 21-23bp small fragments by the enzyme Dicer. Dicer is the enzyme involved in the initiation of RNAi. It is a member of Rnase III family of dsRNA specific endonuclease that cleaves dsRNA in ATP dependent, processive manner to generate siRNA duplexes of length 21-23 bp with characteristic 2 nucleotide overhang at $3'$- OH termini and $5'$ PO_4.

3.2 Effector step: Entry of si RNA into RISC

The siRNAs generated in the initiator step now join a multinuclease effector complex RISC that mediates unwinding of the siRNA duplex. RISC is a ribonucleoprotien complex and its two signature components are the single-stranded siRNA and Argonaute family protien. The active components of an RISC are endonucleases called argonaute proteins, which cleave the target mRNA strand complementary to their bound siRNA, therefore argonaute contributes "Slicer" activity to RISC. As the fragments produced by dicer are double-stranded, they could each in theory produce a functional siRNA. However, only one of the two strands, which is known as the guide strand, binds the argonaute protein and directs gene silencing. The other anti-guide strand or passenger strand is degraded during RISC activation. The process is actually ATP-independent and performed directly by the protein components of RISC. Although it was first believed that an ATP-dependent helicase separated these two strands, the process is actually ATP-independent and performed directly by the protein components of RISC (Senapedis et al., 2011).

3.3 Step 3: Sequence specific cleavage to targeted mRNA

The active RISC further promotes unwinding of siRNA through an ATP dependent process and the unwound antisense strand guides active RISC to the complementary mRNA. The targeted mRNA is cleaved by RISC at a single site that is defined with regard to where the 5' end of the antisense strand is bound to mRNA target sequence. The RISC cleaves the complimentary mRNA in the middle, ten nucleotides upstream of the nucleotide paired with the 5' end of the guide siRNA. This cleavage reaction is independent of ATP. The target RNA hydrolysis reaction requires Mg^{2+} ions. Cleavage is catalyzed by the PIWI Domain of a subclass of Argonaute proteins. This domain is a structural homolog of RNase H, a Mg^{2+} dependant endoribonuclease that cleaves the RNA strand of RNA- DNA hybrids. But each cleavage- competent RISC can break only one phosphodiester bond in its RNA target. The siRNA guide delivers RISC to the target region, the target is cleaved, and then siRNA departs intact with the RISC.

Thus the two important conditions to be fulfilled for the success of silencing by RNAi are established as: 5' phoshorylation of the antisense strand and the double helix of the antisense target mRNA duplex to be in the A form. The A-form helix is required for the stabilization of the heteroduplex formation between the siRNA antisense strand and its target mRNA.

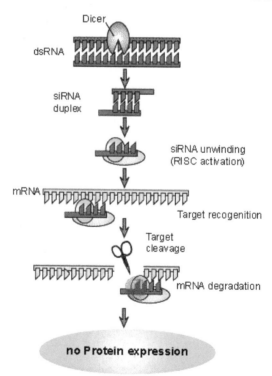

Fig. 2. Mechanism of action of RNAi. Double stranded RNA is introduced into a cell is chopped bydicer to form siRNA, which binds to the RISC complex and is unwound. The anitsense RNA complexed with RISC binds to its corresponding mRNA which is the cleaved by the enzyme slicer rendering it inactive (Christine, 2008).

4. Other forms of RNA interference

In addition to naturally occurring and manufactured siRNAs (Sigoillot and King, 2011), there have been recent publications of alternative forms of RNA, these are

4.1 Micro (mi)–RNAs

These are an abundant class of short (19–25 nt) single-stranded RNAs that are expressed in all higher eukaryotes. They are encoded in the host genome and are processed by Rnase III nuclease Dicer from 70nt hairpin precursors). They can silence gene activity through destruction of homologous mRNA in plants or blocking its translation in plants and animals (Cullen, 2004; Novina and Sharp 2004). Recent work has identified their specific roles in the regulation of early haematopoiesis and lineage commitment). They are sometimes referred to as small temporal RNAs as a reflection of their importance in the regulation of developmental timing (Chen et al., 2004; Medema et al., 2004).

4.2 Piwi-interacting (pi) RNAs

These are single-stranded 25–31 nt RNAs which have recently been detected in mouse, rat, and human testes. They have been shown to associate with Piwi protein (a subclass of Argonaute proteins) and the human RecQ1 protein to form a Piwi-interacting RNA complex (piRC). These complexes are thought to regulate the genome within developing sperm cells (Carthew, 2006).

4.3 Short-Hairpin (sh) RNAs

Short hairpin RNAor shRNA or is a synthetically manufactured RNA molecule of 19–29 nucleotides that contains a sense strand, antisense strand, and a short loop sequence between the sense and antisense fragments. Due to the complementarity of the sense and antisense fragments in their sequence, such RNA molecules tend to form hairpin-shaped double-stranded RNA (dsRNA). shRNA is cloned into a vector, allowing for expression by a pol III type promoter. The expressed shRNA is then exported into the cytoplasm where it is processed by dicer into siRNA which then get incorporated into the siRNA induced silencing complex (RISC)(Medema, 2004).

4.4 Small modulatory (sm)RNAs

These are short, double-stranded RNAs which are found in the nucleus of neural stem cells of mice. They play a critical role in mediating neuronal differentiation through dsRNA/protein interaction (Kuwabara etal., 2004).

5. RNAi application

RNA interference occurs in plants, animals, and humans. It is of great importance for the regulation of gene expression, participates in defense against viral infections, and keeps jumping genes under control. Presently RNA interference has become attractive tool for various researchers and widely used in basic science as a method to study the function of genes and it may lead to novel therapies in the future.

Over the past decade "RNA interference" has emerged as a natural antiviral mechanism for protecting organisms from viruses. It blocks infection by RNA viruses especially in plants and lower animals. For instance, replication of plant viruses, many of which produce dsRNA replication intermediates, very effectively cause a type of RNA silencing called VIGS (Virus induced gene silencing). When viruses or transgenes are introduced into plants, they trigger a post transcriptional gene silencing response in which double stranded RNA molecules, which may be generated by replicative intermediates of viral RNAs or by aberrant transgene coded RNAs. The dsRNAs are then digested into 21-25 nt small interfering RNAs or siRNAs. The siRNAs subsequently assemble into a nuclease complex called RISC, guiding the complex to bind and destroy homologous transcripts. PTGS is believed to be an anti-viral response. Viral RNAs not only trigger PTGS, but they also serve as targets. Cleavages of viral RNA results in reduce virus titers in local and distant leaves and a plant recovery phenotype. In response numerous plant viruses have evolved proteins to suppress PTGS (Elbashir et al., 2001). The results by Silhavy and his colleagues suggest that tombavirus p19 protein suppresses local PTGS by binding to 21-25nt siRNAs. Therefore, siRNAs and not the longer dsRNAs, act as mobile silencing element since p19 can in hibit systemic silencing and p19 can only bind to siRNAs (Sui et al., 2002).

RNA interference besides being working as a genetic immune system against virus, it holds a promising key for maintaining the genome stability by suppressing the movement of mobile genetic elements such as transposons. Jumping genes, also known as transposons, are DNA sequences that can move around in the genome. They are present in all organisms and can cause deleterious effects in the genome if they end up in the wrong place. Many transposons operate by copying their DNA to RNA, which is then reverse-transcribed back to DNA and inserted at another site in the genome. Part of this RNA molecule is often double-stranded and can be targeted by RNA interference. . Thus in this way, foreign elements in the genome (viruses and transposons) can be kept silent.

Moreover, RNAi can be used greatly in both the mammer both "forward genetic" experiments (identifying the gene responsible for a given phenotype) and "reverse genetic" experiments (identifying the function of a known gene). From an application point of view, RNAi may also be useful as a therapy for diseases arising from aberrant gene expression. Typical reverse genetic experiments involve designing siRNAs (chemically or enzymatically synthesized) expressing constructs targeting a gene of interest. Following transient transfection of siRNAs, the phenotype of the cells is assessed using appropriate functional assays.

5.1 Functional genomics

The technology considerably bolsters functional genomics to aid in the identification of novel genes involved in disease processes. Genome sequencing projects have generated wealth of information regarding gene sequences but still clarity on functional role of all genes is missing. The use of small interfering RNA (siRNA) to knock down/ knockout expression of specific gene have opened up exciting possibilities in the study of functional genomics. The ability to easily and economically silence genes promises to elucidate numerous signaling, developmental, metabolic, and related disease pathways. Various studies have been undertaken to elucidate the role of specific genes in basic cellular processes like DNA damage response and cell cycle control (Brummelkamp et al., 2002), general cell metabolism, signaling, the cytoskeleton and its rearrangement during mitosis, membrane trafficking, transcription and, DNA methylation.

Heterochromatin is composed of highly repetitive sequences interspersed with transposons and is non-coding. The condensation pattern is determined by both DNA and histone modification. Recently it has been found to produce RNAi which appear to be the key factor in epigenetic regulation of gene expression, chromosome behaviour and evolution. It may be the mechanism underlying genome imprinting whereby chromosomal condensation pattern is determined by parent-of-origin. Even the phenomenon of hybrid dysgenesis may be explained if siRNA pools that are largely maternal, do not match polymorphic repeats from the paternal chromosome it may result in mobilisation of transposons and consequent chromosomal disruption.

RNAi has been adapted with high-throughput screening formats in *C. elegans*, for which the recombination-based gene knockout technique was not established. Recently, a large-scale functional analysis of 19,427 predicted genes of *C. elegans*, was carried out with RNA interference. This study identified mutant phenotypes for 1,722genes. RNAi technology has been similarly used in the identification of several genes in *D. melanogaster* involved in biochemical signaling cascade as well as embryonic development.

In plants, gene knockdown-related functional studies are being carried out efficiently with transgenes present in the form of hairpin (or RNAi) constructs. Plant endotoxins could be removed if the toxin biosynthesis genes are knocked out. SiRNA results in partial knockout, which is an advantage over complete knockout in that it helps in investigating the effect of various phenotypes. Thus the method holds a great potential to become the most commonly used technique for gene annotation in the near future.

5.2 Genetic improvement of crop plants

Prior to the discovery of RNAi, scientists applied various methods such as insertion of T-DNA elements, transposons, treatment with mutagens or irradiation. These approaches are very cumbersome and the above methods did not always work adequately. For instance, transposons and T-DNA elements were found to occasionally insert randomly in the genome resulting in highly variable gene expression. Furthermore, in many instances the particular phenotype or a trait could not be correlated with the function of a gene of interest. At the same time to improve crop plants transgenes are mainly introduced into the genomes of most model plant species using *Agrobacterium tumefaciens*, a common soil bacterium, and the mechanism of which relies on T-DNA (transfer DNA), that is carried on a resident plasmid. Single T-DNAs can integrate into the genome, but it is very common for multiple copies to integrate in variously permuted head-to-head, tail-to-tail and head-to tail arrays. As we all know till now the most effective genetic approach to pest control has been to make plants that produce a protein called Bt toxin, which causes insects to slow down, then stop eating crops, then die. More than 120,000 square miles of crops genetically engineered to produce Bt were grown last year. But Bt isn't effective against many pests, including corn rootworm, which can cause such extensive damage to corn plants' root systems that the plants blow over in the wind and researchers are concerned that insect pests are becoming resistant to Bt. Here RNAi play a vital role. Now a day, researchers are trying to create plants that kill insects by disrupting their gene expression. The crops, which initiate a gene-silencing response called RNA interference, are a step beyond existing genetically modified crops that produce toxic proteins. Because the new crops target particular genes in particular insects, some researchers suggest that they will be safer and less likely to have unintended effects than other genetically modified plants. Moreover, the quality of crop plants can be improved by RNAi for example Kusaba M 2004 have made significant contribution by applying RNAi to improve rice plants. They were able to reduce the level of glutenin and produced a rice variety called *LGC-1* (low glutenin content 1).The rice mutant line LGC-1 (Low Glutenin Content-1) was the first commercially useful cultivar produced by RNAi. It is low-protein rice and is useful for patients with kidney disease whose protein intake is restricted. This dominant mutation produces hairpin RNA (hpRNA) from an inverted repeat for glutenin, the gene for the major storage protein glutenin, leading to lower glutenin content in the rice through RNAi. Rice down regulation can also be achieved through mutation-based reverse genetics and a gene targeting system (Terada et al., 2002; Shinozuka et al., 2003). However, RNAi has some advantages over these systems. One of these is its applicability to multigene families and polyploidy (Lawrence and Pikaard 2003), as it is not straightforward to knockout a multigene family by the accumulation of mutations for each member of the family by conventional breeding, particularly if members of the family are tightly linked. Another advantage of RNAi lies in the ability to regulate the degree of suppression. Agronomic traits are often quantitative, and a particular degree of suppression of target genes may be required. Control of the level of expression of dsRNA through the choice of promoters with various strengths is thought to be useful in regulating the degree of suppression. However, for wider application of transgene- based RNAi to the genetic improvement of crop plants further feasibility studies are needed.

Engineering of food plants that produce lower levels of natural plant toxins also possible through RNAi. Such techniques take advantage of the stable and heritable RNAi phenotype in plant stocks. For example, cotton seeds are rich in dietary protein but naturally contain the toxic terpenoid product gossypol, making them unsuitable for human consumption. RNAi has been used to produce cotton stocks whose seeds contain reduced levels of delta-cadinene synthase, a key enzyme in gossypol production, without affecting the enzyme's production in other parts of the plant, where gossypol is important in preventing damage from plant pest (Kumar et al., 2006). Similar efforts have been directed toward the reduction of the cyanogenic natural product linamarin in cassava plant (Siitunga and Sayre 2003). Although no plant products that use RNAi-based genetic engineering have yet passed the experimental stage, development efforts have successfully reduced the levels of allergens in tomato plants and decreased the precursors of likely carcinogens in tobacco plant (Gavilano et al., 2006). Other plant traits that have been engineered in the laboratory include the production of non-narcotic natural products by the opium poppy (Allen et al., 2004) resistance to common plant viruses (Zadeh and Foster 2004) and fortification of plants such as tomatoes with dietary antioxidants (Niggeweg et al., 2004). In plants, gene knockdown-related functional studies are being carried out efficiently when transgenes are present in the form of hairpin (or RNAi) constructs. Plant endotoxins could also be removed if the toxin biosynthesis genes are targeted with the RNAi constructs. Therefore, RNAi soon caught the world-wide attention and became a powerful and useful tool for molecular breeders to produce improved crop varieties.

5.3 RNA interference as a novel therapeutic agent

The ability to tap this native RNAi pathway has been recognized as one the most exciting biotechnology advances in the last decade. Given the gene-specific features of RNAi, it is conceivable that this method will play an important role in therapeutic applications and possibly of most commercial interest in the use of RNAi as a therapeutic agent. Indeed, RNAi has revolutionized biology research, including drug target discovery, by allowing for rapid identification and validation of gene function. There are three main time points at which a disease can be stopped. These are transcriptional, post-transcriptional, and post-translational intervention. Before the discovery of antisense RNA and RNAi, most of the drug targets have been proteins, and therefore, post- translational intervention. RNAi is a way to control the development of a disease earlier on in the process. Furthermore, the gene-specific features and potential of RNAi for knocking out a protein without harming a cell has established its most believable role into therapeutic applications. The inhibitory action of siRNAs has been documented for numerous diseases. Some of the examples are highlighted below:

5.3.1 Cancer protection

Gregory Hannon and colleagues have used RNAi to silence expression of p53 — the 'guardian of the genome', which protects against any tumour-associated DNA damage — by introducing several p53-targeting shRNAs into stem cells and looking at the effect in mice (Hemann et al., 2003). The shRNAs produced a wide range of clinical effects, ranging from benign to malignant tumours, the severity and type of which correlated with the extent to which the shRNA had silenced p53. As tumour suppressors such as p53 usually work as part of a complex and finely regulated network, the ability to dampen these networks to varying degrees in these libraries- which the authors term an epi-allelic series of hypomorphic mutations -will be of enormous value when it comes to investigating the early stages of disease. The success of these modified stem cells also gives hope that this could treat diseases in which stem cells can be modified *ex vivo* and then re-introduced into the affected individual.

Researchers at the charity Cancer Research UK and the Netherlands Cancer Institute have recently announced that they intend to generate a large library of human cells, each containing a silenced gene. They initially want to silence 300–8,000 cancer genes, and hope to eventually cover the entire human genome. Their aim is to uncover all the genes that become overexpressed in human cancers and to find out precisely what needs to be taken away from a cancerous cell in order to make it normal again (Hoffman et al., 2010).

5.3.2 HIV protection

HIV infection can be blocked by targeting either viral genes (for example, *gag, rev, tat* and *env*) or human genes (for example, *CD4*, the principal receptor for HIV) that are involved in the HIV life cycle.The strategy used was to silence the main structural protein in the virus, p24, and the human protein *CD4*, which the virus needs to enter the cells. This impairs the virus in infected cells and limits its spread into healthy cells (Paddison et al., 2002). Hence the production of virus is inhibited either by blocking new infections or blocking the production of new viral particles in infected cells (Lohman et al., 1999). The concept of silencing genes in HIV is straightforward: Hit the virus where it counts by eliminating a protein it needs to reproduce or cause infection. siRNA molecules (shorter than 30 base pairs) are added to the cells, where the cell recognizes and degrades mRNA corresponding to the target sequence. As a result little or no protein is produced.

What makes RNAi so exciting to the researchers is its potential for knocking out a protein without harming a cell. By comparison chemotherapy kills tumors by destroying cancerous as well as healthy cells. RNAi strategy includes multiple targets to kill HIV. These could be the targets that block entry into the cells and disrupts the virus life cycle inside the cells. This technology will help researchers dissect the biology of HIV infection and design drugs based on the information. Researchers at City of Hope Cancer Centre in Duarte have developed a DNA-based delivery system in which human cells are generated that produce siRNA against REV protein, which is important in causing human disease (Yu et al., 2002).

5.3.3 Hepatitis protection

This has provided the first tangible evidence for RNAi as a therapy for diseases in live animals. Early RNAi studies noted that RNA silencing was prominent in the liver, which made this organ an attractive target for therapeutic approaches. Many immune-related liver diseases are characterized by apoptosis, which is mediated by a protein called Fas. So Judy Lieberman's group injected siRNA targeting Fas intravenously into two models of autoimmune hepatitis in mice. This decreased Fas mRNA and protein levels in hepatocytes and protected the cells against liver injury from apoptosis, even when siRNA was administered after the induction of injury. Extending these findings to other liver diseases looks hopeful, but the authors concluded that other strategies, such as viral vectors, might be required to target organs in which RNA silencing is less effective than in the liver.

6. Conclusion and future outlook of RNAi

The field of RNAi is moving at an impressive pace, generating exciting results and has established a novel archetype with far-reaching consequences in the field of transcription regulation. The RNA silencing has practical use because of the ability to reduce gene expression in a manner that is highly sequence specific as well as technologically facile, economical and having potential in finding out the function of genes at a faster speed and in agriculture specifically for nutritional improvement of plants and the management of mascotous plant diseases. In addition it has kindled hope for the treatment of several diseases, which have bothered mankind as untreatable by providing an innovative technology for development of therapeutics. However, the major obstacles hindering its immediate applications include selection of targeting sequences and in the delivery of siRNA. The key issues are 1) how to select silencing targets for a particular disease and 2) how to efficiently deliver siRNAs into specific cell types *in vivo*? Besides, RNAi technology can be considered an eco-friendly, biosafe ever green technology as it eliminates even certain risks associated with development of transgenic and it has already added new dimensions in the various field of science. However, a better and comprehensive understanding of RNAi should allow future researchers to work effectively and efficiently in order to manage the phenomenon.

7. References

Allen, R., Millgate, A., Chitty, J., Thisleton, J., Miller, J., Fist, A., Gerlach, W., & Larkin, P. (2004) RNAi-mediated replacement of morphine with the nonnarcotic alkaloid reticuline in opium poppy". *Nat. Biotechnol.* 22: 1559-1566.

Almeida, R. & Allshire, R. C. (2005) RNA silencing and genome regulation. *Trends Cell Biol.* 15(5):251-258.

Bernstein, E. A., Caudy, A., Hammond, S. M. & Hannon, G. J. (2001) Role for a bidentate ribonuclease in the initiation step of RNA interference. *Nature.* 409:363-366.

Brichler, J. A. (2009) Ubiquitous RNA-dependent RNA polymerase and gene silencing. *Genome Biol.* 10:243.1-243.3.

Brummelkamp, T. R., Bernards, R. & Agami, R. (2002) A system for stable expression of short interfering RNAs in mammalian cells. *Science.* 296:550-553.

Carrington, J. C. & Ambros, V. (2003) Role of Micro RNAs in plant and animal development. *Science* 301: 336-338.

Carthew, R. W. (2006) A new RNA dimension to genome control. *Science.* 313: 305-306.

Catalanotto, C., Azzalin, G., Macino, G. & Cogoni, C. (2000) Gene silencing in worms and fungi. *Nature.* 404:245.

Chen, C., Li, L., Lodish, H. F. & Bartel, D. P. (2004) MicroRNAs modulate haematopoeitic lineage differentiation. *Science.* 303: 83-86.

Christine, A. (2008) Antisense RNA *The Science Creative Quarterly*, Issue Three 25-30.

Cogoni, C. & Macino, G. (1999) Homology dependent gene silencing in plants and fungi : a number of variations on the same theme. *Curr. Opin. Microbiol.* 2(6): 657-662.

Corrêa, R. L., Steiner, F. A., Berezikov, E. & Ketting, R. F. (2010) MicroRNA–Directed siRNA biogenesis in *Caenorhabditis elegans. PLoS Genet.* 6(4): e1000903.

Cotta-Ramusino, C., McDonald, E. R. 3rd, Hurov, K., Sowa, M. E., Harper, J. W. & Elledge, S. J. (2011) A DNA damage response screen identifies RHINO, a 9-1-1 and TopBP1 interacting protein required for ATR signaling. *Science.* 332:1313-1317.

Cullen, B. R. (2004) Derivation and function of small interfering RNAs and microRNAs. *Virus Res.* 102:3-9.

Elbashir, S. M., Lendeckel, W. & Tuschl, T. (2001) RNA interference is mediated by 21- and 22-nucleotide RNAs. *Genes Dev.* 15:188-200.

Fagard, M., Boutet, S., Morel, J. B., Bellini, C. & Vaucheret, H. (2000) AGO1, QDE-2, and RDE-1 are related proteins required for post-transcriptional gene silencing in plants, quelling in fungi, and RNA interference in animals. *Proc. Natl. Acad. Sci.* 97:11650-11654.

Fire, A., Xu, S., Montgomery, M. K., Kostas, S. A., Driver, S. E. & Mello, C. C. (1998) Potent and specific genetic interference by double-stranded RNA in Caenorhabditis elegans. *Nature*. 391:806-811.

Gavilano, L., Coleman, N., Burnley, L., Bowman, M., Kalengamaliro, N., Hayes, A., Bush, L. & Siminszky, B. (2006) Genetic engineering of Nicotiana tabacum for reduced nornicotine content. *J. Agric. Food Chem*. 54: 9071-9078.

Golden, T. A., Schauer, S. E., Lang, J. D., Pien, S., Mushegian, A. R., Grossniklaus, U,. Meinke, D. W. & Ray, A. (2002) Short Integuments/ Suspensor1/ Carpel Factory, a Dicer homolog, is a maternal effect gene required for embryo development in *Arabidopsis. Plant Physiol.* 130:808-822.

Guo, S. & Kemphues, K. J. (1995) Par-1, a gene required for establishing polarity in C. elegans embryos, encodes a putative Ser/Thr kinase that is asymmetrically distributed. *Cell* 81:611-620

Hammond, S., Bernstein, E., Beach, D. & Hannon, G. J. (2000) An RNA-directed nuclease mediates post-transcriptional gene silencing in Drosophila cells. *Nature*. 404: 293-296.

Hemann, M. T., Fridman, J. S., Zilfou, J. T., Hernando, E., Paddison, P. J., Cordon-Cardo, C., Hannon, G. J. & Lowe, S.W. (2003) An epi-allelic series of p53 hypomorphs created by stable RNAi produces distinct tumor phenotypes in vivo. *Nature Genet.* 33: 396-400.

Hoffman, G. R., Moerke, N. J., Hsia, M., Shamu, C. E. & Blenis, J. (2010) A high-throughput, cell-based screening method for siRNA and small molecule inhibitors of mTORC1 signaling using the in cell western technique. *Assay Drug Dev. Technol.* 8(2):186-99.

Ketting, R. F., Fischer, S. E. J., Bernstein, E., Sijen, T., Hannon, G. J. & Plasterk, R. H. A. (2000) Dicer functions in RNA interference and in synthesis of small RNA involved in developmental timing in C. *Elegans. Genes Dev.* 15: 2654-2659.

Kuwabara, T., Hsieh, J., Nakashima, K., Taira, K. & Gage, F. H. (2004) A small modulatory dsRNA specifies the fate of adult neural stem cells . *Cell*. 116: 779-93.

Lawrence, R. J. & Pikaard, C. S. (2003) Transgene-induced RNA interference: a strategy for overcoming gene redundancy in polyploids to generate loss-of-function mutations. *Plant J.* 36:114-121.

Lee, N. S., Dohjima, T., Bauer, G., Li, H., Li, M. J., Ehsani, A., Salvaterra, P. & Rossi, J. (2002) Expression of mall interfering RNAs targeted against HIV-1 rev transcripts in human cells. *Nat. Biotechnol.* 20:500-505.

Lohmann, J. U., Endl, I. & Bosch, T. C . (1999) Silencing of developmental genes in Hydra. *Dev. Biol.* 214:211-214.

Martinez, J., Patkaniowska, A., Urlaub, H., Luhrmann, R. & Tuschl, T. (2002) Singlestranded antisense siRNAs guide target RNA cleavage in RNAi. *Cell.* 110:563-574.

Medema, R. H. (2004) Optimizing RNA interference for application in mammalian cells. *Biochem J.* 380:593-603.

Napoli, C., Lemiex, C. & Jorgenson, R. A. (1990) Introduction of a chimeric chalcone synthase gene into Petunia results in reversible co-suppression of homologous genes in trans. *Plant Cell.* 2:279-289.

Nicholson, A. W. (1999) Function, mechanism and regulation of bacterial ribonucleases. *FEMS Microbiol. Rev,* 23:371-390.

Niggeweg, R., Michael, A. & Martin, C. (2004) Engineering plants with increased levels of the antioxidant chlorogenic acid. *Nat Biotechnol.* 22:746-754.

Novina, C. D. & Sharp, P. A. (2004) The RNAi revolution. *Nature.* 430:161-165.

Nykanen, A., Haley, B. & Zamore, P. D. (2001) ATP requirements and small interfering RNA structure in the RNA interference pathway. *Cell.* 107: 309-321.

Paddison, P. J., Claudy, A. A., Bernstein, E., Hannon, G. J. & Conklin, D. S. (2002) Short hairpin RNAs (shRNAs) induce sequence specific silencing in mammalian cells. *Genes Dev.* 16: 948-958.

Senapedis, W. T., Kennedy, C. J, Boyle, P. M. & Silver, P. A. (2011) Whole genome siRNA cell-based screen links mitochondria to Akt signaling network through uncoupling of electron transport chain. *Mol. Biol. of the Cell* 22:1791-1805.

Shinozuka, Y., Onosato, K. & Hirochika, H. (2003) Target site specificity of the Tos 17 retrotransposon shows a preference for insertion within genes and against insertion in retrotransposon-rich regions of the genome. *Plant Cell.* 15:1771-1780.

Sigoillot, F. D., King RW (2011) Vigilance and Validation: Keys to Success in RNAi Screening. *ACS Chem. Biol.* 21:47-60.

Sijen, T., Fleenor, J., Simmer, F., Thijssen, K. L., Parish, S., Timmons, L., Plasterk, R. H. A. & Fire, A. (2001) On the role of RNA amplification in dsRNA-triggered gene silencing. *Cell.* 107: 465-476.

Siritunga, D. & Sayre, R. (2003) Generation of cyanogen-free transgenic cassava. *Planta.* 217: 367-373.

Sui, G., Sohoo, C., Affar, E. B., Gav, F., Shi, Y., Forrester, W. C. & Shi, Y. A. (2002) DNA vector based RNAi technology to suppress gene expression in mammalian cells. *Proc. Natl. Acad. Sci.* 99: 5515-5520.

Kumar, S. G., Campbell, L., Puckhaber, L., Stipanovic, R. & Rathore, K. (2006) Engineering cottonseed for use in human nutrition by tissue-specific reduction of toxic gossypol. *Proc. Natl. Acad. Sci.* 103: 18054–18059.

Tabara, H., Sarkissian, M., Kelly, W. G., Fleenor, J., Grishok, A., Timmons, L., Fire, A. & Mello, C. C. (1999) The rde-1 gene, RNA interference and transposon silencing in C. *elegans. Cell.* 99:123-132.

Terada, R., Urawa, H., Inagaki, Y., Tsugane, K. & Iida, S. (2002) Efficient gene targeting by homologous recombination in rice. *Nat Biotechnol.* 20: 1030-1034.

Thakur, A. (2003) RNA interference revolution. *Electronic J Biotechnol.* 6: 39-49.

Yu, J. Y., Deruiter, S. L. & Turner, D. L. (2002) RNA interference by expression of short interfering RNAs and hairpin RNAs in mammalian cells *Proc. Natl. Acad. Sci.* 99:6047-6052.

Zadeh, A. & Foster, G. (2004) Transgenic resistance to tobacco ringspot virus. *Acta Viro.*, 48: 145-152.

4

Silicon the Non-Essential Beneficial Plant Nutrient to Enhanced Drought Tolerance in Wheat

Mukhtar Ahmed[1], Muhammad Asif[2,*] and Aakash Goyal[3]

[1]Department of Agronomy, PMAS Arid Agriculture University Rawalpindi,
[2]Agricultural, Food and Nutritional Science, 4-10 Agriculture/Forestry Centre,
Univ. of Alberta, Edmonton, AB,
[3]Bayer Crop Science, Saskatoon,
[1]Pakistan
[2,3]Canada

1. Introduction

Present water scarcity is a severe problem and cause of deterioration in quality and productivity of crops to reduce crop yield in arid and semi-arid regions. Silicon is known to better the deleterious effects of drought on plant growth and development. Silicon (Si) found to be an agronomically important fertilizer element that enhances plant tolerance to abiotic stresses (Liang et al., 2005). Silicon also known to increase drought tolerance in plants by maintaining plant water balance, photosynthetic efficiency, erectness of leaves and structure of xylem vessels under high transpiration rates due to higher temperature and moisture stress (Hattori et al., 2005). Similarly, Gong et al., (2003 and 2005) observed improved water economy and dry matter yield of water under application of silicon. A number of possible mechanisms were proposed through which Si may increase salinity tolerance in plants, especially improving water status of plants, increased photosynthetic activity and ultra-structure of leaf organelles. The stimulation of antioxidant system and alleviation of specific ion effect by reducing Na uptake were also drought tolerance mechanisms in plants exposed to silicon application (Liang et al., 2005).

2. Silicon accumulation and its uptake in plants

Silicon (Si) is most abundant in soil next to oxygen and comprises 31% of its weight. It is taken up directly as silicic acid (Ma et al., 2001). It primarily accumulated in leaves because it is distributed with the transpiration stream. In dried plant parts the silica bodies are located in silica cells below the epidermis and in epidermal appendices (Dagmar et al., 2003). Being a dominant component of soil minerals the silicon has many important functions in environment. Many studies have suggested the positive growth effects of silicon, including increased dry mass and yield, enhanced pollination and most commonly

* Corresponding author

increased disease resistance (Rodrigues, et al., 2004). Silicon can also alleviate imbalances between zinc and phosphorus supply. Gypsum is known to improve the productivity of dispersive soils and Sodium silicate has shown to maintain root activity under waterlogged conditions (Ma et al., 1989). Water stress is common problem in the rainfed regions of the world now a day, which have caused deviation of plant functions from normal to abnormal. Therefore, it's necessary to provide plants such type of nutrition which can maintain water balance in the plants. Silicon is considered to be important element under stress because it increased drought tolerance in plants by maintaining leaf water potential, assimilation of CO_2 and reduction in transpiration rates by adjusting plant leaf area (Hattori et al., 2005). Maintenance of higher leaf water potential under stress is one of remarkable feature which silicon nutrition does for plants as reported by Lux et al., (2002). Silicon was reported to enhance growth of many plants particularly under biotic and abiotic stresses (Epstein, 1999). A number of possible mechanisms have been proposed by which Si would increase resistance of plants against salinity stress which is a major yield limiting factor in arid and semiarid areas. (Al-aghabary et al., 2004).

3. How silicon can coup biotic and abiotic stresses

Silicon (Si) is reported to have beneficial effects on the growth, development and yield of plants through protection against biotic and abiotic stresses. Silicon has not yet been considered a generally essential element for higher plants, partly because its roles in plant biology are poorly understood and our knowledge of silicon metabolism in higher plants lags behind that in other organisms (such as diatoms). However, numerous studies have demonstrated that silicon is one of the important elements of plants, and plays an important role in tolerance of plants to drought stress. Agarie et al., (1998) reported that silicon could decrease the transpiration rate and membrane permeability of wheat (*Triticum aestivum* L.) under water deficit induced by polyethylene glycerol. In Pakistan, little work has been done on silicon applications mostly on wheat crop being staple food crop of the country, cultivated under a wide range of climatic conditions. Its contribution towards value added in agriculture and GDP is 13.1 % and 2.7 % respectively. Wheat was cultivated on an area of 8.81 million hectares with the production of 24.2 m. tons and yield of 2750 Kg ha^{-1} (Economic Survey of Pakistan, 2010-11). It is estimated that rainfed area contributes only about 12 percent of the total wheat production. The Punjab province contributes over 71 percent of the national wheat production while the Punjab barani tract contributes 25 percent of the wheat production in the province. The yield of the crop can be increased at least two times with proper management of production factors. The crop also suffers from severe moisture stress that plays a major role in lowering the yield under rainfed areas. So, present study was conducted to evaluate the response of two wheat varieties and two lines under different levels of potassium silicate which will be source of silicon in this study. The specific objectives of proposed study were to evaluate the performance of different wheat varieties and lines under silicon enhanced drought tolerance and to verify that silicon may be useful to enhance the drought tolerance.

3.1 Methodology

The experiment relating to the study of Silicon on wheat growth, development and drought resistance index was conducted in Department of Agronomy, Pir Mehr Ali Shah, Arid

Agriculture University, Rawalpindi. Pakistan. Seeds of two varieties Chakwal-50, GA-2002 and two lines NR- 333 and NR- 372 were taken from National Agricultural Research Center (NARC). The experiment was laid out in glass oven sterilized Petri dishes lined with two layers of Whattman filter paper and one layer of toilet roll. The filter paper and toilet rolls were irrigated with respective solutions at their saturation point and excess solution was discarded. Ten seeds of each variety were sown in total 60 Petri dishes which were set in a complete randomized block design. The lid covered Petri dishes were placed in a germinator under constant darkness at a temperature of 20°C and 30-40 % relative humidity. The solution treatments were applied as T1= control (water only), T2= 5 % Potassium silicate, T3= 10 %, Potassium silicate without irrigation, T4= 5% Potassium silicate T5= 10% Potassium silicate with irrigation. In the experiment II earthen pots of dimensions (25cm length 20cm diameter) with an area of 500cm^2 covered with aluminum foils to prevent an increase in soil temperature caused by solar radiation. Pots were irrigated before adding soil. Each pot was filled with 10 kg of well pulverized soil. Fertilizer was added on the basis of soil weight in the pots. Two wheat cultivars and two wheat advanced lines with three replications were used as plant material in the present study. Ten seeds of each cultivars and advanced lines were sown per pot. At three leaf stage, all the treatments were applied and potassium chloride was applied to the control pots to yield the same total potassium as in Si treatment. The pH in both solutions was adjusted to pH 5.5 with HCl prior to application. Plastic sheets coated with aluminum film were placed on the soil surface to prevent evaporation from the pots. The treatments of the study were:T_1: Control, T_2: 5 % level of Potassium silicate (5: 95ml) without irrigation, T_3: 10 % level of Potassium silicate (10: 90ml) without irrigation, T_4: 5 % level of Potassium silicate (5: 95ml) with irrigation. (300mm) and T_5: 10 % level of Potassium silicate (10: 90ml) with irrigation. (300mm). Data was collected about crop growth rate (CGR), relative growth rate (RGR), net assimilation rate (NAR) and leaf area index (LAI) using the formula by Gardner et al. (1985) while leaf area was measured with the help of CI-202 area meter by averaging the value taken from three plant samples. However, leaf area duration (LAD) was calculated by formula proposed by Power et al., (1967). Similarly, physiological parameters like photosynthesis rate (A) (μ mole/m^2/second), transpiration rate (E) (mole/m^2/s), stomatal coductance (gs) (mol m^2 s^{-1}) were measured by Infrared Gas Analyzer (IRGA) at flag leaf stage (Long & Bernacchi, 2003). Leaf membrane stability index (LMSI) was determined according to method described by Chandrasekar et al., (2000). Meanwhile leaf succulence (mg/m^2) was measured by leaves taken randomly from each plant. Fresh leaf weight was taken and their area was measured and leaves were dried at 70°C for one week and dry weight was taken. Succulence was calculated by using formula (Succulence = fresh weight-dry weight/leaf area). Relative water content (RWC) was measured from fully expended leaves. The leaves were excised and fresh weight (FW) was immediately recorded, then leaves were soaked for 4 hours in distill water at room temperature, and turgid weight (TW) was recorded. After drying for 24 hours at 80 °C total dry weight (DW) was recorded. Relative water content was measured according to that formula (Barrs & Weatherly, 1962). Meanwhile epicuticular Wax (mg m^{-2}) were measured from leaves (0.5 g) randomly taken from the plant and their area was measured. Three leaf samples were washed three times in 10 ml carbon tetrachloride for 30 sec per wash. The extract was filtered, evaporated to dryness and the remaining wax was weighed. Wax content was expressed on the basis of leaf area only, i.e. wax content mg cm^{-2}

(Silva Fernandus et al., 1964). SPAD chlorophyll meter was used to measure chlorophyll contents at flag leaf. Drought resistance index is define as DC multiplied by variety minimum yield and then divided by average minimum yield of the varieties used in experiments (DRI =DC× (Ɏa / Ya) where DC =Drought resistance coefficient=Ya /Ym, Whereas Ya is the average yield of all the varieties with no irrigation, Ɏa is yield of the variety without irrigation and Ym is maximum yield of variety under irrigation). Proline content (μg g^{-1}) was estimated spectrophotometerically following the ninhydrine method (Bates et al. 1973). The silicon concentration in leaves (mg) were measured at flag leaf stage according to Lux et al., (2002) The dried powdered plant sample was ashed in a muffle oven at 500 ^{0}C for 5 h. The plant ash was dissolved in diluted HCl (1: 1; 10 ml) at 100 ^{0}C. The process of dissolving in HCl and evaporation to dryness was repeated three times. Then, diluted HCl (1: 1; 15ml) was added and sample was heated at 100 ^{0}C, filtered placed into a ceramic crucible and ashed again in the oven at 540 ^{0}C for 5 h. After cooling, the weight of silicon was determined gravimetrically. At the end observations collected were analyzed for variance by STATISTIX 8.1.

4. Effect of Silicon on growth kinetics

4.1 Crop Growth Rate (CGR) (g m^{-2} day^{-1})

Crop growth rate is unit increase in drymatter of crop on daily basis. Temperature and moisture are the main determinant factors which affects growth and many other physiological processes of plants. Significant findings were observed for CGR at flag leaf stage under present study. The results revealed that maximum CGR was observed for NR-333 line under 10 % silicon application with irrigation followed by Chakwal-50 and NR-372 (Table 1). Whereas, NR-333 under controlled conditions exhibited minimum crop growth rate. Application of silicon enhanced crop growth rate under stressed conditions. Our results were in the line with the findings of Hattori et al., (2005)) who reported an increase in drymatter and growth rate of crop by silicon application under drought stress conditions.

4.2 Relative Growth Rate (RGR) (g g^{-1} day^{-1})

The pattern of relative growth rate for different wheat genotypes under various silicon treatments presented in table 2 .The results of current study depicted that maximum relative growth rate was observed for Chakwal-50 under 10 % silicon application with irrigation, followed by NR-333 under same treatment. While, minimum relative growth rate was exhibited by NR-333 under controlled conditions (without silicon application). This variation in relative growth rate might be due to difference in genetic potential and adaptability measures to cope stress using silicon enhanced treatments. Our results were similar to the findings that RGR decreased under stress conditions and negatively related to plant age.

4.3 Net Assimilation Rate (NAR) (g cm^{-2} day^{-1})

Remarkable increase in net assimilation rate (NAR) was recorded due to silicon treatments. The maximum NAR was recorded for Chakwal-50 (6.54 g cm^{-2} day^{-1}) with 10% of silicon application which might increase water conversion capacity toward assimilates by boosting photosynthesizing machinery. However, under control conditions, NAR was 1.99 g cm^{-2} day^{-1}

for NR-333 which further confirmed the significant effect of silicon application on NAR (Table 3). Drought tolerant genotypes have maximum NAR under silicon nutrition as compared to other genotypes. The effect of silicon enhanced treatments to further elucidate effect of silicon on NAR of crop revealed that silicon application has positive effect on crop NAR.

4.4 Leaf area (cm^2)

Leaf area might be an important index in determining crop growth as it directly involved in many plant processes. Photosynthesis, transpiration and stomatal aperture take place in leaves of plants. Leaf area attributed toward good uptake of water and translocation of photoassimilates from source to sink. Results demonstrated significant variation for leaf area among different cultivars under various silicon treatments (Table 4). The results demonstrated maximum leaf area for Chakwal-50 (201.67 cm^2) under 10 % silicon application with irrigation. While, minimum leaf area was observed for NR-333(173.00 cm^2) under controlled condition. Leaf area significantly contributed toward physiological indices, which boosted up crop growth and accumulation of more photoassimilates from source to sink and consequently, it led to higher grain yield (Fig.1). Our results were in the line with the findings of Ahmed et al., (2011a) who reported significant impact of stress conditions upon leaf area of crop.

4.5 Leaf area index

Leaf area index (LAI) at flag leaf stage among four wheat cultivars under five different concentrations of silicon application revealed that silicon nutrition has depicted significant effect upon crop growth. However, this effect was more significant for Chakwal-50 as compared to other genotypes which are drought resistance exhibiting maximum leaf area and ultimately higher leaf area index as compared to other genotype (Table. 5). While, minimum leaf area index was noted down for GA-2002 under controlled treatment. The increment in LAI due to silicon nutrition was considerable (Ahmed et al., 2011a).

4.6 Leaf area duration

Leaf area duration for all four genotypes under different silicon application was shown in Table.6. Leaf area duration increased from tillering to flag leaf stage and reached maximum at flag leaf stage as it is the critical stage of wheat. Maximum physiological attributes take place at this stage which determines crop productivity. In current study, Chakwal-50 showed maximum leaf area duration (116.00) under 10% application of silicon, whereas, NR-333 exhibited minimum leaf area duration (87.00) under control conditions.

Treatments	Chakwal-50	GA-2002	NR-333	NR-372	Mean
T1	10.10ijkl	10.00jkl	9.23l	9.56l	9.72E
T2	11.53ghi	9.83kl	10.10ijkl	11.23hijk	10.68D
T3	12.10fgh	11.36hij	12.85efg	13.50ef	12.45C
T4	16.23bc	13.90de	15.20cd	16.93ab	15.57B
T5	18.10a	16.77ab	18.20a	18.17a	17.80A
Mean	13.61AB	12.73C	13.12B	13.88A	

LSD for Genotypes= 0.64, Treatments=0.72 and Genotypes x Treatments= 1.43

Table 1. Crop growth rate (g m^{-2} day^{-1}) of wheat genotypes under different silicon treatments

Treatments	Chakwal-50	GA-2002	NR-333	NR-372	Mean
T1	87.37fg	81,13k	73.69l	84.00hij	81.55D
T2	92.23e	86.23gh	83.33jk	86.07ghi	86.97C
T3	93.73de	89.30f	96.53bc	89.03f	92.15B
T4	85.37ghi	83.50ijk	83.17jk	94.90bcd	86.73C
T5	103.00a	93.93cde	96.80b	94.27bcde	97.00A
Mean	92.34A	86.82C	86.70C	89.65B	

LSD for Genotypes= 1.18, Treatments= 1.32 and Genotypes x Treatments= 2.64

Table 2. Relative growth rate (g m^{-2} day^{-1}) of wheat genotypes under different silicon treatments

Treatments	Chakwal-50	GA-2002	NR-333	NR-372	Mean
T1	5.46bc	3.90de	1.99g	2.14fg	3.37C
T2	5.78ab	4.30d	2.75fg	2.91f	3.93B
T3	6.58a	4.70cd	2.88fg	3.02ef	4.30B
T4	5.59ab	4.33d	2.87fg	3.87de	4.25B
T5	6.54a	5.48bc	5.42bc	5.39bc	5.71A
Mean	6.06A	4.54B	3.47C	3.18C	

LSD for Genotypes= 0.41, Treatments= 0.46 and Genotypes x Treatments= 0.91

Table 3. Net assimilation rate (g m^{-2} day^{-1}) of wheat genotypes under different silicon treatments

Treatments	Chakwal-50	GA-2002	NR-333	NR-372	Mean
T1	184.33def	184.00ef	173.00g	183.00ef	181.08D
T2	193.33bc	186.00de	179.33fg	186.00de	186.17C
T3	197.67ab	187.67cde	183.00ef	189.00cde	189.33BC
T4	198.00ab	187.00cde	187.00cde	193.33bc	191.33AB
T5	201.67a	183.00ef	190.67cd	198.67ab	193.50A
Mean	195.00A	185.53C	182.60D	190.00B	

LSD for Genotypes= 2.86, Treatments= 3.19 and Genotypes x Treatments= 6.39

Table 4. Leaf area (cm^2) of wheat genotypes under different silicon treatments

Treatments	Chakwal-50	GA-2002	NR-333	NR-372	Mean
T1	1.16d	0.55k	0.85g	0.68j	0.815D
T2	1.13d	0.72ij	0.81gh	0.72ij	0.845D
T3	1.93b	0.94e	0.92ef	0.85g	1.16D
T4	1.83c	0.85fg	0.77hi	0.78ghi	1.06C
T5	2.32a	1.14d	1.11d	0.95e	1.38A
Mean	1.68A	.85C	.89B	0.79D	

LSD for Genotypes= 0.03, Treatments= 0.04 and Genotypes x Treatments= 0.07

Table 5. Leaf area index (LAI) of wheat genotypes under different silicon treatments

Treatments	Chakwal-50	GA-2002	NR-333	NR-372	Mean
T1	101.67h	98.00i	87.00k	96.00i	95.67E
T2	108.67de	103.00gh	91.00j	101.00h	100.92D
T3	114.00ab	107.33ef	97.67i	105.67fg	106.17C
T4	115.00a	111.67bc	101.00h	110.33cd	109.50B
T5	114.00ab	116.00a	105.00fg	115.00a	112.50A
Mean	110.67A	107.20B	96.33D	105.60C	

LSD for Genotypes= 1.23, Treatments= 1.37 and Genotypes x Treatments= 2.74

Table 6. Leaf area duration (LAD) of wheat genotypes under different silicon treatments

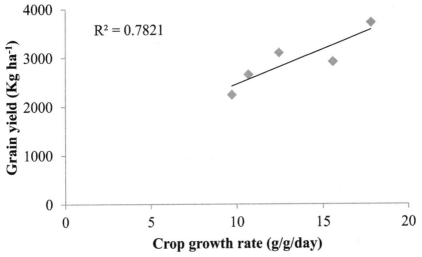

Fig. 1. Crop growth rate and grain yield of wheat genotypes under different silicon treatments

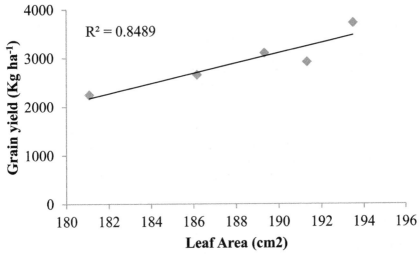

Fig. 2. Leaf area and grain yield of wheat genotypes under different silicon treatments

5. Effect of silicon on physiological parameters

5.1 Photosynthesis (A) (μ mole/m^2/second)

Photosynthesis is a determinant factor for crop growth and development as maximum photosynthesis contributes toward more yield and production. Results demonstrated significant variation for photosynthetic efficiency for various genotypes. Genotypes from diverse climatic regions behaved differently for photosynthetic efficiency. Chakwal-50 (17.47μ mole/m^2/second) performed well over all other genotypes for photosynthesis followed by NR-372 (15.33 μ mole/m^2/second) under 10 % silicon application with irrigation. While, minimum photosynthesis observed in NR-333 (8.23 μ mole/m^2/second) which ultimately led to reduced yield (Table 7). Chakwal-50 genotype has some adaptability characteristics which promoted physiological attributes and ultimately better production due to silicon application. Our results were in the line with the findings of Ahmed et al., (2011b) who reported significant impact of drought variation upon photosynthetic efficiency of wheat crop. Flag leaf stage is crucial stage in crop growth and development as crop produces maximum photosynthate using all available resources and it can only be achieved if suitable genotype sown at optimum time. Increased temperature and reduction in moisture promote photosynthetic efficiency up to an optimum. A linear relationship between photosynthetic rate and grain yield was observed (Fig. 3) which depicted that more photosynthesis led to maximum accumulation of photoassimilates from source to sink and ultimately maximum yield. Chakwal-50 genotypes performed very well due to efficient translocation of photoassimilates from source to sink using available thermal units under drought stress conditions.

5.2 Transpiration rate (E) (mole/m^2/s)

Transpiration is the removal of moisture from plant parts and it has significant impact on yield of crops and a major constituent to measure water use efficiency of agricultural crops. Crop yield is inversely related to transpiration rate. It depends upon climatic variants like solar radiation, temperature, water vapor pressure deficit, wind speed and the water status of the plants (Ahmed et al., 2011a). Present study depicted significant variation among different genotypes under changing silicon concentration which enhanced transpiration rate (Table 8). Resistant cultivars have defensive mechanism to cope with stress and gaining yield potential under limited available resources. The results of present study described maximum transpiration rate for GA-2002 under control and 10% application of silicon, while, minimum transpiration was showed by NR-333 with 5% application of silicon followed by NR-372 under the same treatment. This variation in transpiration might be due to genetic potential of genotypes and various silicon treatments impact. Stomatal aperture plays a vital role in leaves as water evaporates through them. Stomatal conductance and transpiration were positively correlated and stomatal closure led to reduce transpirational losses and ultimately good production. Our results were at par with the outcomes of Ahmed et al., (2011b).

5.3 Stomatal conductance (mol m^2 s^{-1})

Stomata plays important role in plant physiological indices, as water and nutrients enter in plant through stomata. Stomatal conductance is the speed of passage of water and nutrients through stomata. Optimum conductance led to maximum uptake of water and CO_2 for photosynthetic efficiency and ultimately more yield. Stomatal conductance depends upon favourable climatic conditions. Drought resistant genotypes have good conductance of stomata and adaptation strategies to cope stress. The results revealed maximum stomatal conductance for Chakwal-50 (0.42 mol m^2 s^{-1}) with 10% application of silicon as compared to other genotypes and silicon applications. Whereas, NR-333 showed minimum stomatal conductance (0.17 mol m^2 s^{-1}) under controlled conditions (Table 9). Silicon application boosted up physiological attributes under stressed conditions and resistant cultivars showed better response under enhanced silicon concentrations. A positive and linear relationship was observed between stomatal conductance and grain yield which support our results (Fig.4).

5.4 Leaf Membrane Stability Index (LMSI)

Results regarding leaf membrane stability index (Table 10) showed significant variation. It was observed that in control treatment, all the varieties showed little variation in LMSI but all varieties had greater membrane stability as compared to stress conditions. Chakwal-50 showed maximum value of LMSI (84.2%) followed by NR-333 (81.9%) under the 10% application of silicon. On the other hand NR-333 showed minimum value (60.33%) Perhaps this was the distinction which made some varieties able to perform better under water stress condition and others not. Similarly Plants having more LMSI value show less membrane injury by accumulating more saturated fatty acids, which is very important mechanism of plants to resist drought. So the selection of wheat varieties like Chakwal-50 with high LMSI value under drought condition is very essential to increase production.

5.5 Leaf succulence

The outcomes of the current study demonstrated that in controlled condition, Chakwal-50 (15.80 mg/m^2) showed maximum leaf succulence followed by NR-372 (14.70 mg/m^2) (Table 11). On the other hand NR-333 showed minimum value of 6.48 mg/m^2 preceded by GA-2002 (7.42 mg/m^2). It was noted that with increase in water stress, value of leaf succulence was also reduced. Different varieties showed different behaviors, Chakwal-50 and NR-372 showed maximum leaf succulence under enhanced silicon application followed by GA-2002 and NR-333. Some varieties showed less reduction in leaf succulence and some varieties showed much reduction. Here the distinction was developed among drought resistant and drought susceptible varieties. Leaf succulence is important adaptive mechanism of plants to resist drought. Varieties with high LS are considered to be more drought resistant. It has also been observed by many scientists that increased level of leaf succulence under drought conditions is a key adaptive mechanism to resist the drought. Ahmed et al., (2011a) also found that decreasing water potential cause reduction in leaf succulence value. So selection of wheat varieties, having more LS under water stress is very important for water deficit conditions.

5.6 Relative water content

The maximum relative water contents were recorded for drought tolerant genotype due to silicon nutrition. This elaborate that silicon entered inside the plants and might followed an active transport pathway and through xylem it reached inside the leaves in order to maintain water potential under water stress. The results of present study investigated that Chakwal-50 being drought resistant genotype, exhibited maximum relative water contents under enhanced silicon application (91.83%). While, minimum RWC was noted down for NR-372 (71.43) (Table 12). Control treatment (without silicon application) depicted less water contents as compared to silicon enhanced treatments. The findings of Ma et al., (2001) supported our results as they were of the opinion that silicon improved crop relative water potential.

5.7 Epicuticular wax

Epicuticular wax might be an important attribute in drought tolerant genotypes. Maximum epicuticular wax (8.62 mg) was observed in Chakwal-50 (Table 13). Whereas, minimum epicuticular wax was measured for NR-333 (4.36 mg). Drought resistant genotypes develop epicuticular wax on leaves which prohibited loss of water from plant leaves which then be used by plants under stress conditions. Silicon application played a crucial role in development of waxy layer on resistant varieties. The similar result reported by earlier researcher who documented that cuticle wax accumulation increase the drought tolerance in plants and silicon holds a vital place under such circumstances.

5.8 Chlorophyll contents (SPAD)

Crop growth could be related to rate of photosynthesis which is directly proportional to chlorophyll contents in leaves. Plants use chlorophyll to trap light from sun for photosynthesis and green colour of plants is due to absorption of all visible colours instead green by photosynthesizing pigments. In this experiment maximum chlorophyll contents were measured for Chakwal-50 (51.26) followed by GA-2002 (45.60), while, minmum for NR-372(21.33) under control conditions. Our findings were in close agreement with Paknejad et al., (2007) who found that chlorophyll contents in different wheat cultivars could be reduced more than 25% due to drought stress. A linear relationship was observed between chlorophyll contents and grain yield (Fig. 5) which described that increase in chlorophyll contents led to increased photosynthesis and consequently grain yield. So it can be concluded that selection of wheat varieties having more chlorophyll contents under drought stress is very important to increase production.

5.9 Drought resistance index for selected wheat cultivars

Drought resistance index is define as DC multiplied by variety's minimum yield, and then divided by average minimum yield of the varieties used in experiment. Cultivars showing greater value of DRI are considered to have more resistance against drought. However, the cultivars having less value of DRI were considered to be less resistant to drought. The findings of current study highlighted that Chakwal-50 showed maximum Drought Resistance Index (0.58) whereas GA-2002 showed minimum value (0.34) for drought resistance index. Our results were at par with the findings of previous researcher who calculated the DRI values for seven wheat cultivars grown under irrigated and non-irrigated conditions and found that cultivars having more DRI values were more resistant to drought.

5.10 Proline contents (μg g^{-1})

Proline, accumulates in plants under environmental stresses is proteinogenic amino acid with an exceptional conformational rigidity, and is essential for primary metabolism. It acts as a signal to triggers specific gene action which may be essential for crop recovery from stress.In the present study, proline contents differed significantly among wheat genotypes. Chakwal-50 being a drought resistance genotype, exhibited maximum proline accumulation (52.23 μg g^{-1}) followed by NR-372 (51.70μg g^{-1}). Whereas, minimum proline accumulation observed for GA-2002 (42.70 μg g^{-1}) (Table 16). Similar findings have been documented that proline contents reduced under increased temperature and moisture stress; however, it increased in resistant cultivars which led to higher yield. Grain yield of wheat found to be positively related to proline contents under different silicon regimes (Fig.6).

5.11 Correlation coefficients among physiological attributes with grain yield

Correlation coefficients among physiological attributes with grain yield of various wheat genotypes under silicon enhanced treatments showed variable response. Grain yield of wheat genotypes under stressed conditions and silicon application found to be positively correlated with photosynthesis (A), stomatal conductance (gs), chlorophyll contents (cc), proline contents (PC), relative water contents (RWC) and drought resistance index (DRI). However, significant negative correlation was observed between transpiration rate (E), leaf membrane stability (LMS) and grain yield. Significant and positive relationship led to conclusion that grain yield increase with increase in these physiological indices and application of silicon in this regard holds a vital place. Silicon concentration enhanced resistivity of some genotypes which had adaptability measures to cope stress.

Treatments	Chakwal-50	GA-2002	NR-333	NR-372	Mean
T1	11.74fgh	8.56jk	8.23k	9.20ijk	9.42E
T2	13.35def	10.34hij	10.53hi	11.23gh	11.36D
T3	14.34cd	12.49efg	12.5efg	13.47def	13.21C
T4	17.47a	15.15bcd	14.27cde	14.73bcd	14.99B
T5	15.79abc	16.23ab	14.77bcd	15.33bc	15.95A
Mean	14.54A	12.55B	12.06B	12.8B	

LSD for Genotypes= 0.81, Treatments= 0.90 and Genotypes x Treatments= 1.81

Table 7. Net photosynthesis (An) of wheat genotypes under different silicon treatments

Treatments	Chakwal-50	GA-2002	NR-333	NR-372	Mean
T1	5.95bc	6.67a	5.87bcde	5.73cdef	6.05A
T2	5.90bcd	6.13b	6.00bc	5.70def	5.93A
T3	5.90bcd	6.67a	5.73cdef	5.73cdef	6.00A
T4	5.87bcde	5.22gh	4.87i	5.23gh	5.30B
T5	5.47fg	5.60ef	5.30gh	5.17h	5.39B
Mean	5.82B	6.06A	5.55C	5.51C	

LSD for Genotypes= 0.12, Treatments= 0.13 and Genotypes x Treatments= 0.27

Table 8. Transpiration rate (E) of wheat genotypes under different silicon treatments

Treatments	Chakwal-50	GA-2002	NR-333	NR-372	Mean
T1	0.26i	0.24ij	0.37cd	0.24ij	0.28D
T2	0.32fg	0.26i	0.17k	0.28hi	0.26E
T3	0.35def	0.31gh	0.21j	0.33efg	0.30C
T4	0.38bcd	0.37cd	0.26i	0.36cde	0.34B
T5	0.42a	0.39abc	0.30gh	0.41ab	0.38A
Mean	0.35A	0.31B	0.26C	0.32B	

LSD for Genotypes= 0.02, Treatments= 0.02 and Genotypes x Treatments= 0.03

Table 9. Stomatal conductance (Gs) of wheat genotypes under different silicon treatments

Treatments	Chakwal-50	GA-2002	NR-333	NR-372	Mean
T1	80.57bc	76.7efg	80.67abc	60.80i	74.68B
T2	80.03bcde	75.67fgh	77.90c-g	61.27i	73.71BC
T3	80.17bcde	76.90defg	77.93c-g	63.00i	74.5B
T4	78.83b-f	74.80gh	74.83gh	60.37i	72.20C
T5	84.2a	80.33bcd	81.9ab	73.06h	79.88A
Mean	80.70A	76.80C	78.64B	63.70D	

LSD for Genotypes= 1.58, Treatments= 1.77 and Genotypes x Treatments= 3.54

Table 10. Leaf membrane stability index (LMSI) of wheat genotypes under different silicon treatments

Treatments	Chakwal-50	GA-2002	NR-333	NR-372	Mean
T1	15.80a	14.33abc	12.81b-f	14.70ab	14.41A
T2	15.40a	12.59cdef	10.40ghi	12.60cdef	12.74B
T3	14.36abc	11.39defg	9.20hij	12.90bcde	11.96B
T4	13.26bcd	9.24hij	8.53ij	10.80fgh	10.46C
T5	10.86efgh	7.42jk	6.48k	10.40ghi	8.79D
Mean	13.94A	10.99C	9.48D	12.28B	

LSD for Genotypes= 0.91, Treatments= 1.02 and Genotypes x Treatments= 2.03

Table 11. Leaf succulence of wheat genotypes under different silicon treatments

Treatments	Chakwal-50	GA-2002	NR-333	NR-372	Mean
T1	73.00hij	77.66k	74.76ghi	71.93jk	72.59E
T2	77.03g	72.66ijk	74.80ghi	71.50jk	74.00D
T3	81.66f	76.67g	74.96gh	71.43jk	76.18C
T4	85.20de	81.34f	82.96ef	87.76bc	84.31B
T5	91.83a	86.66cd	88.20bc	89.40b	89.05A
Mean	81.74A	77.60C	79.14B	78.40BC	

LSD for Genotypes= 1.02, Treatments= 1.14 and Genotypes x Treatments= 2.29

Table 12. Relative water content (RWC) of wheat genotypes under different silicon treatments

Treatments	Chakwal-50	GA-2002	NR-333	NR-372	Mean
T1	7.90abc	8.20abc	7.10cdef	8.13abc	7.83A
T2	7.83abc	7.60abcd	6.10fghi	8.16abc	7.42AB
T3	8.26ab	7.33bcde	5.30hij	6.33def	6.88B
T4	5.90ghi	6.36efgh	5.10ij	6.56defg	5.98C
T5	8.66a	5.80ghi	4.36j	5.53ghi	6.09C
Mean	7.71A	7.06B	7.00B	5.60C	

LSD for Genotypes= 0.51 Treatments= 0.57 and Genotypes x Treatments= 1.15

Table 13. Epicuticular wax (EW) of wheat genotypes under different silicon treatments

Treatments	Chakwal-50	GA-2002	NR-333	NR-372	Mean
T1	29.83hi	25.60kl	22.50m	21.33m	24.18E
T2	35.91de	29.67hi	26.63jk	23.10lm	28.82D
T3	45.06b	40.50c	34.13ef	33.06ef	38.19B
T4	39.30c	34.86ef	31.41gh	28.50ij	33.52C
T5	51.26a	45.60b	38.26cd	35.30ef	42.60A
Mean	40.27A	35.24B	30.59C	28.60D	

LSD for Genotypes=0.95 Treatments=0.10 and Genotypes x Treatments= 0.21

Table 14. Chlorophyll contents (cc) of wheat genotypes under different silicon treatments

Treatments	Chakwal-50	GA-2002	NR-333	NR-372	Mean
T1	0.46hi	0.36k	0.38k	0.37k	0.39D
T2	0.52cd	0.48fgh	0.43j	0.49efg	0.48C
T3	0.54bc	0.50def	0.45ij	0.52cde	0.50B
T4	0.5800a	0.525bcd	0.47ghi	0.54bc	0.53A
T5	0.55b	0.54bc	0.49fg	0.55b	0.53A
Mean	0.53A	0.48B	0.44C	0.49B	

LSD for Genotypes= 0.13 Treatments= 0.14 and Genotypes x Treatments= 0.24

Table 15. Drought Resistance Index (DRI) of wheat genotypes under different silicon treatments

Treatments	Chakwal-50	GA-2002	NR-333	NR-372	Mean
T1	50.16def	51.53bcd	50.50cde	51.23bc	51.05B
T2	48.23gh	49.56efg	41.83k	48.00h	46.90D
T3	51.20bcd	44.20i	43.96ij	49.43efgh	47.20CD
T4	48.90fgh	42.70jk	52.00b	48.00h	47.90C
T5	52.23a	51.50bcd	54.06a	51.70bc	52.12A
Mean	49.94A	47.90B	48.47B	49.82A	

LSD for Genotypes=0.66 Treatments= 0.74 and Genotypes x Treatments= 1.4886

Table 16. Proline contents (pc) of wheat genotypes under different silicon treatments

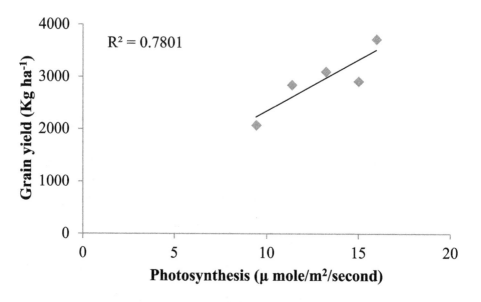

Fig. 3. Photosynthesis and grain yield of wheat genotypes under different silicon treatments

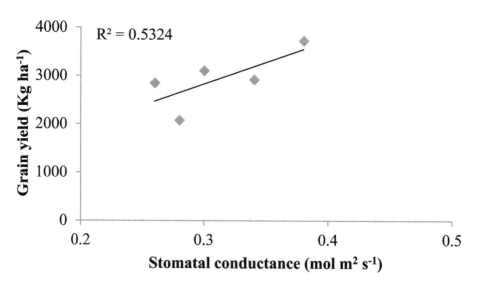

Fig. 4. Stomatal conductance and grain yield of wheat genotypes under different silicon treatments

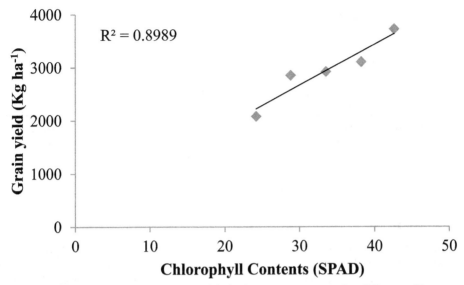

Fig. 5. Chlorophyll contents and grain yield of wheat genotypes under different silicon treatments

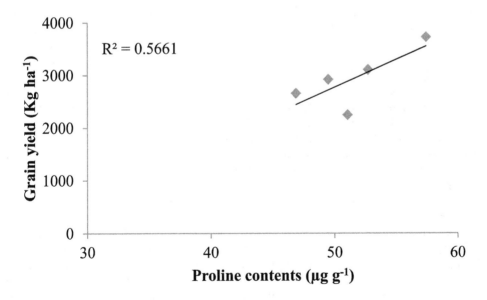

Fig. 6. Proline contents and grain yield of wheat genotypes under different silicon treatments

	GY	An	E	GS	CC	PC	RW	DRI
An	0.52***							
E	-0.28**	-0.34***						
GS	0.35***	0.61***	-0.23*					
CC	0.12ns	0.41***	0.39***	0.26**				
PC	0.28**	0.11ns	-0.20ns	0.29**	0.09ns			
RW	0.16ns	0.69***	-0.57***	0.58***	0.13ns	0.16ns		
DR	0.51***	0.81***	-0.22*	0.65***	0.46***	-0.02ns	0.54***	
LM	-0.25**	0.07ns	0.10ns	-0.04ns	0.43***	0.02ns	0.23*	-0.06ns

(GY= Grain yield, An= Net photosynthesis rate, E= Transpiration rate, GS= Stomatal conductance, PC= Proline contents (PC), RWC= Relative water content, DRI= drought resistance index (DRI) ***, **, * = Significant at 1 %, 5 % and 10 % respectively, ns = Non-significant)

Table 17. Correlation coefficients among physiological attributes of wheat genotypes with grain yield under different silicon applications

6. Conclusion

Drought is the major threat to agriculture in the world and Pakistan. Amongst different approaches being used to combat the drought stress, exogenous application of nutrients is much more important e.g. k+ the addition of Silicon in the growth medium is also beneficial to enhance the crop growth affected due to drought. Si is reported to accumulate in the plant body of various crops like rice that enables the plants to tolerate the drought stress. Likewise wheat an important serial in Pakistan has been designated as a Si accumulator. In this regard, the present study was undertaken in pots at PMAS, Arid Agriculture University Rawalpindi. Seeds of two varieties i.e. GA-2002, Chakwal-50 and two lines i.e. NR-333 and NR- 372 were taken from National Agricultural Research Center (NARC). The source of silicon, potassium silicate was used in the silicon applied treatments (+Si) and potassium chloride in the silicon deficient treatment (-Si). Effect of potassium silicate at 5 % and 10 % was investigated for germination and compared with control. Secondly two wheat cultivars and two advanced lines with three replications were sown per pot. At three leaf stage, 5 % and 10 % of potassium silicate solution were applied to the pots of the +Si treatments and it was compared with control. Potassium chloride solution was applied to the pots of the –Si treatments to yield the same total potassium as in the +Si treatments. Parameters like leaf membrane stability index, epicuticular wax, crop growth rate, relative water content, stomatal conductance, transpiration rate, photosynthetic rate, leaf area, leaf area index, chlorophyll contents, leaf succulence, relative leaf water contents, silicon concentration in leaves, proline contents, spikelets per spike, no. of grains per spike and weight of 100 grains were measured. The outcomes of the study highlighted maximum crop growth, physiological attributes and yield parameters for Chakwal-50 under 10 % silicon application with irrigation as compared to other genotypes under other levels of silicon concentrations. Drought resistant genotype showed responsive behaviour toward silicon and adaptability measures to cope stress condition. From all the discussion, it can be concluded that A single trait can not make a plant resistant to water stress, rather complex combinations of different traits make a plant able to survive in the drought conditions. Application of silicon in this regard would be beneficial for screening of drought resistant genotypes.

7. References

Agarie, S., Hanaoka, N., Ueno, O., Miyazaki, A., Kubota, F., Agata W. & Kaufman P.B. (1998). Effects of silicon on tolerance to water deficit and heat stress in rice plants (*Oryza sativa* L.) monitored by electrolyte leakage. *Plant Production Science*. 96–103.

Ahmed, M., F.U. Hassen, U. Qadeer & M.A. Aslam. (2011a). Silicon application and drought tolerance mechanism of sorghum. *African Journal of Agricultural Research*. 6(3): 594-607. Available from http://www.academicjournals.org/AJAR.

Ahmed, M., Hassan, F. U.., & Y. Khurshid. (2011b). Does silicon and irrigation have impact on drought tolerance mechanism of sorghum? *Agricultural Water Management*. 98(12): 1808-1812.

Al-aghabary, K., Zhu Z.J. & Q.H. Shi. (2004) Influence of silicon supply on chlorophyll content, chlorophyll fluorescence, and antioxidative enzyme activities in tomato plants under salt stress, *Journal of Plant Nutrition* 27, pp. 1–15.

Barrs, H. D. & P. E. Weatherley. (1962). Are-examination of relative turgidity technique for estimating water deficit in leaves. *Aust. J. Biol. Sci.*, 15: 413-428.

Bates, L.S., Waldren R.P., & I.D. Teare. (1973). Rapid determination of free proline for water stress studies, *Plant and Soil*, 39, 205-207.

Chandrasekar, V., Sarium R.k. & G.C.Srivastava. (2000). Physiology and biological response of hexapoloid and tetraploid wheat to stress. *J. Agron. Crop Sci.*, 18: 219-227.

Dagmar, D., Simone, H., Wolfgang, B., Rüdiger, F., Bäucker, E., Rühle, G., Otto, W. & M. Günter.(2003). Silica accumulation in Triticum aestivum L., and Dactylis glomerata L. *Analytical and Bioanalytical Chemistry*. 376(3): 399-404.

Epstein, E., (1999). Silicon. *Annl. Rev. Plant Physiol. Plant Mol. Biol.*, 50:641-664.

Gardner, F. P., Pearce, R. B. & R. L. Mitchell. (1985). Physiology of Crop Plants. 2nd ed. Lowa stat univ. press, *Ames, IA.*, 200-205.

Gong, H. J., Chen, K. M., Chen, G. C., Wang, S. M. & C. L. Zhang. (2003). Effect of silicon on growth of wheat under drought. *Journal of Plant Nutrition*. 26(5): 1055-1063.

Gong, H. J., Chen K. M., Chen G. C., Wang S. M. & C. L. Zhang. (2005). Silicon alleviates oxidative damage of wheat plants in pots under drought. *Plant.Sci.* 169: 313–321.

GOP. (2011). Economic Survey of Pakistan, 2010-11. Economic advisory wing, Finance Division, Islamabad.

Hattori, T., Inanaga, S., Tanimoto, E., Lux, A., Luxova M. & Y. Sugimoto. (2005). Silicon-induced changes in viscoelastic properties of sorghum root cell walls. *Plant Cell Physiol.*, 44: 743–749.

Hattori, T., Inanaga S., Araki H., An P., Mortia S., Luxova M. & A. Lux. (2005). Application of silicon enhanced drought tolerance in sorghum bicolor. *Physiolgia Plantarum*. 123: 459-466.

Liang, Y., Si, J., & V. Romheld. (2005). Silicon uptake and transport is an active process in *Cucumis sativus*. *New Phytologist*. 167: 797–804.

Long, S.P. & C.J. Bernacchi. (2003). Gas exchange measurements, what can they tell us about the underlying limitation to photosynthesis? Procedure and sources of error, *J. Exp. Bot.* 54: 2393-2401.

Lux, A., Luxova, M., Hattori, T., Inanaga, S., & Y., Sugimoto. (2002). Silicification in sorghum (*Sorghum bicolor*) cultivars with different drought tolerance. *Physiol. Plant*, 115: 87–92.

Ma, J., Nishimura K. & E. Takahashi. (1989). Effect of silicon on the growth of rice plant at different growth stages. *Soil Sci. Plant Nutr.*, 35: 347-356.

Ma, J.F., Miyake Y. & E. Takahashi. (2001). Silicon as a beneficial element for crop plants. Elsevier Science, Amsterdam, pp. 17-39.

Paknejad, F., Nasir M., Moghadam H. R. T., Zahidi H.,& M. J. Alahmadi. (2007). Effect of drought stress on cholorophyll fluorescense parameters, chlorophyll contents and grain yield of wheat cultivars. Dept. of Agric. Islamic Azad Uni. Karaj Branch Iran. *J. of Biol. Sci.* 7(6): 841-847.

Power, J.F., Willis W.O., Grunes D.L. & G.A. Riechman. (1967). Effect of soil, temperature, phosphorous and plant age on growth analysis of barley. *Agron. J.,* 59: 231-234.

Rodrigues, F.Á., McNally, D.J., Datnoff, L.E., Jones, J. B., Labbe, C., Benhamou, N., Menzies, J.G., & R.R. Bélanger. (2004). Silicon enhances the accumulation of diterpenoid phytoalexins in rice: A potential mechanism for blast resistance. *Phytopathology.* 94: 177-183.

Silva Fernandus, A.M., Baker E.A. & J.T. Martin. (1964). Studies of plant cuticle vi. The isolation and fractionation of cuticular waxes. *Ann. Appl. Biol.,* 53: 43-58.

5

Boron Deficiency in Soils and Crops: A Review

Waqar Ahmad[1], Munir H. Zia[2],
Sukhdev S. Malhi[3], Abid Niaz[4] and Saifullah[5,6]
[1]Faculty of Agriculture, Food, and Natural Resources, The University of Sydney,
[2]Research & Development Section, Fauji Fertilizer Company Ltd, Rawalpindi,
[3]Agriculture and Agri-Food Canada, Melfort, Saskatchewan,
[4]Soil Chemistry Section, Institute of Soil Chemistry & Environmental Sciences,
Ayub Agricultural Research Institute, Faisalabad,
[5]Institute of Soil and Environmental Sciences, University of Agriculture Faisalabad,
[6]School of Earth and Environment, Faculty of Natural and Agricultural Sciences,
The University of Western Australia, Crawley, Perth,
[1,6]Australia
[2,4,5]Pakistan
[3]Canada

1. Introduction

Boron (B) is a unique non-metal micronutrient required for normal growth and development of plants. In 1923, it was first time reported that B is essential for cell structure of plants (Warington, 1923). The possible roles of B include sugar transport, cell wall synthesis, lignification, cell wall structure integrity, carbohydrate metabolism, ribose nucleic acid (RNA) metabolism, respiration, indole acetic acid (IAA) metabolism, phenol metabolism, and as part of the cell membranes (Parr & Loughman, 1983; Welch, 1995; Ahmad et al., 2009). In soils, concentration of total B is reported to be in the range of 20 to 200 mg B kg^{-1} (Mengel & Kirkby, 1987), and its available concentrations also vary greatly from soil to soil.

Boron is absorbed by roots as undissociated boric acid [B (OH)$_3$ or H$_3$BO$_3$] (Mengel & Kirkby, 1982; Marschner, 1995) which has a strong ability to form complexes with diols and polyols, particularly with cis-diols inside the plant system (Loomis & Durst, 1992). Among the elements required by plants that are taken up from the soil, B is the only element that is taken up by plants not as an ion, but as an uncharged molecule (Marschner, 1995; Miwa & Fujiwara, 2010). The factors affecting B uptake include soil type (texture, alkalinity/calcareousness, pH, organic matter content), B concentration, moisture, and plant species (Welch et al., 1991). Boron absorption by plant roots is closely related to pH and B concentration in the soil solution; and is probably a non-metabolic process (Brown & Hu, 1998). The supplying mechanism of B to plant roots is primarily through mass flow, while its distribution in plants is governed by the transpiration stream through the xylem (Raven, 1980). Boron is relatively immobile in plant, and thus its availability is essential at all stages of growth, especially during fruit/seed development. However, recent physiological studies have revealed the presence of channel-mediated facilitated diffusion and energy-dependent active transport against concentration gradients in B transport systems (Dannel et al., 2000, 2001; Stangoulis et al., 2001).

Boron deficiency is one of the major constraints to crop production (Sillanpaa, 1982), and has been reported in more than 80 countries and for 132 crops over the last 60 years (Shorrocks, 1997). Boron deficiency has been realized as the second most important micronutrient constraint in crops after that of zinc (Zn) on global scale. Boron deficiency has been reported to result considerable yield reduction in annual [fiber (cotton - *Gossypium hirsutum* L.), cereal (rice – *Oryza sativa* L., maize/corn *Zea mays* L.- , wheat – *Triticum aestivum* L.), legume/pulse (soybean – *Glycine max* L.), oilseed (groundnut – *Arachis hypogaea* L., oilseed rape/canola – *Brassica napus or B. rapa* L.)] and perennial [citrus fruit orchards, alfalfa – *Medicago sativa* L.] crops (Arora et al., 1985; Patil et al., 1987; Sakal et al., 1988; Ali & Monoranjan, 1989; Takkar et al., 1989; Dwivedi et al., 1990; Sinha et al., 1991; Borkakati & Takkar, 2000; Niaz et al., 2002, 2007; Rashid et al., 2005; Johnson, 2006; Zia et al., 2006). Rashid (2006) estimated a substantial potential net economic benefit from the use of B fertilizers in B-deficient crops.

Boron bioavailability decreases under drought condition because of reduced mobility of B from soil by mass flow to roots (Chiu & Chang, 1985; Chang et al., 1992; Chang, 1993; Barber, 1995). Boron can move relatively long distances by mass flow and diffusion to roots. Soil drying reduces B diffusion by reducing the mobility of soil solution and increasing the diffusion path length (Scott et al., 1975). The lack of moisture in soil reduces transpiration rate, thereby reducing B transport to shoots (Lovatt, 1985). Wetting and drying cycles and increasing soil temperature (25 to 45 °C) also increased B fixation by montmorillonite and kaolinite clays (Biggar & Fireman, 1960). Low temperature in spring and autumn season of temperate regions reduced availability of B to forage legumes while increased temperature enhanced B concentration for sugarcane (Gupta, 1993).

Boron deficiency has been commonly reported in soils which are highly leached and/or developed from calcareous, alluvial and loessial deposits (Takkar et al., 1989; Razzaq & Rafiq, 1996; Borkakati & Takkar, 2000). Several soil factors and conditions render soils deficient in B. For example, low soil organic matter content, coarse/sandy texture, high pH, liming, drought, intensive cultivation and more nutrient uptake than application, and the use of fertilizers poor in micronutrients are considered to be the major factors associated with the occurrence of B deficiency (Dregne & Powers, 1942; Elrashidi & O'Connor, 1982; Takkar et al., 1989; Goldberg & Forster, 1991; Rahmatullah et al., 1999; Eguchi & Yamada, 1997; Rashid et al., 1997, 2005; Mengel & Kirkby, 2001; Niaz et al., 2002, 2007; Rashid & Rayan, 2004). This paper reviews the roles of B in plant nutrition and the factors affecting B availability in soils in general, while focusing on a number of case studies related to diagnosis and correction of B deficiency in soils and crops.

2. Factors affecting boron availability in soils

Boron concentrations in soil vary from 2 to 200 mg B kg⁻¹, but generally less than 5-10% is in a form available to plants (Diana, 2006). Boron concentration and its bioavailability in soils is affected by several factors including parent material, texture, nature of clay minerals, pH, liming, organic matter content, sources of irrigation, interrelationship with other elements,   and environmental conditions like moderate to heavy rainfall, dry weather and high light intensity (Moraghan & Mascagni, 1991). Therefore, knowledge of these factors affecting B uptake is essential for the assessment of B deficiency and toxicity under different conditions.

Upon mineralization from organic matter or B addition to soils through irrigation or fertilization, a proportion of it remains in the soil solution while left of it is adsorbed by soil particles and other soil constituents. Tourmaline is a mineral which contains B in a very insoluble form while hydrated B minerals are the most soluble form of B minerals. These minerals do not usually determine the solubility of B in the soil solution (Goldberg, 1993), which is governed by B adsorption reactions mainly. The equilibrium exists between the soil solution and adsorbed B (Russell, 1973). Plants obtain B from the soil solution (Hatcher et al., 1959), and buffering against abrupt changes in the level of B in the soil solution is controlled by the adsorbed pool of B (Hatcher et al., 1962). Therefore, it is important to know the distribution of B between the solid and the liquid phases of the soil. Factors affecting the amount of B adsorbed by soils and the B bioavailability in soils include soil pH, texture, moisture, temperature, and management practices such as liming (Evans & Sparks, 1983).

2.1 Parent material

Parent material is considered a dominant factor affecting supply of B from the soil. Soils are quite variable in their B and clay forming minerals contents, and therefore have a fundamental effect on the availability of B. In general, soils derived from igneous rocks, and those in tropical and temperate regions of the world, have much lower B concentrations than soils derived from sedimentary rocks, and those in arid or semi arid regions (Ho, 2000). High B concentrations are usually found in the soils that have been formed from marine shale enriched parent material. Soils derived from acid granite and other igneous rocks, fresh-water sedimentary deposits, and in coarse textured soils low in organic matter have been reported with low B concentrations (Liu et al., 1983). Boron bioavailability is also reduced in soils derived from volcanic ash (Sillanpaa & Vlek, 1985) and in soils rich in aluminum (Al) oxides (Bingham et al., 1971). Soils along the sea shore as well as those derived from mudstone are usually B enriched. Conversely, lateritic soils, and soils derived from sandstone, slate or crystalline limestone do not contain much B. The levels of total B in common rocks are presented in Table 1.

Rock type	Minerals	B (mg B kg^{-1})^z	B (mg B kg^{-1})^y
Igneous	Basic: gabbro, basalt	5-20	5
	Intermediate: diorite	9-25	15
	Acid: granite, rhyolite	10-30	
Metamorphic	Gneiss	10-30	-
Sedimentary	Shale	120-130	100
	Sandstone	30	35
	Limestone dolomite	20-30	20

[z]Kabata-Pendias & Pendias, 1992; [y]Sillanpaa & Vlek, 1985.

Table 1. Total B concentrations in major rock types

2.2 Soil reaction (pH)

Soil pH is one of the most important factors affecting the availability of B in soils. Its bioavailability becomes less at the higher solution pH. Consequently, application of lime to acid soils, in excessive amounts, can sometimes render plants deficient in B. There is a close association with the pH of the soil solution and the level of soluble B in soils (Elrashidi & O'Connor, 1982; Takkar et al., 1989; Rashid et al., 1994; Niaz et al., 2002, 2007). Boron uptake by plants growing in soil, with the same water soluble B concentration, was noticed to be higher where pH of the soil solution was lower (Wear & Patterson, 1962). The adsorption of B by soils is much dependent on pH of the soil solution. Boron adsorption by soils increased when the pH rose from 3 to 9 (Bingham et al., 1971; Mezuman & Keren, 1981; Keren & Bingham, 1985; Barrow, 1989), and it decreased when the pH was increased further in the range 10 to 11.5 (Goldberg & Glaubig, 1986). In several studies, highest levels of B adsorption by soil depicted close correlation with the pH of the soil solution (Okazaki & Chao, 1968; Evans, 1987; Shafiq et al., 2008).

2.3 Soil texture and clay minerals

Coarse-textured soils often contain less available B than fine-textured soils (Takkar et al., 1989; Raza et al., 2002; Malhi et al., 2003). This might be one of the reasons that B deficiencies in crop plants have often been observed on sandy soils (Gupta, 1968; Fleming, 1980). Niaz et al. (2002) concluded from a study in Punjab, Pakistan that B concentrations of coarse- and medium-textured soils and plants grown in such soils were lower than their respective critical levels, because these soils were well drained and had good leaching. Besides aluminum and iron oxides, calcium carbonate and organic matter, clay minerals are considered to be amongst the primary B adsorbing surfaces in soils (Goldberg, 1997). The mechanism of B adsorption on these surfaces is considered to be ligand exchange with reactive surface hydroxyl groups leading to strong specific adsorption (Goldberg & Chunming, 2007). Boron adsorption in fine-textured soils is higher compared with the coarse- and medium-textured soils at the same equilibrium concentration (Table 2). The level of native B is also closely related to the clay content of the soil (Elrashidi & O'Connor, 1982; Raza et al., 2002). At the same time, water soluble B concentration and B uptake are reported to be higher in plants grown in coarse-textured soils (Wear & Patterson, 1962). The level of B adsorbed by the soil thus largely depends on soil texture in addition to pH of soil solution. It increases with an increase in clay content (Bhatnager et al., 1979; Wild & Mazaheri, 1979; Mezuman & Keren, 1981; Elrashidi & O'Connor, 1982).

More B adsorption is commonly found in illite as compared with kaolinite or montmorillonite clay types. . In fact, kaolinite adsorbs B the least (Hingston, 1964; Fleet, 1965). Frederickson & Reynolds (1959) proposed that most of the B in the clay mineral fraction of sedimentary rocks is contained in the illite fraction. Sims & Bingham (1967, 1968a, 1968b) found that B adsorption was greater for iron (Fe) and Al coated kaolinite or montmorillonite than for uncoated clays. It was concluded that hydroxyl of Fe and Al compounds present in the layer as silicates or as impurities dominate over clay mineral species per se in determining B adsorption characteristics. Bingham et al. (1971) and Schalscha et al. (1973) also inferred that B adsorption by certain soils was primarily due to their Al oxide content.

Sand dune		Sandy loam		Black clay	
Equilibrium conc.	B adsorbed	Equilibrium conc.	B adsorbed	Equilibrium conc.	B adsorbed
3.8	1.2	3.0	2.0	1.2	3.8
7.8	2.2	6.4	3.6	3.8	6.2
16.5	3.5	14.8	5.2	8.5	11.5
25.0	5.0	24.0	6.0	14.0	21.0
34.0	6.0	33.0	7.0	14.0	21.0
43.0	7.0	42.2	8.0	25.0	25.0

[z]Source: Gupta, 1979a.

Table 2. Boron adsorption (mg B kg $^{-1}$) in soils as affected by texture[z]

2.4 Organic matter

Organic matter (OM) is the storehouse for most nutrients in soil and is known to improve soil health and availability of plant nutrients. Many researchers have suggested that the level of soil organic matter (SOM) influences the nutrient bioavailability (Sarwar & Mubeen, 2009). Boron may bind with OM or with carbohydrates released during humification. Boron associated with humic colloids is the principal B pool for plant growth in most of the agricultural soils (Jones, 2003). However, there is limited information on the role of OM in B nutrition. The strongest evidence that OM affects the availability of soil B is derived from studies that show a positive correlation between levels of SOM and the amount of hot-water-soluble B (Kao & Juang, 1978; Chang et al., 1983; Takkar et al., 1989; Niaz et al., 2002; 2007; Raza et al., 2002; Shafiq et al., 2008).

The association between B and SOM is said to be caused by the assimilation of B by soil microbes (Gupta et al., 1985). Albeit, B present in SOM is not immediately available to plants, it seems to be a major source of available B when it is released through mineralization (Gupta et al., 1985). It is well documented, that the interaction of dissolved organic matter (DOM) with soil is affected by the presence of OM and hydroxides in the clay fraction particles (McDowell & Likens, 1988; Jardine et al., 1989; Donald et al., 1993). The role of DOM to affect B availability/adsorption has already been reported by Mackin (1986) from pore waters of marine sediments and recently by Communar & Keren (2008) for soil-plant system. Further the B solution concentration assessment may be driven through interaction of effluent DOM with native soil OM, B complexation with DOM, and adsorption of B and B–DOM complexes by soil. Correlations of total dissolved boron (TDB) and ratios of B to chloride with DOM, in organic-rich sediments, predispose that the fraction of dissolved boron (DB) that is complexed by OM is a function of dissolved organic matter concentration (Mackin, 1986). It can be inferred that DB concentrations equilibrium is highly related with organic-B complexes. Further, such potentially useful approximation should also be applied for determining concentrations of organic-B complexes in marine waters and sediments. Both deep understanding of the mechanisms of these relations and parameterization according to the local conditions permit to improve the model for B transport in soil (Communar et al., 2004; Communar & Keren, 2005, 2006). But all these investigations call for more extensive research on role of DOM pertaining to B desorption/release.

2.5 Sources of irrigation water

There are two common sources of water to irrigate crops, i.e., canal water and tube well water. The soil B status, and its availability and toxicity to plants also depend on the source of irrigation water. Underground water used for irrigation purpose has been reported to contain toxic amounts of B in many parts (Uttar Pradesh, Rajasthan, Haryana, Punjab, and Gujrat) (Chauhan & Asthana, 1981) of India. This toxicity reduces growth, particularly of shoots, and causes chlorosis starting at the leaf tip and margins of mature leaves (Nable et al., 1997; Reid et al., 2004; Reid & Fitzpatrick, 2009). Similarly, underground water for irrigation in the western desert of Egypt was also shown to be high in B (Elseewi, 1974). Boron toxicity has been reported in many crops irrigated with high-B water in Spain (Salinas et al., 1981), Arizona (Ryan et al., 1977), northern Greece (Sotiropoulos, 1997) and Philippines (Dobermann & Fairhurst, 2000). Ahmad et al. (2004) conducted a survey to determine the B concentrations in canal and ground waters used for irrigation in different villages of Faisalabad (Pakistan). The results showed that B in the tube well waters collected during February-March ranged from 0.14-0.65 mg B L^{-1} [standard deviation (SD) = 0.16] with a mean of 0.38 mg B L^{-1}, and those collected in July-August ranged from 0.52 to 0.66 mg B L^{-1} (SD = 0.28), with a mean 0.56 mg B L^{-1}. Boron in river water samples collected during February-March ranged from 0.11 to 0.43 mg B L^{-1} (SD = 0.10), with a mean of 0.21 mg B L^{-1}. The authors concluded that tube well waters contain higher B concentrations compared to the canal waters, so farmers should get their water samples analyzed prior to irrigation and should consider these B concentrations in order to adjust B fertilizer doses to crops. This suggests that farmers using these waters for irrigation of their crops should pay attention to this potential source of B availability. This recommendation is of prime importance, because B is the unique element in the sense that there is a very narrow range between its deficient and toxic levels (< 0.5 mg B kg^{-1} and > 5 mg B kg^{-1}, respectively). Farmers can calculate the amounts of B being added to their fields through irrigations of canal and tube well water. These results are also in line with those of Sillanpaa (1982) and Keren & Bingham (1985).

2.6 Interactions of boron with other nutrients

Some functions of B interrelate with those of nitrogen (N), phosphorus (P), potassium (K) and calcium (Ca) in plants (US Borax, 2009). Its interaction (synergistic, antagonistic) with most of the nutrients (N, P, K, Ca, Mg [magnesium] Al [aluminum] and Zn) may be sometimes influential in regulating B availability to plants in soil. Application of B may improve the utilization of applied N in cotton plants by increasing the translocation of N compounds into the boll (Miley et al., 1969). Smithson and Heathcote (1976) found that when B deficiency occurred in cotton, the application of 250 kg N ha^{-1} reduced the yield. However, when B was applied, crop biomass escalated with the same dose of N.

Graham et al. (1987) found that B uptake by barley (*Hordeum vulgare* L.) was lower when Zn was applied compared to in its absence. Further, they also showed that rate of B accumulation in plants is increased even at low levels of Zn and high levels of P. Therefore, Zn fertilization may reduce B accumulation, and lessen the risk of toxicity in plants (Ahmed et al., 2008). A significant relationship has been found between K and B fertilizers regarding their assimilation/uptake by crop plants as well as crop produce (Hill & Morrill, 1975). At heavy applications of K and other intensive production practices B may need to be applied to prevent reduction in corn yield (Woodruff et al., 1987). , Yang & Gu (2004) studied the effect of B on Al toxicity on seedlings of two soybean cultivars. The results showed that high B was found to ameliorate Al toxicity by significantly increasing the growth characters including root length under 2 mM Al stress, and epicotyl length and fresh weight under 5 mM Al stress of the two cultivars. Similar kind of study was conducted by Hossain and Hossain (2004) which confirmed the relationship of B with Al. The ratio between Ca and B in the plant is sometimes used to identify B deficiency. In a recent study, application of the both Ca and B to four cultivars of maize significantly enhanced shoot dry matter production (Kanwal et al., 2008). Nevertheless, B concentration in the shoot of maize cultivars was antagonized with Ca application. A curvilinear relation was exhibited between Ca/B ratio in shoot and relative shoot dry matter. In this regard, further work is warranted on Ca/B utilizing association for ameliorating B deficient calcareous soils (Rashid et al., 1997).

3. Sensitivity of crop species/cultivars to boron deficiency

Crop species differ in their capacity to take up B, even when they are grown in the same growth medium. These differences generally reflect different B requirements for growth. In general, dicots (cotton and leguminous plants) have 4-7 times higher B requirement (20-70 mg B kg^{-1}) than monocots (graminae family), 5-10 mg B kg^{-1} (Bergmann, 1988, 1992; Marschner, 1995). As the most important functions of B in plants are thought to be its structural role in cell wall development and stimulation or inhibition of specific metabolism pathways (Gupta, 1979a, 1979b, 1993; Ahmad et al., 2009), thereby, differences in the B demand of graminaceous and dicotyledonous species are probably related to the differences in their cell wall composition, and cis-diol configuration in the cell walls, such as pectic substances. A meager amount of pectic material is constituted in the cell walls of graminaceous species (wheat and rice). Such species also have much lower Ca requirements. In fact, these two plant categories also differ in their capacity for silicon (Si) uptake, which is usually inversely related to B and Ca requirements (Loomis & Durst, 1992). All the three elements are located mainly in the cell walls. Brown & Shelp (1997) and Brown & Hu (1998) concluded that knowledge of the relative mobility of B within a particular species determines the optimum fertilization strategy and the same can also be used in partial understanding of the causes and consequences of B deficiency.

In summary, B deficiency is commonly induced under the following soil conditions; 1) soils which are inherently low in B, such as those derived from the parent material made from acid granite and other igneous rocks, and freshwater sedimentary deposits, 2) leaching impacted naturally acid soils from which native B has been removed, 3) light-textured sandy soils and gravelly soils, 4) alkaline and calcareous soils, 5) irrigated soils having low B concentration in irrigation water, and where salt or carbonate has been deposited, and 6) soils low in OM.

4. Diagnosis and correction of boron deficiency

4.1 Identification of boron deficiency

Boron is very vulnerable to leaching, so its deficiency can temporarily be expected in countries like Pakistan and India during and after monsoon rains, especially in coarse-textured soils. However, its major source mineral (i.e., tourmaline) is highly insoluble. In Pakistan, Alfisols appear to be the soil group most likely to produce B-deficient crops (Zia et al., 2006). Singh (2001) explored that out of 36,825 soil samples collected throughout India, 33% were deficient in B. In India, laterite and lateritic soils (Ferralsols and Dystric Nitisols) have been widely reported for the deficiency of B. Boron deficiencies are also more pronounced during drought periods when root activity is restricted. Once B has accumulated in a particular organ, it has restricted mobility in most plant species but not all. Boron is immobile in plants, so its deficiency symptoms develop firstly, and are more severe, on young leaves with marginal, dull yellow chlorosis at the tip of young leaves. Because B plays an important role in the elongation of stems and leaves, stems of B deficient plants are short and stout. If B deficiency is severe, many tillers can die before maturity, or whole plant may die before producing heads. Boron deficiency also manifests itself in poorly developed stamens, blast of pear blossoms, inadequate fruit set, bark necrosis of apple, corking in the fruit, and cracking of fruit. When leaf B levels are in the range of 20 to 25 mg B kg^{-1} (desired is 35 mg B kg^{-1}) on a dry-weight basis, supplemental B is needed. Boron is taken up from the soil only at higher soil temperatures than are other elements. Most values of the critical concentration for B deficiency range from 0.15 to 0.50 mg kg^{-1} soil (HWE – hot water extractable). However, in highly sensitive crops and alkaline clay soils, these values can double. This is because, B sorption increases to a maximum between pH 7.5-9.5. Hence, the critical range of extractable B is generally higher in alkaline soils. For example Bell (1997) reported that for wheat grown on alkaline clay soils in northern China, a critical range of 0.32-0.38 mg B kg^{-1} (HWE) was proposed, whereas on loams in northern Thailand the figure was 0.12-0.15 mg B kg^{-1}. The critical concentration of B (HWE) in soils which is considered deficient to most crops in Pakistan was 0.45-0.50 mg B kg^{-1} until revised recently to 0.65 mg B kg^{-1} (Rashid et al., 1994; Rashid, 2006). Singh (1994) concluded that depending upon groundnut genotypes and soil, the critical limits of B may vary from 0.2-0.4 mg B kg^{-1}.

4.2 Sources, rates, methods and timing of boron application

There are eight different sources of B [borax ($Na_2B_4O_7.10H_2O$ with 11% B), solubor - $Na_2B_8O_{13}.4H_2O$ (20% B), sodium borate ($Na_2B_4O_7.5H_2O$ with 20% B), sodium tetraborate ($Na_2B_4O_2.5H_2O$ with 14% B), boric acid (H_3BO_3 with 17% B), Colemanite ($Ca_2B_6O_{11}.5H_2O$ with 10% B), B frits containing 2-6% B, and boronated superphosphate being used to prevent/correct B deficiency in crops. Borax, solubor, sodium borate and sodium tetraborate have been most commonly used for soil application. Boric acid, colemanite and B frits are considered to be more promising on highly leached sandy soils as well as for long duration field crops including perennial forages and fruit plants owing to their low solubility and slow release of B. Boronated superphosphate has also been tried to correct B deficiency in crops (Patil et al., 1987).

Among these B fertilizer sources, borax is the most commonly used B fertilizer to prevent and/or correct B deficiencies in crops. Because of the narrow margin between B sufficiency and toxicity, an excess dose can easily occur and harm plant growth (Gupta, 1972; Marschner, 1995). Therefore, extreme care is needed to apply the correct dose of B fertilizer and to distribute it uniformly. Boron application rates generally range from 0.25 to 3.0 kg B ha^{-1}, depending on crop requirement and the method of application (Arora et al., 1985; Nuttall et al., 1987; Patil et al., 1987; Sakal et al., 1988; Ali & Monoranjan, 1989; Dwivedi et al., 1990; Sinha et al., 1991; Mortvedt & Woodruff, 1993). Higher rates are required for broadcast applications than for banded soil applications or foliar sprays. Because B is immobile in plants, B deficiency in crops growing in soils with marginal B levels can occur during peak growing periods (vegetative, flowering, and seed development stages), so a steady supply of B throughout the growing season is essential for optimum growth and seed yield. Foliar fertilization is also an effective way to supply B to plants, especially when root activity is restricted and B deficiency in crop appears under dry soil conditions in the growing season (Mortvedt, 1994). Experiments regarding the effect of B on yield, mobility and stress tolerance in different crop species revealed that B significantly enhanced yield and it was attributed to the significant increase in the panicle fertility. In extreme cases, crops on low B soils grow well until flowering when floral abortion or seed set failure can result in severe yield losses. Boron deficiency at critical stages of reproductive development has been shown to cause pod abortion with poor seed setting in wheat in Western Australia (Wong, 2003). Boron application at the onset of reproductive phase was found to be more effective, most likely due to its immobile nature in the plants depending upon the photosynthetic efficiency of the plants (Anonymous, 2007a). These findings are in agreement with the recent work of Ahmad et al. (2009).

Relatively small amounts of B that are normally required to make significant improvements in B status of annual crops, namely 1–2 kg B ha[-1], are in broad accord with such recovery rates. For many crops, absorption of 100–200 g B ha[-1] of applied B could be expected to be sufficient (Shorrocks, 1997). Application of 10 kg of boric acid ha[-1] (1.7 kg B ha[-1]) or 18 kg of borax ha[-1] (2.0 kg B ha[-1]) proved to be effective for 4-5 years in order to cure B deficiency in rice, wheat and cotton soils.. It was found that in case of cotton, 0.1% solution of B would be economical if used with insecticides foliar sprays. Value cost ratios (VCR) for B use in these crops have been very good , particularly in the case of cotton, where it ranged from 5:1 to 13:1 by soil application and 20:1 by foliar application of B. It was revealed that application of B significantly boosted rice yield, mainly because of increase in the panicle fertility (Anonymous, 2007b). Application of B fertilizers up to 2.5 kg B ha[-1] is recommended for major crops like cotton, rice and wheat in Pakistan (Anonymous, 1998). Boron may safely be applied to orchard crops at a rate of 0.56 kg B ha[-1] as a maintenance dose and at a rate of 1.12 kg B ha[-1] as a deficiency dose (Zia et al., 2006) and its residual effect has generally been reported for at least two years. In the case of borax, application rates should not exceed 90 g borax per orchard tree (Zia et al., 2006).

In India, soil application of B at 20 kg sodium tetraborate to supply 2.8 kg B ha[-1] as well as two foliar sprays with 0.2% solution of this salt proved equally effective in increasing soybean grain yield and the residual effect of soil applied B on subsequent wheat crop was significantly higher as compared with direct foliar B application (Table 3). Since B undergoes less leaching in fine-textured soils, single application may produce residual effect. In view of very sharp and narrow difference between optimum and toxic levels of B, more precaution is needed in its repeat application, particularly in medium- to fine-textured soils. Boron deficiency is also invariably corrected by its soil application depending upon soil type (Arora et al., 1985; Sakal et al., 1988; Ali & Monoranjan, 1989). In calcareous soils of Bihar, the rate varying between 1.0 to 2.5 kg B ha[-1] has been found to be optimum for different crops (Sakal et al., 1988; Sinha et al., 1991).

Treatment	Grain yield (Mg ha[-1])	
	Soybean	Wheat
Soil Application (20 kg sodium tetraborate ha[-1])	1.41	1.46
Foliar application (0.2% sodium tetraborate soil)	1.33	1.35
Control	0.89	0.66
LSD (p=0.05)	0.1	0.11

[z]Source: Dwivedi et al., 1990.

Table 3. Effect of mode of B application on grain yield of soybean and wheat[z]

5. Yield response of selected crops to boron fertilization

5.1 Cotton

Cotton (*Gossypium hirsutum* L.) is an important fiber crop grown in many countries of the world. There are several factors responsible for low yields of cotton, and micronutrient deficiency is one of them. Boron has been recognized as the most important micronutrient for cotton production in some countries. Its deficiency inhibits petiole and peduncle cell development and reduces growth of cotton (De Oliveira et al., 2006). In a number of studies, application of B fertilizer has been shown to increase cotton yield (Murphy & Lancaster, 1971; Rashid, 1995, 1996; Anonymous, 1998). Research has shown that as little as 1.1 kg of B ha[-1] can increase cotton seed yield by more than 560 kg ha[-1] (US Borax, 2002). In Pakistan, 50% cotton fields have been reported to be deficient in B (Anonymous, 1998). Cotton is very responsive to B fertilization on B deficient soils. For example, in a study in Pakistan with 30 field experiments, B application has been reported to increase cotton yield in the range of 2 to 30%, with an average value of 14%, over the zero-B control (Malik et al., 1992; Rashid, 1995, 1996; Anonymous, 1998). The value cost ratio (VCR) data indicated that by spending one rupee on B fertilizer, crop yield increase was worth Rs. 5 to 20 (average Rs. 16) in cotton (Anonymous, 1998). Use of B and Zn fertilizers proved highly profitable, benefit cost ratio being 15:1 for soil application and 30:1 for foliar spray (Rashid & Akhtar, 2006). Niaz et al. (2002) conducted field experiments on cotton at 13 different sites in Punjab, Pakistan; five were medium-textured (clay loam), two were silty clay, one was loam, and five were coarse-textured (sandy loam or loamy sand). Of the 13 soils, 12 were found deficient in B (less than 0.5 mg B kg[-1] 0.05M HCl extractable). Boron concentration in younger leaves, at flowering stage and harvest, ranged from 7.8 to 23.8 mg B kg[-1] with an average of 11.4 mg B kg[-1], whereas only one of the 13 samples had adequate B concentration (15 mg B kg[-1]). Similar results have also been reported from Australia (Reuter & Robinson, 1986; Shorrocks, 1997), Egypt (Ibrahim et al., 2009), Turkey (Gormus, 2005) and USA (Zhao & Oosterhuis, 2003). In Taiwan, Smithson & Heathcote (1976) found that when B deficiency occurred in cotton, the application of 250 kg N ha[-1] reduced the yield. However, if B was applied, the same application of N increased the crop yield. In pot experiments, application of 0.06 g of borax to 40 kg soil, deficient in B, was sufficient to overcome B deficiency problem in cotton. In field experiments, 10 to 30 kg ha[-1] of applied borax (to supply 1.1 to 3.3 kg B ha[-1]) was enough to prevent B deficiency. Since B is essential for the transfer and assimilation of sugars and N into complex carbohydrates (fiber) and protein, demand for this element is the greatest during lint and seed development (Lancaster et al., 1962).

5.2 Rice

Rice (*Oryza sativa* L.) is grown worldwide, but it is one of the most important cereal grains especially in Asia. Severe B deficiency has been reported in 10-45% rice fields in Pakistan (Tahir et al., 1990; Zia, 1993) and 1-69% (average 33%) rice fields in India (Singh, 2001). Average increase in rice paddy yield with B application in 22 field experiments was 14% over the zero-B control (Anonymous, 1998). Results of recent research have shown 15-25% increase in seed yield over N, P and Zn, coupled with appreciable improvement in grain/cooking quality (more recovery and less breakage of kernels during milling, greater grain elongation, less bursting and less stickness upon cooking) with application of B (Rashid et al., 2009). The authors also found that the B use in rice was highly profitable. Similarly, Mehmood et al. (2009) worked on three rice cultivars [viz., KS-282 (salt-tolerant), BG-402-4 (mixed behavior) and IR-28 (salt-sensitive)] to investigate the ameliorative nutritional aspects of B. Boron was applied at 25, 50, 100, 200, 400 and 800 ng B mL-1 in the presence (80 mol m-3) and absence (0 mol m-3) of NaCl salinity, whereas in solution culture B was applied at 1.5, 3.0 and 6.0 kg B ha-1 to saline [electrical conductivity of saturated paste extract (ECe) 9.0 dS m-1, sodium adsorption ratio (SAR) 5.46, and pH 7.8), and saline-sodic soils (ECe 9.0 dS m-1, SAR 28.2, pH 8.2]. Application of B improved all growth parameters, i.e., tillering capacity, shoot and root length, and shoot and root weight at external B application rates of 200-400 ng B mL-1 in solution culture in the presence and absence of NaCl salinity. Moreover, rice cultivars have shown differential response to B application (Table 4). In contrast, a marked increase in the paddy rice yield with the application of B was also reported in a non-saline soil (Chaudhry et al., 1976). Increasing supply of B increased the accumulation of B in roots and shoots (Nable et al., 1990; Akram et al., 2006). Vasil (1987) reported that the stigma, style and ovary often contain high concentration of B, and this B occurs in pollen at about 0.7 mg B kg-1 dry weight.

SN.	Cultivar	B concentration (mg B kg-1) in leaves		Remarks
		Control (no B)	B applied	
1	Super Basmati	7.58	11.24	
2	Basmati-6129	9.36	16.19	
3	DR-83	9.07	14.70	B concentration in
4	KS-282	9.86	16.67	leaves of different
5	Basmati-385	7.14	8.42	cultivars increased
6	Pakhal	9.29	11.56	with B application
7	Basmati-370	7.43	9.79	
8	IR-6	8.62	11.31	

[z]Adapted from Rashid et al., 2005.

Table 4. Response of B concentration in leaves of different rice cultivars to B application[z]

5.3 Wheat

Wheat (*Triticum aestivum* L.) is amongst the major cereal crops grown in almost every part of the world. Boron deficiency in wheat field was first observed almost concurrently on different sides of the world following the spread of semi-dwarf wheat in the 1960s (Rerkasem & Jamjod, 2004). Its deficiency has been reported to cause grain set failure and considerable yield losses in the wheat belt of the world's wheat growing countries (Rerkasem & Jamjod, 2004; Rerkasem et al., 2004). Since introduction of green revolution cases of severe B deficiency have been reported from several wheat growing countries of the world (Li et al., 1978; da Silva & de Andrade, 1980; Misra et al., 1992). Bangladesh, Brazil, Bulgaria, China, Finland, India, Madagascar, Nepal, Pakistan, South Africa, Sweden, Tanzania, Thailand, USA, USSR, Yugoslavia and Zambia are amongst the countries where B fertilization response based B deficiency, in wheat has been reported (Shorrocks, 1997). The B deficiency pronedregionsare believed to be in adjoining areas of eastern Nepal, northeastern India and northwestern Bangladesh, through to southwestern China (Rerkasem & Jamjod, 2004; Bhatta & Ferrara, 2005). Boron application on such fields (B-deficient soils) can make profound contributions to grain yield in wheat (Chakraborti & Barman, 2003; Soylu & Topal, 2004). In Pakistan, Chaudhry et al. (2007) conducted a study to identify the wheat response to micronutrients (B, Fe, Zn) in rainfed areas. The authors observed an increase in the yield of wheat and other crops (rice, maize and cotton) in a number of field experiments in response to B application. Summary of 16 field experiments revealed that application of B contributed 16% increase in grain yield, and also increase in value to cost ratio (VCR) over the zero-B control (Table 5; Anonymous, 1998). Further, genotypic differences were observed among wheat cultivars for their response to B application (Table 6).

Crop	Field experiments	Control yield (Mg ha^{-1})	Yield increase (%)	VCR
Wheat	16	3.286	14	4:1
Rice	19	3.081	14	5:1
Maize	9	2.512	20	7:1
Cotton				
Soil	30	2.377	14	16:1
Foliar	13	2.156	12	33:1

[z]Source: Anonymous, 1998.

Table 5. Yield responses and crop value to cost ratios (VCR) of four major crops to B fertilizer application in field experiments, Pakistan[z]

SN.	Cultivar	B concentration (mg B kg^{-1}) in leaves		Remarks
		Control (no B)	B added	
1	Rohtas-90	5.2	11.0	
2	Sindh-81	9.0	17.0	All cultivars showed a positive response to B application as depicted from the increase in B contents of the leaves
3	Faisalabad-85	8.0	10.3	
4	Rawal-87	9.3	18.0	
5	Pak-81	8.7	15.8	
6	Sariab-92	10.0	19.5	
7	Inqalab-91	7.2	20.7	
8	Bakhtawar	11.0	21.0	

[z]Adapted from Rashid et al., 2005.

Table 6. Response of B concentration in leaves of different wheat cultivars to B application[z]

5.4 Oilseed *Brassica spp.*

Canola or oilseed rape (*Brassica spp.* L.) is one of the major oilseed crops grown worldwide. Canola is considered to have high requirements for B. In addition, a steady supply of B during the peak vegetative, flowering, pod production and seed development stages is needed for optimum seed yield of canola. Deficiency of B at any growth stage in canola can severely affect its seed yield (US Borax, 1996). Research has shown that application of B fertilizers can be most effective if incorporated into the soil, seed and band placement may have toxic effects, and foliar application may be very effective to supply B to plants when deficiency is noted in the growing season. Foliar fertilization is an effective way to supply B to plants, especially when root activity is restricted by dry soil (Mortvedt, 1994). In a field study comparing rapeseed (*Brassica campestris* L.), barley (*Hordeum vulgare* L.) and potato (*Solanum tuberosum* L.) test crops, rapeseed showed the largest response to B fertilization (Wooding, 1985). In that study, symptoms of B deficiency on rapeseed plants did not appear until upper parts of the plants formed pods, with seed development limited to only those pods located on the lower parts of the plant. Also, B deficiency delayed maturity and kept the plants in an indeterminate stage of growth with flowers forming up to the time of the first killing frost. In other study in Alberta, Canada, B-deficient oilseed rape appeared normal in early growth stages, showed red margins and/or inter-veinal mottling at bloom stage and had reduced seed set (Nyborg & Hoyt, 1970).

In China, on a clayey soil with 0.7 mg B kg^{-1}, application of B fertilizer to *Brassica napus* L. improved plant height, pod-bearing branches and pod number per plant, seed number per pod, seed yield and oil content (Hu et al., 1994). Recently, Shi & Wang (2009) reported decrease in seed yield in oilseed rape (*Brassica napus* L.) with B deficiency. In Pakistan, based on various nutrient indexing field experiments on rapeseed-mustard, Rashid (1993, 1994) reported that 65% of the tested sites were deficient in B under the agro-ecological conditions. Oilseed crops responded well to B application for the reported B-deficient sites in Pakistan (Anonymous, 1998). In other studies, canola yield was not affected by B fertilization, although B concentration in plants was significantly increased and 20-30 mg B kg^{-1} in plant tissue was considered adequate for optimum yield (Bullock & Sawyer, 1991). In the Parkland region of western Canada, canola grown on Gray Luvisol soils has sometimes shown failure of flower bud development and poor seed set, more often on sandy soils. Deficiency of B was suspected to be responsible for these conditions because the symptoms match B deficiency symptoms (Grant & Bailey, 1993). In an earlier study on the Parkland region soils of Saskatchewan, B fertilization was observed to enhance rapeseed yield in a greenhouse experiment but its effect was not consistent in field experiments (Nuttall et al., 1987). In another field study in Saskatchewan, Canada, application of B fertilizer did not have any consistent influence on seed yield increase of canola, grown on soils ranging between 0.11 to 0.82 mg plant-available B kg^{-1} (Malhi et al., 2003).

5.5 Maize/corn

Maize (*Zea mays* L.) belongs to Graminae family. It ranks second (after wheat) in the world cereal production. Contribution to world corn/maize production is 2% from India, while it is 10% from China, and U.S. contribution to the total maize production of the world is known to be 43%. Approximately 80% of the maize production in Pakistan is concentrated in North West Frontier Province and Northern and Central Punjab. Maize in Pakistan is cultivated as a multipurpose food and forage crop, therefore the economic potential of this important crop is overwhelming (Khan et al., 2008). According to National Fertilizer Development Centre, (NFDC), forty percent (40%) maize fields in Pakistan, surveyed for fertility status, have been reported to be deficient in B (Anonymous, 1998). Boron application on nine fields exhibiting B deficiency in maize has shown to be very effective for yield increases, ranging from 12% to 35%, with an average increase of 20% over zero-B control. Maize yield increases worth of 5-15 Rs. (mean Rs. 7) have been documented after spending one rupee on B fertilizer (Anonymous, 1998). In a study in India, Mishra & Shukla (1986) reported considerable increase in plant height, metabolic rate, content of photosynthetic pigment and all dry weight fractions measured after the application of B containing amendment to maize.

5.6 Groundnut/peanut

Groundnut (*Arachis hypogaea* L.) belongs to Leguminosae family and is known as arachide in France, mani or cacahuete in Spain, pistachio di terra in Italy, erdnuss in Germany and amendoim in Portugal. Farmers in Asia and Africa grow 90% of the world's total groundnut production. The leading groundnut growing countries include India, China and USA. Groundnut is also one of the important cash crops of the Potohar plateau in the Punjab province of Pakistan. The crop is grown under rainfed conditions on relatively poorly fertile alkaline-calcareous soils with no adequate fertilization history (Rashid et al., 1997). Its average yield in Pakistan is reported to be 921 kg ha[-1] (Anonymous, 2009) and is much less as compared with the average yields of some other countries, e.g., in China, 2180 kg ha[-1] (Luo et al., 1990). As in alkaline and calcareous soils B is deficient (Tisdale et al., 1993), its deficiency is suspected in a highly sensitive crop species like groundnut (Katyal & Randhawa, 1983; Luo et al., 1990) when grown over such a soils. Application of B on such soils has shown positive results throughout the world.

In Pakistan, 50% B-deficient test sites have been reported in farm fields with groundnut, based on multi-locations field trials by Rashid & Qayyum (1991) and Rashid (1993, 1994). The value cost ratio (VCR) data of NFDC (Anonymous, 1998) indicated that by spending one rupee on B fertilizer, crop yield increase was worth Rs. 11 in groundnut. Seed yield increases in groundnut have been reported from 9 to 12% by borax application in B deficient Chinese soil (0.3-0.5 mg B kg[-1]) by Zhang et al. (1986). A 10% increase in pod yield of groundnut after B fertilization over the control) was obtained with 1 kg B ha[-1] (Rashid et al., 1997) in Pakistan. Encouraging responses of groundnut to B application have also been recorded in India, with average pod yield increase of 180 kg ha[-1] (Takkar & Nayyar, 1984). In China, Zhang et al. (1986) and Luo et al. (1990) indicated that 1 kg B ha[-1] (borax) can be the optimal B fertilizer requirement of groundnut. Foliar application of B is also very effective and it can be used with herbicides for groundnut. Nonetheless, B use in Pakistan is not a promising practice in groundnut as it is a low-input high-risk rainfed crop. Since internal B requirements of various groundnut genotypes vary greatly; a viable and practical solution of managing B deficiency in groundnut could be the screening of the available germplasm with respect to its sensitivity to B deficiency (Rashid et al., 1997).

5.7 Alfalfa

Alfalfa (*Medicago sativa* L.; also called lucerne) is one of the most important forage crops globally. It is well adapted to a wide range of growing conditions on soils of varied fertility. Boron deficiency caused nutritional disorders are quite common (Shorrocks, 1997; Dell & Huang, 1997). Its deficiency in alfalfa is causative of leaf yellowing, reddening of the upper leaves, shorten internodes and rosette appearance of the plant. At this stage the growing point becomes dormant or dies, flowering is reduced and the flower falls before setting seed (Bell, 1997; Shorrocks, 1997). Boron and other micronutrients applications on Indian soils for alfalfa have shown positive results in the form of increase in forage and seed yield (Kormilitsyn, 1992; Hazra & Tripathi, 1998; Patel & Patel, 2003). In a study on Chinese soils, B application along with other micronutrients increased yield and crude protein content in alfalfa (Wang & Chen, 2003; Liu & Zhang, 2005). Rammah & Khedr (1984) reported positive response of alfalfa to B application in some Egyptian soils.

Alfalfa is sometimes grown on the Coastal Plain of southern United States, but poor soil fertility status is one of the production problems in these areas. Field-scale demonstrations have shown a considerable increase in alfalfa forage yields (3.9 Mg ha^{-1} or 159%) with B application. The sustainable economic production is possible under rainfed conditions on selected, limed Coastal Plain soils of US with improved methods of site selection, adequate fertility and management guidelines (Haby & Leonard, 2000, 2005).

In a field study (Greece), foliar B application helped to increase the percentage of pods formed per inflorescence up to 52% as compared with the control. However, no significant difference between the different rates of B application was observed. The seed yield was increased by an average of 37% compared with the zero-B control during the second year at both locations. Moreover, foliar application of B improved seed germination and increased seed vigor which was increased by 27% in 2003 and up to 19% in 2004 as compared with the control (Dordas, 2006). Recently in a field study on calcareous soils in eastern Turkey, Turan et al. (2010) have also reported positive responses to B application. The authors concluded that lucerne production requires B addition to alleviate natural B deficiency problem in soils. This study warrants further studies with different soils and initial soil test B levels needed to conclude critical soil and tissue values for wider application across the region.

5.8 Soybean

Soybean (*Glycine max* L.) belongs to Leguminosae family. China, India and Indonesia are the leading soybean growing countries after USA. The occurrence of B deficiency based on responses at farmers fields have been reported for many countries like Australasia (China, India, Korea, Thailand), Europe (USSR) and in South America (Shorrocks, 1997). Generally, B deficiency is a common problem for this crop, especially when grown on alkaline calcareous soils of the world. The alkaline, silt, and sandy loam soils in Northeast Arkansas are also known to suffer from B deficiency (Anonymous, 2007a). Soybean is known to respond positively to B application on deficient sites of the world (Wu, 1986; Kirk & Loneragan, 1988). The increase in oil content and other quality parameters in soybean with combined application of B and sulfur in India have been noticed by Dinesh & Sudkep (2009), and Kumar & Sidhu (2009). In another study, Eguchi (2000) found a depressing effect of B deficiency on growth, yield, and protein and fat contents in the grains of soybean.

Boron application was also found to ameliorate Al toxicity by increasing growth characters (Yang & Gu, 2004). Furthermore, the genotypic variations in responses to B and other micronutrients (Zn, Mn) deficiencies have been observed by Graham & Heavner (1993) at the cellular level. In other field experiments the susceptibility of soybean cultivars to B deficiency was examined on "Typic Tropaqualf soils" in Northern Thailand, where B deficiency depressed seed yield by 60% in different cultivars. Sometimes, B deficiency also induced a localized depression on the internal surface of cotyledons in soybean seeds resembling to the symptom of 'hollow heart' in peanut. However, addition of B either decreased or eliminated such symptoms. In a comparative study of 19 soybean cultivars, the incidence of hollow heart symptoms in seeds at control (zero-B) appeared to be 75% but by the addition of only 1% B it reduced from none to 36%. The results suggested that susceptibility to B deficiency is sufficiently important and variable among soybean genotypes to warrant its inclusion as a selection criterion when breeding cultivars for areas with low soil B (Rerkasem et al., 1993).

5.9 Potato

Potato (*Solanum tuberosum* L.) belongs to Solanaceae family. Potatoes were probably brought to the Indian sub-continent hundred years ago by the Portuguese, and its cultivation expanded under British colonial rule in the 19th century (Geddes et al., 1989). The occurrence of B deficiency in potato based on responses at farmers fields have been reported in Australasia (Australia, China, India, Pakistan), Europe (Belgium, Czechoslovakia, Finland, Germany, Hungary, Sweden, USSR), and USA (Shorrocks, 1997). Despite favorable diversity of soils, climate and agricultural practices for potato cultivation, the average yield of potato has been reported to be 20.3 Mg ha^{-1} in Pakistan (Anonymous, 2008). Imbalanced use of fertilizers is one of the main reasons for this low yield (Nazli, 2010). Thus, the balanced use of micro and macronutrients (B, Zn, N, P and K) can considerably increase the yield. Lora (1978) obtained B responses in potato crop on Andosols soils in Colombia. Boron has also been identified to play a key role in forming abscission layers such as scar tissues at the stem end of potato at maturity that seals the tuber and thereby preventing it from diseases and bacterial infection on its storage. In a study, the growth of potatoes in sand cultures with zero-B resulted in poor vigour, yield and quality of tuber. Application of even 1 μg mL^{-1} B resulted in normal growth (Hill, 1936). Potatoes have shown positive response to B application if drilled in the row with fertilizer (Midgley & Dunklee, 1947). Higher rates of B were suggested if it was broadcasted and worked into the soil. In Pakistan, application of B fertilizer on three fields exhibiting B deficiency in potato has shown to be very effective for yield increases, ranging from 15% to 30%, with an average increase of 21% over zero-B control (Anonymous, 1998). Potato yield increases worth of Rs. 9 to Rs. 40 have been documented after spending one rupee on B fertilizer.

5.10 Citrus fruits

Citrus production is one of the world's largest agricultural industries. It is sown in more than 125 countries in the belt within 35º latitude north and south of equator (Duncan and Cohn, 1990). In addition to other factors, micronutrient deficiency (including B) is also considered among constraints that are currently hampering citrus yield (Johnson, 2006). In Pakistan, B deficiencies have been exhibited in citrus and other deciduous fruits (Tariq et al.,

2004; Zia et al., 2006). Its deficiency decreases growth and photosynthesis, and increases starch and hexoses in leaves of citrus seedlings (Han and Chen, 2008). Soil and plant analysis showed that > 50% of the cultivated soils of Pakistan were unable to supply sufficient B to meet the needs of many crops (Khattak, 1995) including citrus. Keeping in view the export potential for this crop, in a survey, 1250 citrus orchard growers were interviewed and soil samples were selected from their respective orchards in district Sargodha of Punjab to investigate B application trend and its suspected deficiency. The results revealed that out of the 1250 citrus growers majority (58.8%) never used the B since the establishment of their orchards, 18.4% farmers were in a practice of using recommended doses of macro and micronutrients depending upon their current financial position, while 11.6% citrus growers seasonally applied B (Table 7). Only 140 (11.2%) farmers were in a habit to use recommended doses of B. This percentage of the growers (11.2%) is the main contributor to the foreign exchange while exporting citrus.

In Pakistan (citrus belt, Sargodha), soil samples from varied depths (0-22.5 and 22.5-45 cm) were analyzed for B concentrations by Azomethine-H method (Ponnamperuma et al., 1981). Forty eight percent (600) orchards (samples) were found to be deficient in B. Since, in the past, farmers used to fertilize their citrus orchards without any soil testing, nutrient problem were common. Other reasons for B deficiency in the orchards might be due to the effect of sampling time. As the samples were taken during the monsoon season, this sampling time could be one of the reasons to affect B concentrations. These results are in agreement with the findings of Zia et al. (2006).

Growers category	Number of citrus growers	Percentage	Remarks
Never used	735	58.8	Low income, low education
Seasonal users	145	11.6	Lack of interest
Conditional users	230	18.4	Low income from the other sources of livelihood
Regular users	140	11.2	Education level, accessible extension and advisory services, good income

[z]Source: Ahmad et al., unpublished data.

Table 7. Boron use matrix in Bhalwal, Sillanwali and Sahiwal tehsil of district Sargodha[z]

As far as B concentrations in leaves and fruits of citrus are concerned, 38% of the samples were also found to be B deficient. At the appearance of B deficiency symptoms during drought year , B concentration was below 10 mg B kg[-1] in fruit peel and leaves, but at abundant precipitation B was 20 mg B kg[-1] in leaves and 14 mg B kg[-1] in peel, and no B deficiency symptoms were observed. In an orchard where fruit had deficiency symptoms, 0.5 M HCl-extractable-B concentrations were 0.15 mg B kg[-1] in surface soil (0-22.5 cm) and 0.10 mg B kg[-1] in the subsoil (22.5-45 cm). The reasons for this deficiency could be, low B status in soils of the orchards, less than recommended use of B containing fertilizers and moisture stress during the drought periods. This aspect of B deficiency induced by drought might be due to the restricted mineralization of organically bound soil B (Evans & Sparks, 1983; Flannery, 1985). Research has also shown B deficiency to be responsible for diseases and/or application of B fertilizers to correct those diseases in vegetables such as brown heart of turnips, heart rot of beets, browning of cauliflower (Hill, 1936; Greenhill, 1938), and fruits such as corky-core, blotchy cork and drought spot of apple (Hill, 1936; Greenhill, 1938; McLarty, 1940; Fritzsche, 1955), die-back of apricot (Fitzpatrick & Woodbridge, 1941) and deformed mandarin fruits of citrus (Chiu & Chang, 1985, 1986).

6. Boron deficiency and crop diseases

As discussed earlier, B deficiency is widespread in Pakistan (Rashid et al., 2009), India (Gupta, 1983, 1984; Sillanpaa & Vlek, 1985; Sakal & Singh, 1995), China (Liu Zheng et al., 1980, 1982a, 1982b, 1983, 1989) and Western Australia (Wong, 2003) and many other countries (Sillanpaa, 1982). Boron deficiency has been reported to be associated with internal tissue breakdown in root crops, groundnut/peanut, and some bean cultivars. Warncke (2005) observed the amelioration of internal black spot in cranberry bean seed with B application. Boron nutrition of cereal crops in connection to chilling tolerance has also been demonstrated by Huang & Ye (2005). Problems like stunning of cotton growth, wilting of the plants and reddening of cotton leaves have been observed in the major cotton growing areas of Sindh province in Pakistan. At present, integrated plant nutrient management including B with the best management practices seems to be the only solution of this lethal problem (Abid Niaz - personal communication). However, no clear relationships have been established between above mentioned symptoms and B nutrition. Similar relation of B application in curing the problem of rust in wheat has also been reported by Nuclear Institute for Agriculture and Biology (NIAB), Faisalabad, Pakistan. Application of B also minimized the adverse effects of drought stress in crop plants during dry spell. Severe attack of rust was observed on B-deficient wheat plants exposed to drought stress, whereas no incidence of disease was fond in B treated plants (Anonymous, 2007a).

7. Boron extractants and their comparative efficiency

Several extractants used for soil B extraction have been employed over time for example, hot water for plant-available B (Berger & Truog, 1939; Parker & Gardner, 1981; Mahler et al., 1984; Rahmatullah et al., 1999), 0.05 M HCl for plant-available B (Ponnamperuma et al., 1981), 0.018 M $CaCl_2$ for non-specifically adsorbed/readily soluble B on soil surfaces (Iyenger et al., 1981; Aitken & McCallum, 1988; Spouncer et al., 1992; Hou et al., 1996; Rahmatullah et al., 1999), 1 M NH_4OAc for multi-element extraction (Gupta & Stewart, 1975; Chaudhary & Shukla, 2004), 0.25 M sorbitol-DTPA for bioavailable B (Goldberg, 1997; Miller et al., 2000; Gloldberg et al.,, 2002; Shiffler et al., 2005), 0.05 M mannitol prepared in 0.01 M $CaCl_2$ for B in soil solution and its nonspecifically adsorbed forms to assess regenerative power of soil for B (Cartwright et al., 1983; Aitken et al., 1987; Jin et al., 1988; Rahmatullah et al., 1999;Vaughan & Howe, 1994), and 0.005 M AB-DTPA for multi-element extraction (Gestring & Soltanpour, 1984, 1987; Matsi et al., 2000).

There are a number of methods for extracting available B from soils. The colorimetric and other methods of determining B in the soil extract remain the same for testing on acid and alkaline soils (Bingham, 1982; Gupta, 2006). The most common extractant is hot water (Berger & Truog, 1939) because soil solution B is most important with regard to plant uptake. Li and Gupta (1991) compared hot water, 0.05 M HCl, and hot 0.01 M $CaCl_2$ solutions as B extractants in relation to B accumulation by soybean, red clover, alfalfa, and rutabaga. The authors concluded that 0.05 M HCl was the best extractant (r=0.82) followed by hot water, and hot 0.01 M CaCl$_2$. Tsadilas et al. (1994) in a study using diverse soils concluded that HWS-B was a valuable measure of available soil B and it correlated strongly with 0.05 M mannitol in 0.1 M CaCl$_2$ extractable, 0.05 M HCl-soluble B. Another extractant, ammonium bicarbonate-diethylenetriaminepentaacetic acid (AB-DTPA) has been suggested for effective B determination in alkaline soils (Gestring & Soltanpour, 1984, 1987, Gupta, 2006). Studies involving 31 US soils (Kaplan et al., 1990) and 100 Dutch soils (Novozamsky et al., 1990) have also confirmed that B values of cold extraction using 0.01 M CaCl$_2$ were highly associated with those of hot extraction (hot 0.01 M CaCl$_2$). However, for hot water B extraction method, several researchers have marked some problems of significance such as problematic comparability of the basic soil parameters determined routinely, precision, time consumption etc. (Shiffler et al., 2005).

Using irrigated rice soils (n=53, pH 3.5-8.0) Ponnamperuma et al. (1981) recommended 0.05 M HCl as equally good extractant as HWS-B method (r=0.96) of Berger and Truog (1939). Cartwright et al. (1983) in a study concluded that extraction of wide range of soils (pH 5.4-10.1, CaCO$_3$ 0-85%) with 0.01 M CaC1, + 0.05 M mannitol was found to be a more convenient soil test for plant-available B than the standard HWS-B method, and to be as good in predicting the response in B uptake by plants. With a cold 0.01 M CaCl$_2$ extraction (n=100, pH 3.9-6.5) equally valuable soil B values can be obtained as with the more difficult to standardize hot water extraction procedure (Novozamsky et al., 1990). Vaughan and Howe (1994) suggested sorbitol (prepared in a buffered solution of I N ammonium acetate and 0.1 M triethanolamine) as an alternate for HWS-B test in determining available soil B. The amounts of B recovered by HWS, 0.05 M mannitol in 0.01 M CaCl$_2$ extractable B, 0.05 M HCl soluble B methods were strongly correlated with each other, the highest correlation obtained being between HWS-B and HC1-B. Plant B was highly correlated to the B recovered (n=50, pH 6.1-8.2, CaCO$_3$ 0-9.2%) by all the three extractants (Tsadilas et al., 1997). Amounts of extractable B with AB-DTPA and with hot water were similar (r=0.84) for ten soils (pH 5.8-7.8, CaCO$_3$ 0-61 g/kg) studied by Matsi et al. (2000). There were highly significant positive correlations between the amounts of B extracted through hot water-soluble, 1:1 soil: distilled water and 1:2 soil:distilled water, ammonium acetate, calcium chloride - mannitol, and DTPA - sorbitol extractants (Goldberg et al., 2002). Latter, Chaudhary & Shukla (2004) also accentuated the advantageous features of sorbitol + NH$_4$OAc + TEA and mannitol + NH$_4$OAc + TEA extractants. Further, the simplicity of these extractants has also been compared to hot water and hot 0.01 M CaCl$_2$ methods..These extractants have the tendency to demarcate the available B status of arid soils on a routine basis where a large number of samples are to analyze. DTPA-Sorbitol has been recommended as a replacement to these cumbersome hot water extraction procedures (Shiffler et al., 2005).

8. Advances in boron analysis

The spectrophotometric technique using a colorimetric reaction with azomethine-H has been the most extensively tested B determination method for soil and plant samples (Ogner, 1980; Parker & Gardner, 1981; Porter et al., 1981; Lohse, 1982; Garcia et al., 1985; Lee et al., 1987; Chen et al., 1989; Kaplan et al., 1990; Banuelos et al., 1992; Campana et al., 1992; Nogueira et al., 1993). In this type of determination, hot water and 0.5 M HCl have commonly been used as extractants, both for acidic and alkaline soils. The use of these extractants is attributed with certain merits and demerits. For example, the HWE procedure embodies several potential sources of error like difficulty to standardize (Novozamsky et al., 1990), time consuming and tedious for routine and reproducible usage (Deabreu et al., 1994). Further, the amount of B extracted is affected by the reflux time (McGeehan et al. 1989), extraction time and temperature (Spouncer et al., 1992). The coloured hot water extracts in some soils may affect B determination. The HWE method has limitations for some soils and B extracted by this method did not correlate with crop responses under some management conditions (Gestring & Soltanpour, 1987; Offiah & Axley, 1988; Mustafa et al., 1993). The use of 0.05 M HCL has eliminated the problems of extraction with hot water. Overall, the colorimetric methods, in general, suffer several interferences, such as sample pH in the range of 6.4 to 7.0 (Carrero et al., 1993), sample colour (McGeehan et al., 1989; Evans & Krahenbuhl, 1994a), nitrate complexes in the wet HNO_3 acid digests of plants (Gestring & Soltanpour, 1981a) and the presence Fe, Al, Cu, Zn and Mo (Arruda & Zagatto, 1987). These interferences and lack of sensitivity limit the application of these methods for the samples with low B concentrations and complex matrices.

The reliability of B measurements has improved in the last decade with better instrumentation and analytical methodology (Sah & Brown, 1997). After spectrophotometry, B has been determined utilizing potentiometer, chromatography, flame atomic emission and absorption spectrometry, inductively coupled plasma (ICP) optical emission (OES) and mass spectrometry (MS), and neutron activation analysis using neutron radiography and prompt-activation analysis. The extraction with 0.05 M HCl is concerned; it has also worked well for predicting B availability to crop plants in acid soils (Ponnemperuma et al., 1981; Renan & Gupta, 1991). However, Fe extracted with the acid extractant often interferes in B determination by spectrophotometric and ICP–OES methods (Evans & Krahenbuhl, 1994a; Pougnet & Orren, 1986a, 1986b).

There are reports on the use of plasma-source OES for assaying B (Pritchard & Lee, 1984; Nilsson & Jennische, 1986; Lee et al., 1987; Jeffrey & McCallum, 1988; Novozamsky et al., 1990; Goto et al., 1992; Spouncer et al., 1992; Ferrando et al., 1993; Evans & Krahenbuhl, 1994a). Reported detection limits for B are 10 to 15 mg B L^{-1} in soil solutions and plant digests (Spiers et al., 1990). Boron determination by ICP–OES is also affected by other interfering species, for example, Si (Owens et al., 1982; Din, 1984), Ni, Cr, Al, V, Mn, Ti, Mo and high concentrations of Na (Pougnet & Orren, 1986b; Kavipurapu et al., 1993). Nevertheless, recently, Mehlich-3 has been promoted as a "universal" extractant in a wide variety of soils. But ICP analysis of B following extraction with Mehlich- 3 chemicals has proven difficult because of B contamination within the ICP unit. Secondly, the effects of distilled water, nitric acid and sorbitol solutions used between samples for correcting B contamination has also been diagnosed by Allen et al. (2005). Sorbitol solution is found as the most effective solution to rectify the contamination problem. The contamination problem unique to Mehlich-3 has the tendency to limit the development of Mehlich-3 as the widely accepted extractant. However, no such problem was observed with ICP analysis of B with hot water, pressurized hot water or DTPA-Sorbitol extractions. (Allen et al., 2005). Consequently, pressurized hot water or DTPA-Sorbitol extractions have been proposed as replacement.

Overall, mostly all the above mentioned soil B extractants, provide good correlaton with plant B contents under controlled conditions. However, the efficacy of these extractants should also be tested under field conditions (Goldberg & Chunming, 2007). Historically, B soil tests have been developed to predict B deficient soils and have not generally been evaluated for their ability to predict soil conditions that produce B toxicity effects in plants. The work reported by Goldberg et al. (2002, 2003) for using shallow groundwater to apply for crops could predict improvement in irrigation efficiency. Such sort of attributes must be incorporated in B investigation techniques/methodologies to take into loop B content of both field grown and crops grown under controlled conditions of potential B toxicity. Boron determination by ICP–MS suffers no spectroscopic interferences, and is considered the most practical and convenient technique for B isotope determination. Among the present technologies, ICP–MS has emerged as the method of choice for determining B concentration and a convenient method for B isotope determination (Sah & Brown, 1997). With the increased use of inductively coupled plasma atomic emission spectrophotometry (ICP) (especially instruments with simultaneous detection capability) soil test laboratories would welcome the need for different extractions for many element analyses. Boron is readily measured with ICP instrumentation. However, the predictions of B deficiencies by soil testing needs to be based on local data and not from broad generalizations from other areas.

9. Summary

Boron (B) is a unique micro mineral nutrient required for normal plant growth and optimum yield of crops. Its deficiency is widespread in alkaline/calcareous, coarse-textured and low organic matter soils in many countries of the world. Annual [fiber (cotton), cereal (rice, maize/corn, wheat), legume/pulse (soybean), oilseed (groundnut/peanut, oilseed rape/canola)], vegetable (potato), and perennial [citrus fruit orchards, alfalfa] crops grown on such soils usually suffer from B deficiency. This paper discusses factors affecting B availability in soils, including parent material, soil pH, texture, clay minerals and organic matter, irrigation sources, nutrient interactions, and plant species. The paper also documents the diagnosis and correction of B deficiency in several important crops in a wide range of soils. Crop yield increases up to 14% each in cotton and wheat, 14-30% in rice, 20% in maize, 58% in soybean, 10% in groundnut, 45% in oilseed rape, 30% in potato, 37% in alfalfa seed and 159% in alfalfa forage are reported with application of B by using appropriate rates, methods (soil or foliar) and sources (such as borax) on B-deficient soils. Application of B fertilizers up to 2.5 kg B ha^{-1} is recommended to prevent/correct B deficiency in major crops depending on the placement method. The paper also reviews comparative efficiency of various boron extracts under different soil conditions in addition to advances in boron analysis. Among the present technologies, ICP–MS has emerged as the method of choice for determining B concentration and a convenient method for B isotope determination.

Prevention and/or correction of B deficiency in crops on B-deficient soils can have a dramatic effect on yield and produce quality of many crops including fibers, cereals, pulses, oilseeds, vegetables, citrus fruits and alfalfa. Source, rate, formulation, time and method of B application and proper balancing of B with other nutrients in soil all affect crop yield on B-deficient soils. Both soil and foliar application methods of B are effective in improving crop yield, produce quality, concentration and uptake of B, and economic returns. Soil applied B leaves residual effect for years on succeeding crops grown on B-deficient soils in the same fields. The actual fraction of B fertilizer removed by the crops is only 1-2% of the total applied fertilizer through soil. However, it is very important that research for improving crop yields must move beyond applications of B based on general recommendations, and that deriving methods to predict site-specific deficiencies (e.g., soil or plant tests) are essential since the potential for B toxicity is large and the difference between deficiency and toxicity is very narrow. This could be especially important if B is applied sequentially to fields over a series of years without knowing the residual effects.

10. Conclusions and future research needs

Prevention and/or correction of B deficiency in crops on B-deficient soils can have a dramatic effect on yield and produce quality of many crops including fibers, cereals, pulses, oilseeds, vegetables, citrus fruits and alfalfa. An increase in yield of 14% each in cotton and wheat, 14-30% in rice, 20% in maize/corn, 58% in soybean, 10% in groundnut/peanut, 45% in oilseed rape, 30% in potato, 37% in alfalfa seed and 159% in alfalfa forage was observed with B fertilization. Source, rate, formulation, time and method of B fertilizer application, and proper balancing of B with other nutrients in soil all affect crop yield on B-deficient soils. Both soil and foliar application methods of B are effective in improving crop yield, produce quality, concentration and uptake of B, and economic returns. Application of B to rice on B-deficient soils also enhanced milling recovery and head rice recovery, and improved kernels quality traits like stickiness and cooking quality. Zinc application has been found to neutralize toxic effect of B in some crop plants and produced increase in crop yield.

Soil applied B leaves residual effect for years on succeeding crops grown on B-deficient soils in the same fields. The actual fraction of B fertilizer removed by the crops is only 1-2% of the total applied fertilizer through soil. Research on recycling of crops from rotation system rather than mono-cropping culture can generate useful information for B management. Moreover, adaptive research is also a pre-requisite for B management under efficient irrigation systems, e.g., drip, sprinkler and others. Impact of B nutrient use on product quality is needed, especially for high B requirement crops. Moreover, B efficient genotypes for different crops need to be identified and developed for commercial use.

Effect of B fertilizer use in high input system should also be given priority as a futuristic option for the sustainability of crop production, soil quality and environment. Further, management decisions for use of B fertilizers should consider both immediate and long-term effects of B fertilizer on crop yield, produce quality and economic returns. Research is also required, in different agro-ecological zones, to determine the long-term effects of different sources of B on accumulation and distribution of B and its balanced application with other nutrients to investigate its relationship with disease and insect resistance in different crops. However, it is very important that research for improving crop yields must move beyond applications of B based on general recommendations, and that deriving methods to predict site-specific deficiencies (e.g., soil or plant tests) are essential since the potential for B toxicity is large and the difference between deficiency and toxicity is very narrow. This could be especially important if B is applied sequentially to fields over a series of years without knowing the residual effects. It could become a serious problem if B was applied for several years for a tolerant crop and then change to B sensitive crop. Soil test could potentially determine potential problems of excessive accumulations of B in the soil.

11. References

Ahmad, W.; Ahmad, N.; Ibrahim, M. & Niaz, A. (2004). Boron contents in ground and river waters used for irrigation in district Faisalabad. *In: Abstracts 10th International Congress of Soil Science. March 16-19, 2004,* Sindh Agricultural University, Tandojam, Pakistan. 109 pp.

Ahmad, W.; Niaz, A.; Kanwal, S.; Rahmatullah, & Rasheed, M. K. (2009). Role of boron in plant growth: A review. *Journal of Agricultural Research.* Vol. 47, pp. 329-338.

Ahmed, N.; Abid, M. & Ahmad, F. (2008). Boron toxicity in irrigated cotton (*Gossypium hirsutum* L.). *Pakistan Journal of Botany.* Vol. 40, pp. 2443-2445.

Aitken, R. L.; Jeffrey, A. J. & Compton, B. L. (1987). Evaluation of selected extractants for boron in some Queensland soils. *Australian Journal of Soil Research.* Vol. 25, pp. 263-273.

Aitken, R. L. & McCallum, L. E. (1988). Boron toxicity in soil solution. *Australian Journal of Soil Research.* Vol. 26, pp. 605–610.

Akram, M. S.; Ali, Q.; Athar, H. & Bhatti, A. S. (2006). Ion uptake and distribution in *Panicum antidotale* Retz. under salt stress. *Pakistan Journal of Botany.* Vol.38, pp. 1661-1669.

Ali, S. J. & Monoranjan, R. (1989). Effect of NPK and micronutrient in controlling sterility in wheat. *Fertilizer News.* Vol. 34, pp. 35-36.

Allen, S. K.; Jolley, V. D.; Webb, B. L.; Shiffler, A. K. & Haby, V. (2005). Challenges of Mehlich 3 Extraction for B and Comparison with Other Methods in Boron-Treated Soils. *Brigham Young University Department of Plant and Animal Sciences, University, Texas Agricultural Expt.* Stn., PO Box 200, Overton, TX 75684-0200, U.S.A.

Anonymous. (1998). Micronutrients in Agriculture - Pakistan Perspective. NFDC Pub. No. 4/98. National Fertilizer Development Centre, Islamabad, Pakistan.

Anonymous. (2007a). Soybean response to boron fertilization, 2007 [Online]. Available:http://ipni.net/far/farguide.nsf/$webindex/article=5F51B22A852573F1 0001A460FBF6FC76!opendocument [15 May 2010].

Anonymous. (2007b). National Institute of Agriculture and Biology (NIAB). Seventh five-year report on research and other activities (6 April 2002 to 5 April 2007). Faisalabad, Pakistan. 67 pp.

Anonymous. (2008). Pakistan statistical year book 2008. [Online] Available: http://www.statpak.gov.pk/depts/fbs/publications/yearbook2008/yearbook200 8.html [15 May 2010].

Anonymous. (2009). Pakistan statistical year book 2009. [Online] Available: http://www.statpak.gov.pk/depts/fbs/publications/yearbook2009/yearbook200 9.html [15 May 2010].

Arora, C. L.; Singh, B. & Takkar, P. N. (1985). Secondary and micronutrient deficiency in crops. *Progressive Farming.* XXI 8: 13.

Arruda, M. A. Z. & Zagatto, E. A. G. (1987). A simple stopped-flow method with continuous pumping for the spectrophotometric flow-injection determination of boron in plants. *Analytica Chimica Acta.* Vol. 199, pp. 137–145.

Ba~nuelos, G.S.; Cardon, G.; Pflaum, T. & Akohoue, S. (1992). Comparison of dry ashing and wet acid digestion on the determination of boron in plant tissue. *Communications in Soil Science and Plant Analysis.* Vol. 23, pp. 2383–2397.

Barber, S. A. (1995). Soil nutrient bioavailability: a mechanistic approach. John Wiley and Sons, New York, NY, U.S.A.

Barrow, N. J. (1989). Testing a mechanistic model. X. The effect of pH and electrolyte concentration on borate sorption by a soil. *Journal of Soil Science*. Vol. 40, pp. 427-435.

Bell, R. W. (1997). Diagnosis and prediction of boron deficiency for plant production. *Plant and Soil*. Vol. 193, pp. 149-168.

Berger, K. C. & Truog, E. (1939). Boron determination in soils and plants using the quinalizarin reaction. Industrial and Engineering Chemistry. Analytical Edition. Vol. 11, pp. 540-545.

Bergmann, W. (1988). Ernährungsstörungen bei Kulturpflanzen. Entstehung, visuelle und analytische diagnose. Fischer Verlag, Jena, Germany.

Bhatnagar, R. S.; Attri, S. C.; Mathur, G. S. & Chaudhary, R. S. (1979). Boron adsorption equilibrium in soils. *Annals of Arid Zone*. Vol. 18, pp. 86-95.

Bhatta, M. R. & Ferrara, G. O. (2005). Wheat sterility induced by boron deficiency in Nepal. In: Micronutrients in South and South East Asia, 221-229; P. Andersen, J. K. Tuladhar, K. B. Karki & S. L. Maskey (Eds), International Centre for Integrated Mountain Development (ICIMOD), Kathmandu, Nepal.

Biggar, J. W. & Fireman, M. (1960). Boron adsorption and release by soils. Soil Science Society of America Proceedings. Vol. 24, pp. 115-120.

Bingham, F. T. (1982). Boron. In Methods of Soil Analysis, eds. A. L Page, 431–447. Madison, Wisconsin, U.S.A.: ASA.

Bingham, F. T.; Page, A. L.; Coleman, N. T. & Flach, K. (1971). Boron adsorption characteristics of selected soils from Mexico and Hawaii. *Soil Science Society of America Journal*. Vol.35, 546-550.

Borkakati, K. & Takkar, P. N. (2000). Forms of boron in acid alluvial and lateritic soils in relation to ecosystem and rainfall distribution. In: *Proceedings of International Conference on Managing Resources for Sustainable Agricultural Production in the 21st Century. Better Crops*. Vol. 2, pp. 127-128.

Brown, P. H. & Hu, H. (1998). Boron mobility and consequent management. In: *Better Crops*. Vol. 82, pp. 28-31.

Brown, P. H. & Shelp, B. J. (1997). Boron mobility in plants. *Plant and Soil*. Vol. 193, pp. 85-101.

Bullock, D. G. & Sawyer, J. E. (1991). Nitrogen, potassium and boron fertilization of canola. *Journal of Production Agriculture*. Vol. 4, pp. 550-555.

Campana, A. M. G.; Barrero, F. A. & Ceba, M. R. (1992). Spectrofluorimetric determination of boron in soils plants and natural waters with Alizarin red-S. *Analyst*. Vol. 117, pp. 1189–1191.

Cartwright, B.; Tiller, K. G.; Zarcinas, B. A. & Spouncer, L. R. (1983). The chemical assessment of the boron status of soils. *Australian Journal of Soil Research*. Vol. 21, pp. 321-332.

Chakraborti, S. K. & Barman, P. (2003). Enhancement of yield of wheat genotypes by application of borax in Terai region. *Journal of Interacademicia*. Vol. 7, pp. 256-261.

Chang, S. S.; Hu, N. H.; Chen, C. C. & Chu, T. F. (1983). Diagnosis criteria of boron deficiency in papaya and the soil boron status of Taitung area (Taiwan). *Chinese Journal of Soil Science*.Vol. 32, pp. 238-252.

Chang, S. S. (1993). Nutritional physiology of boron and the diagnosis and correction of boron deficiency and toxicity in crops. *Proceedings of the Symposium on Reclamation of the Problem Soils in the Eastern Taiwan* (S. N. Hwang & G. C. Chiang, eds). *Chinese Society of Plant Nutrition and Fertilizer Science and Hwalian District Agricultural Improvement Station, Taiwan*. pp. 109-122.

Chang, S. S.; Huang, W. T.; Lian, S. & Wu, W. L. (1992). Research on soil testing and leaf diagnosis as guides to fertilization recommendation for the citrus orchards in Taiwan. *Annual Research Reports on Soils and Fertilizers 81*. Published by the Provincial Department of Agriculture and Forestry, Taiwan. pp.167-195.

Chaudhary, D. R. & L. M. Shukla. (2004). Evaluation of extractants for predicting availability of boron to Mustard in arid soils of India. *Communications in Soil Science and Plant Analysis*. Vol. 35, pp. 267-283.

Chaudhry, E. H.; Timmer, V.; Javed, A. S. & Siddique, M. T. (2007). Wheat response to micronutrients in rain-fed areas of Punjab. *Soil & Environment*. Vol. 26, pp. 97-101.

Chaudhry, F. M.; Latif, A.; Rashid, A. & Alam, S. M. (1976). Response of the rice varieties to field application of micronutrient fertilizers. *Pakistan Journal of Industrial and Scientific Research* Vol. 19, pp. 134-139.

Chauhan, R. P. S. & Asthana, A. K. (1981). Tolerance of lentil, barley and oats to boron in irrigation water. *J. Agric. Sci.* Vol. 97, pp. 75-78.

Chen, D.; Lazaro, F.; Decastro, L. & Valcarcel, M. (1989). Direct spectrophotometric determination of total boron in soils with ultrasonic leaching in automatic flow systems. *Analytica Chimica Acta*, Vol. 226, pp. 221–227.

Chiu, T. F. & Chang, S. S. (1985). Diagnosis and correction of boron deficiency in citrus orchard. In: *Seminar on Leaf Diagnosis as a Guide to Orchard Fertilization. Technical Bulletin No. 91*. Food Fertilizer and Technology Center, Taipei, Taiwan. pp. 1-12.

Chiu, T. S. & Chang, S. S. (1986). Diagnosis and correction of boron deficiency in the citrus orchard of Taiwan. *Soils and Fertilizers in Taiwan*, Taipei, Taiwan.

Communar, G. & Keren, R. (2008). Boron adsorption by soils as affected by dissolved organic matter from treated sewage effluent. *Soil Science Society of America Journal*. Vol. 72, pp. 492-499.

da Silva, A. R. & de Andrade, J. M. V. (1980). A cultura do trigo nas varzeas de Minas Gerais-possibilidades e dificuldades. Embrapa, Centro de Pesquisa Agropecuária dos Cerrados (CPAC). *Circular Técnica* No. 2: 69 pp. (in Portuguese).

Dannel, F.; Pfeffer, H.; Romheld, V. (2000). Characterization of root boron pools, boron uptake and boron translocation in sunflower using the stable isotope 10B and 11B. *Australian Journal of Plant Physiology* Vol. 156, pp. 756–761.

Dannel, F.; Pfeffer, H.; Walch-Liu, P. & Romheld, V. (2001). Plant nutrition – food security and sustainability of agro-ecosystems. Dordrecht: Kluwer, 162–163.

De Oliveira, R. H.; Milanez, C. R. D.; Moraes-Dallaqua, M. A. & Rosolem, C. A. (2006). Boron deficiency inhibits petiole and peduncle cell development and reduces growth of cotton. *Journal of Plant Nutrition*, Vol. 29, pp. 2035-2048.

Deabreu, C. A.; Deabreu, M. F, Vanraij, B. & Bataglia, O. C. (1994). Extraction of boron from soil by microwave heating for ICPAES determination. *Communications in Soil Science and Plant Analysis*. Vol. 25, pp. 3321–3333.

Dell, B. &. Huang, L. (1997). Physiological response of plants to low boron. *Plant and Soil*. Vol. 193, pp. 103–120.

Diana, G. (2006). Boron in the soil, from deficit to toxicity. *Informatore Agrario*. Vol. 62, pp. 54-58.

Din, V. K. (1984). The preparation of iron-free solutions from geological materials for the determination of boron (and other elements) by inductively coupled plasma emission spectrometry. *Analytica Chimica Acta*. Vol. 159, pp. 387–391.

Dinesh, K. & Sudkep, S. (2009). Influence of soil applied sulfur and boron on yield and quality parameters of soybean. *Annals of Biology*. Vol. 25, pp. 105-111.

Dobermann, A. & Fairhurst, T. (2000). Rice nutrient disorders & nutrient management [Online] Available: http://books.irri.org/9810427425_content.pdf [May, 15, 2010].

Dordas, C. (2006). Foliar boron application improves seed set, seed yield, and seed quality of alfalfa. *Agronomy Journal*. Vol. 98, pp. 907-913.

Dregne, H. E. & Powers, W. L. (1942). Boron fertilization of alfalfa and other legumes in Oregon. *Journal of American Society of Agronomy*. Vol. 34, pp. 902-12.

Duncan, L. W. & Cohn, E. (1990). Nematode parasites of citrus. In: *Plant parasitic nematodes in sub-tropical and tropical agriculture,* Luc, R.; Sikora, R. A. & Bridge, J. (Eds). CAB International, 321-346, Wallingford, Oxon, UK.

Dwivedi, G. K.; Dwivedi, M. & Pal, S. S. (1990). Mode of application of micronutrients in soybean –wheat crop sequence. *Journal of the Indian Society of Soil Science,* Vol. 38, pp. 458-463.

Eguchi, S. (2000). Effect of boron deficiency on growth, yield and contents of protein and fat in grains of soybean (*Glycine max*). *Journal of soil science and plant nutrition.*(Japan). Vol. 71, pp. 171-178.

Eguchi, S. & Yamada, Y. (1997). Long term field experiment on the application of slow-release boron fertilizer: Part 2. Behaviour of boron in the soil. In: *Proceedings of International Symposium on Boron in Soils and Plants*, R. W. Bell & B. Rerkasem, (Eds.), 49–56, September, 1997, Chiang Mai, Thailand.

Elrashidi, M. A. & O'Connor, G. A. (1982). Boron sorption and desorption in soils. *Soil Science Society of America Journal*. Vol. 46, pp. 27-31.

Elseewi, A. A. (1974). Some observations on boron in water, soils, and plants at various locations in Egypt. *Alexandria Journal of Agricultural Research*. Vol. 22, pp. 463-473.

Evans, C. M. & Sparks, D. L. (1983). On the chemistry and mineralogy of boron in pure and in mixed systems: A review. *Communications in Soil Science and Plant Analysis*. Vol. 14, pp. 827-846.

Evans, L. J. (1987). Retention of boron by agricultural soils from Ontario. *Canadian Journal of Soil Science*. Vol. 67, pp. 33-42.

Evans, S. & Krahenbuhl, U. (1994a). Boron analysis in biological material – microwave digestion procedure and determination by different methods. *Fresenius' Journal of Analytical Chemistry*. Vol. 349, pp. 454–459.

Ferrando, A. A.; Green, N. R.; Barnes, K. W. & Woodward, B. (1993). Microwave digestion preparation and ICP determination of boron in human plasma. *Biological Trace Element Research*. Vol. 37, pp. 17–25.

Fitzpatrick, R. E. & Woodbridge, C. G. (1941). Boron deficiency in Apricots. *Scientific Agriculture*. Vol. 22, No. 4, pp. 271-273.

Flannery, R. L. (1985). Understanding boron needs in crop production. *Fertilizer Progress*. Fertilizer Institute, Washington, U.S.A. Vol. 16, pp. 41-45.

Fleet, M. E. L. (1965). Preliminary investigations into the sorption of boron by clay minerals. *Clay Minerals*. Vol. 6, pp. 3-16.

Fleming, G. A. (1980). Essential micronutrients. I. Boron and molybdenum. In: Applied soil trace elements, B. E. Davies (Ed.), 155-197, John Wiley and Sons, New York, NY, U.S.A.

Frederickson, A. F. & Reynolds, R. C. Jr. (1959). Clays. In: *Clay Minerals Proceedings*, 203-213, 8th Conf.; Pergamon Press, Oxford, England.

Fritzsche, R. (1955). On the cork disease of Glocken Apples. *Schweizerische Zeitschrift fur Obst- und Weinbau*. Vol. 64, No. 11, pp. 193-198.

Garcia, I. L.; Cordova, M. H. & Sanchez-Pedrono, C. (1985). Sensitive method for the spectrophotometric determination of boron in plants and waters using crystal violet. *Analyst*. Vol. 110, pp. 1259.

Geddes, A. M .W.; Khan, S. M.; Naumann-Etienne, K.; Edwards, R. J. A.; Smith, A. E.; Bajwa.; K. & Hussain, A. (1989). *Potato atlas of Pakistan: Information on potato production by agro-ecological zones*. 1st edition. Pakistan Agricultural Research Council, Islamabad, Pakistan.

Gestring, W. D. & Soltanpour, P. N. (1984). Evaluation of the ammonium bicarbonate-DTPA soil test for assessing B availability to alfalfa. *Soil Science Society of America Journal*. Vol. 48, pp. 96-100.

Gestring, W. D. & Soltanpour, P. N. (1987). Comparison of soil tests for assessing boron toxicity to alfalfa. *Soil Science Society of America Journal*. Vol. 51, pp. 1214–1219.

Goldberg, S. (1993). Chemistry and mineralogy of boron in soils. In: U. C. Gupta (Ed.), Boron and its role in crop production, Pages 344, CRC Press, Boca Raton, FL, U.S.A.

Goldberg, S. (1997). Reactions of boron with soils. *Plant and Soil*. Vol. 93, pp. 35-48.

Goldberg, S. & Forster, H. S. (1991). Boron sorption on calcareous soils and reference calcites. *Soil Science*. Vol. 152, pp. 304–310.

Goldberg, S. & Glaubig, R. A. (1986). Boron adsorption on California soils. *Soil Science Society of America Journal*. Vol 50, pp. 1173-1176.

Goldberg, S.; Shouse, P. J.; Lesch, S. M.; Grieve, C. M.; Poss, J. A.; Forster, H. S. & Suarez, D. L. (2002). Soil boron extractions as indicators of boron content of field-grown crops. *Soil Science*. Vol. 167, pp. 720-728.

Gormus, O. (2005). Interactive effect of nitrogen and boron on cotton yield and fiber quality. *Turkish Journal of Agriculture and Forestry*. Vol. 29 pp. 51-59.

Goto, I.; Muramoto, J. & Ninaki, M. T. I. (1992). Application of inductively coupled plasma atomic emission spectrometry (ICP-AES) to soil analysis (Part 5) determination of hot water soluble boron in soils by ICP-AES. *Japanese Journal of Soil Science and Plant Nutrition*. Vol. 63, pp. 53–57.

Graham, M. J. & Heavner, D. L. (1993). Response of soybean genotypes to boron, zinc and manganese deficiency in tissue culture. *Plant and Soil*. Vol. 150, pp. 307-310.

Graham, R. D.; Welch, R. M.; Grunes, D. L.; Cary, E. E. & Norvell, W. A. (1987). Effect of zinc deficiency on the accumulation of boron and other mineral nutrients in barley. *Soil Science Society of America Journal*. Vol. 51, pp. 652-657.

Grant, C. & Bailey, L. D. (1993). Fertility management in canola production. *Canadian Journal of Soil Science*. Vol. 73, pp. 651-670.

Greenhill, A. W. (1938). Boron deficiency in horticultural crops: recent developments. *Scientific Horticulture*. Vol. 6, pp. 191-198.

Gupta, S. K. & Stewart, J. W. B. (1975). The extraction and determination of plant-available boron in soils. *Schweiz. Landwirtsch. Forsch.* Vol. 14, pp. 153–159.

Gupta, U. C. (1968). Relationship of total and hot-water soluble boron and fixation of added boron, to properties of Podzol soils. *Soil Science Society of America Proceedings.* Vol. 32, pp. 45-48.

Gupta, U. C. (1972). Interaction of boron and lime on barley. *Soil Science Society of America Proceedings.* Vol. 36, pp. 332-334.

Gupta, U. C. (1979a). Boron nutrition of crops. *Advances in Agronomy.* Vol. 31, pp. 273-307.

Gupta, U. C. (1979b). Some factors affecting the determination of hot water-soluble boron from Podzol soils using azomethine-H. *Canadian Journal of Soil Science.* Vol.59, pp. 241-247.

Gupta, U. C. (1983). Boron deficiency and toxicity symptoms for several crops as related to tissue boron levels. *Journal of Plant Nutrition.* Vol. 6, pp. 387–395.

Gupta, U. C. (1984). Boron nutrition of alfalfa, red clover, and timothy grown on podzol soils of eastern. *Canadian Journal of Soil Science.* Vol. 137, pp. 16–22.

Gupta, U. C. (1993). Introduction. In: *Boron and its role in crop production,* U. C. Gupta (Ed.), Pages 237, CRC Press, Boca Raton, FL, U.S.A.

Gupta, U. C. (2006). Chapter 8. Boron. In: *Hand book of Plant Nutrition,* Barker, A.V. & D. J. Pilbeam (Eds.), 241-277, CRC Press, Boca Raton, FL, USA.

Gupta, U. C.; Jame, Y. M.; Campbell, C. A.; Leyshon, A. J. & Nicholaichuk, W. (1985). Boron toxicity and deficiency: A review. *Canadian Journal of Soil Science.* Vol. 65, pp. 381-409.

Haby, V. A. & Leonard, A. T. (2000). Alfalfa production on coastal plain soils. In: *Proceedings/Reports of the American Forage and Grassland Council,* 37th North American Alfalfa Improvement Conference, July 16 19, 2000, Madison, WI, U.S.A.

Haby, V. A. & Leonard, A. T. (2005). Sustainable alfalfa production on coastal plain soils of the United States. *Communications in Soil Science and Plant Analysis.* Vol. 36, pp. 47-63.

Han, S. & Chen, L. S. (2008). Boron deficiency decreases growth and photosynthesis, and increases starch and hexoses in leaves of citrus seedlings. *Journal of Plant Physiology.* Vol. 165, pp. 1331-1341.

Hatcher, J. T.; Blair, G. Y. & Bower, C. A. (1959). Response of beans to dissolved and absorbed boron. *Soil Science.* Vol. 88, pp. 98-100.

Hatcher, J. T.; Blair, G. Y. & Bower, C. A. (1962). Adjusting soil solutions to specified B concentrations. *Soil Science.* Vol. 94, pp. 55-57.

Hazra, C. R. & Tripathi, S. B. (1998). Effect of secondary and micronutrients on yield and quality of forages. Fert. News 43: 77-82.

Hill, H. (1936). Minor elements affecting horticultural crops. *Scientific Agriculture.* Vol. 17, pp. 148-153.

Hill, W. E. & Morrill, L. G. (1975). Boron, calcium and potassium interactions in Spanish peanuts. *Soil Science Society of America Proceedings.* Vol. 39, pp. 80-83.

Hingston, F. J. (1964). Reaction between boron and clays. *Australian Journal of Soil Research.* Vol. 2, pp. 83-95.

Ho, S. B. (2000). Boron deficiency of crops in Taiwan. Department of Agricultural Chemistry, National 684 Taiwan University, Taipei 106, Taiwan.

Hopkins, B. G.; Jolley, V. D.; Webb, B. L. & Callahan, R. K. (2010). Boron fertilization and evaluation of four soil extractants: Russet Burbank potato. *Communications in Soil Science and Plant Analysis.* Vol. 41, pp. 527-539.

Hossain, A. K. M. Z. & Hossain, M. A. (2004). Effects of aluminum and boron supply on growth of seedlings among 15 cultivars of wheat (*Triticum aestivum* L.) grown in Bangladesh. *Soil Science and Plant Nutrition.* Vol. 50, pp. 189-195.

Hou, J.; Evans, L. J. & Spiers, G. A. (1996). Chemical fractionation of soil boron. I. Method development. *Canadian Journal of Soil Science.* Vol. 76, pp. 485–491.

Hu, Y. S.; Ma, Y. H.; Sun, Y. L. & Guo, G. (1994). Effect of B application on the agronomic traits, yields and oil contents of a double-row rape (*Brassica napus* L.) cultivar. *Oil Crops (China).* Vol. 16, pp. 43-46.

Huang, L. & Ye, Z. (2005). Boron nutrition and chilling tolerance of warm climate crop species. *Annals of Botany.* Vol. 96, pp. 755-767.

Ibrahim, M. E.; Bekheta, M. A.; El-Moursi, A. & Gaafar, N. A. (2009). Effect of arginine, pro-hexadione-Ca, some macro and micro-nutrients on growth, yield and fiber quality of cotton plants. *World Journal of Agricultural Sciences.* Vol. 5, pp. 863-870.

Iyenger, S. S.; D. C. Martens, & W. P. Miller. (1981). Distribution and plant availability of soil zinc fractions. *Soil Science Society of America Journal.* Vol. 45, pp. 735–739.

Jeffrey, A. J. & McCallum, L. E. (1988). Investigation of a hot 0.01 M CaCl2 soil boron extraction procedure followed by ICP-AES analysis. *Communications in Soil Science and Plant Analysis.* Vol. 19, pp. 663–673.

Jin, JI-Yun, D. C. Martens, & L. W. Zelazny. (1988). Plant availability of applied and native boron in soils with diverse properties. *Plant and Soil.* Vol. 105, pp. 127-132.

Johnson, G. (2006). *Pakistan citrus industry challenges: Opportunities for Australia-Pakistan collaboration in research, development and extension.* Islamabad, Pakistan.

Jones, J. B. (Jr). (2003). Plant mineral nutrition. In: *Agronomic handbook: Management of crops, soils and their fertility,* Pages 325, CRC Pres, Boca Raton, FL, U.S.A.

Kabata-Pendias, A. & Pendias, H. (1992). *Trace elements in soils and plants,* 2nd ed., CRC Press, Boca Raton, FL, U.S.A.

Kanwal, S.; Rahmatullah; Aziz, T.; Maqsood, M. A. & Abbas, N. (2008). Critical ratio of calcium and boron in maize shoot for optimum growth. *Journal of Plant Nutrition,* Vol. 31, pp. 1535-1542.

Kao, M. M. & Juang, T. C. (1978). Comparison of boron test methods for sugarcane soils in Taiwan. In: *Report of Taiwan Sugar Experiment Station,* 15-27, Tainan, Taiwan.

Kaplan, D. I.; Burkman, W.; Adriano, D. C.; Mills, G. L. & Sajwan, K. S. (1990). Determination of boron in soils containing inorganic and organic boron sources. *Soil Science Society of America Journal.* Vol. 54, pp. 708–714.

Katyal, J. C. & Randhawa, N. S. (1983). Micronutrients. *FAO Fertilizer and Plant Nutrition Bulletin. No. 7.* Rome, Italy.

Kavipurapu, C. S.; Gupta, K. K.; Dasgupta, P.; Chatterjee, N. & Pandey, L. P. (1993). Determination of boron in steels by inductively coupled plasma optical emission spectrometry. *Analusis.* Vol. 21, pp. 21–25.

Keren, R. & Bingham, F. T. (1985). Boron in water, soils, and plants. *Advances in Soil Sciences.* Vol. 1, pp. 229-276.

Keren, R. & Communar, G. (2009). Boron sorption on wastewater dissolved organic matter: pH effect. *Soil Science Society of America Journal.* Vol. 73, pp. 2021-2025.

Khan, M. A.; Shaukat, S. S. & Khan, M. A. (2008). Economic benefits from irrigation of maize with treated effluent of waste stabilization ponds. *Pakistan Journal of Botany.* Vol. 40, pp. 1091-1098.

Khattak, J. (1995). Micronutrients in Pakistan agriculture. *Project report, Pakistan Agriculture Research Council,* Islamabad and NWFP Agriculture University Peshawar, Pakistan. 135 pp.

Kirk, G. J. & Loneragan, J. F. (1988). Functional boron requirement for leaf expansion and its use as a critical value for diagnosis of boron deficiency in soybean. *Agronomy Journal.* Vol. 80, pp. 758-762.

Kormilitsyn, V. F. (1992). Effect of rates of boron, manganese and molybdenum on lucerne yield under irrigation in the Volga region. *Agrokhimiya.* Vol. 8, pp. 94-98.

Kumar, D. & Sidhu, S. (2009). Influence of soil applied sulfur and boron on yield and quality parameters of soybean. *Annals of Biology.* Vol. 25, pp. 105-111.

Lancaster, J. D.; Murphy, B. C.; Hurt, B. C. Jr.; Arnold, B. L.; Coats, R. E.; Albritton, R. C. & Walton, L. (1962). *Boron for cotton.* Bull. 635 pp. State College, Mississippi State University, Agricultural Experiment Station, Mississippi, U.S.A.

Lee, J. J.; van-derWalinga, I.; Manyeki, P. K.; Houba, V. J. G. & Novozamsky, I. (1987). Determination of boron in fresh and in dried plant material by plasma emission spectrometry after extraction with HF-HCl. *Communications in Soil Science and Plant Analysis.* Vol. 18, pp. 789-802.

Li, B.; Li, H.; Kui, W. H.; Chao, M. C.; Jern, W. S.; Li, H. P.; Chu, W. J. & Wang, C. L. (1978). Studies on cause of sterility of wheat. *Journal of Northeastern Agriculture College.* Vol. 3, pp. 1-19.

Liu, G. & Zhang, X. (2005). Effects of combined application of B, Mo and Zn fertilizers on yield and crude protein content of lucerne. *Grassland (China).* Vol. 27, pp. 13-18.

Li, R. & Gupta, U. C. (1991). Extraction of soil B for predicting its availability to plants. *Communications in Soil Science and Plant Analysis.* Vol. 22, pp. 1003-1012.

Liu Z.; Zhu, Q. Q. & Tong, L. H. (1980). Boron deficient soils and their distribution in China. *Acta Pedologica Sinica.* Vol. 17, pp. 228-239.

Liu Z.; Zhu, Q. Q. & Tong, L. H. (1982a). On the status of microelements in soils and their role in crop production of China. In: *Proceedings of International Conference of Chemicals and World Food Supplies,* December 1982, Manila, Philippines.

Liu Z.; Zhu, Q. Q. & Tong, L. H. (1983). Microelements in the main soils of China. *Soil Science.* Vol. 135, pp. 40-46.

Liu Z.; Zhu, Q. Q. & Tong, L. H. (1989). Regularities of content and distribution of boron in soils. *Acta Pedologica Sinica.* Vol. 26, pp. 353-361.

Liu Z.; Zhu, Q. Q.; Tong, L. H.; Xu, J. X. & Yen, C. L. (1982b). Geographical distribution of trace elements deficient soils in China. *Acta Pedologica Sinica.* Vol. 19, pp. 209-223.

Lohse, G. (1982). Microanalytical azomethine-H method for boron determination in plant tissue. *Communications in Soil Science and Plant Analysis.* Vol. 13, pp. 127-134.

Loomis, W. D. & Durst, R. W. (1992). Chemistry and biology of boron. *Biofactors.* Vol. 3, pp. 229-239.

Lora, R. S. (1978). Respuesta de los cultivos de clima frio a la aplicación de micronutrientos. Memorias del V coloquio de suelos. *Suelos Ecuatoriales. SOCIEDAD Colombiana de la Ciencia del Suelo.* Vol. 9, pp. 183-191.

Lovatt, C. J. (1985). Evolution of xylem resulted in a requirement of boron in the apical meristems of vascular plants. *New Phytologist.* Vol. 99, pp. 509-523.

Luo, X. Y.; Peng, Y. C. & Wang, B. Y. (1990). Effect of boron fertilization on yield and quality of groundnut. *Journal of Zhejiang Agricultural Science.* Vol. 1, pp. 30-32.

Mackin, J. E. (1986). The free-solution diffusion coefficient of boron: influence of dissolved organic matter. *Marine Chemistry.* Vol. 20, pp. 131-140.

Mahler, R. L.; Naylor, D. V. & Fredrichson, M. K. (1984). Hot water extraction of boron from soils using sealed plastic pouches. *Communications in Soil Science and Plant Analysis.* Vol. 15, pp. 479-492.

Malhi, S. S.; Raza, M.; Schoenau, J. J.; Mermut, A. R.; Kutcher, R.; Johnston, A. M. & Gill, K. S. (2003). Feasibility of B fertilization for yield, seed quality, and B uptake of canola in north eastern Saskatchewan. *Canadian Journal of Soil Science.* Vol. 83, pp. 99-108.

Malik, M. A.; Makhdum, M. I. & Shah, S. I. H. (1992). Cotton response to boron fertilizer in silt loam soils. In: *Proceedings of* soil health for sustainable agriculture, 331-336, *3rd National Congress of Soil Science,* March 20-22, 1990, Lahore, Pakistan, Soil Science Society of Pakistan.

Marschner, H. (1995). *Mineral nutrition of higher plants.* 2nd ed. Academic Press. London. UK. 889 pp.

Matsi, T.; Antoniadis, V. & Barbayiannis, N. (2000). Evaluation of the NH_4HCO_3-DTPA soil test for assessing boron availability to wheat. *Communications in Soil Science and Plant Analysis.* Vol. 31, pp. 669-678.

McGeehan, S. L.; Topper, K. & Naylor, D. V. (1989). Sources of variation in hot water extraction and colorimetric determination of soil boron. *Communications in Soil Science and Plant Analysis.* Vol. 20, pp. 1777–1786.

McLarty, H. R. (1940). British Columbia uses boron for fruit. In: *Better Crops.* Vol. 24, No. 4, pp. 8-11; 37-38.

Mehmood, E. H.; Kausar, R.; Akram, M. & Shahzad, S. M. (2009). Is boron required to improve rice growth and yield in saline environment? *Pakistan Journal of Botany.* Vol. 41, pp. 1339-1350.

Mengel, K. & Kirkby, E. A. (1982). *Principles of plant nutrition.* 3rd ed. International Potash Institute, Worblaufen-Bern, Switzerland.

Mengel, K. & Kirkby, E. A. (1987). *Principles of plant nutrition.* 4th ed. International Potash Institute, Worblaufen-Bern, Switzerland.

Mengel, K. & Kirkby, E. A. (2001). Boron. In: *Principles of plant nutrition.* 621-638, Kluwer Academic Publishers (5th ed.) Dordrecht/ Boston/ London, Netherlands.

Mezuman, U. & Keren, R. (1981). Boron adsorption by soils using a phenomenological adsorption equation. *Soil Science Society of America Journal.* Vol. 45, pp. 722-726.

Midgley, A. R. & Dunklee, D. E. (1947). Boron for Vermont soils and crops. *Exp. Sta. Bull.* 539. 20 pp. Agricultural Experiment Station, University of Vermont and State Agricultural College, Burlington, VT, U.S.A.

Miley, W. N.; Hardy, G. W. & Sturgis, M. B. (1969). Influence of boron, nitrogen and potassium on yield, nutrient uptake and abnormalities of boron. *Agronomy Journal.* Vol. 61, pp. 9-13.

Miller, R. O.; B. Vaughan, & J. Kutoby-Amacher. (2000). Extraction of soil boron with DTPA-sorbitol. *Soil and Plant Analysis Spring:* pp. 4–10.

Mishra, L. C. & Shukla, K. N. (1986). Effects of fly ash deposition on growth, metabolism and dry matter production of maize and soybean. *Environmental Pollution Series A: Ecol. Biol.* Vol. 42, 1-13.

Misra, R.; Munankarmi, R. C.; Pandey, S. P. & Hobbs, P. R. (1992). Sterility work in wheat at Tarahara in the eastern Terai of Nepal. In: *Boron deficiency in wheat, Wheat Special Report No. 11. CIMMYT*, 67-71, C. E. Mann & B. Rerkasem (Eds.), Mexico.

Miwa, K. & Fujiwara, T. (2010). Boron transport in plants: co-ordinated regulation of transporters. *Annals of Botany.* Vol. 105, pp. 1103–1108.

Moraghan, J. T. & Mascagni. H. J. (1991). Environmental and soil factors affecting micronutrient deficiencies and toxicities. In: *Micronutrients in agriculture*, 371-425, R. J. Luxmoore (Ed.), Soil Sci. Soc. Am.; Madison, WI, U.S.A.

Mortvedt, J. J. (1994). Boron diet essential for crops. *Farm Chemicals.* February, 1994. 2 pp.

Mortvedt, J. J. & Woodruff, J. R. (1993). Technology and application of boron fertilizers for crops. In: Boron and its role in crop production, U. C. Gupta (Ed.), 156-176, CRC, Press, Boca Raton, FL, U.S.A.

Murphy, B. C. & Lancaster, J. J. D. (1971). *Response of cotton to boron.* Agronomy Journal. *Vol.* 63, pp. 539-540.

Mustafa, W. A.; Falatah, A. M. & El. Shall, A. A. (1993). Effect of excess boron fertilization on status and availability of boron in calcareous soils. *Fertilizer Research.* Vol. 36, pp. 71–78.

Nable, R. O.; Banuelos, G. S. & Paull, J. G. (1997). Boron toxicity. *Plant and Soil.* Vol. 193, pp. 181–198.

Nable, R. O.; Lance, R. C. M. & Cartwright, B. (1990). Uptake of boron and silicon by barley genotypes with differing susceptibilities to boron toxicity. *Annals of Botany.* Vol. 66, pp. 83–90.

Nazli, F.; Bibi F.; F.; Gul, M. & Hannan, A. (2010). Effect of graded level of potassium on quality of potato crop. In: *Abstracts 13th International Congress of Soil Science*, 13 pp., March 24-27, 2010, Faisalabad, Pakistan.

Niaz, A.; Ibrahim, M.; Nisar, A. & Anwar, S. A. (2002). Boron contents of light and medium textured soils and cotton plants. *International Journal of Agriculture and Biology.* Vol. 4, pp. 534–536.

Niaz, A.; Ranjha, A. M.; Rahmatullah; Hannan, A. & Waqas, M. (2007). Boron status of soils as affected by different soil characteristics–pH, $CaCO_3$, organic matter and clay contents. *Pakistan Journal of Agricultural Sciences.* Vol. 44, pp. 428-435.

Nilsson L G & Jennische P. (1986). Determination of boron in soils and plants. *Swedish Journal of Agricultural Research.* Vol. 16, pp. 97–103

Nogueira, A. R. A.; Brienza, S. M. B.; Zagatto, E. A. G.; Lima, J. L. F. C. & Araujo, A. N. (1993). Multi-site detection in flow analysis: Part 2. Monosegmented systems with relocating detectors for the spectrophotometric determination of boron in plants. *Analytica Chimica Acta.* Vol. 276, pp. 121–125.

Novozamsky, I.; Barrera, L. L.; Houba, V. J.; Vanderlee, J. J. & Eck, R. (1990). Comparison of a hot water and cold 0.01 M $CaCl_2$ extraction procedures for the determination of boron in soil. *Communications in Soil Science and Plant Analysis.* Vol. 21, pp. 2189–2195.

Nuttall, W. F.; Ukrainetz, H.; Stewart, J. W. B. & Spurr, D. T. (1987). The effect of nitrogen, sulphur and boron on yield and quality of rapeseed (*Brassica napus* L. and *Brassica compestris* L.). *Canadian Journal of Soil Science.* Vol. 67, pp. 545-559.

Nyborg, M. & Hoyt, P. B. (1970). Boron deficiency in turnip rape grown on gray wooded soils. *Canadian Journal of Soil Science.* Vol. 50, pp. 87-88.

Offiah, O. & Axley, J. H. (1988). Improvement of boron soil test. *Communications in Soil Science and Plant Analysis.* Vol. 19, pp. 1527-1542.

Ogner, G. (1980). Automatic determination of boron in water samples and soil extracts. *Communications in Soil Science and Plant Analysis.* Vol. 11, pp. 1209-1219.

Okazaki, E. & Chao, T. T. (1968). Boron adsorption and desorption by some Hawaiian soils. *Soil Science.* Vol. 105, pp. 255-259.

Owens, J. W.; Gladney, E. S. & Knab, D. (1982). Determination of boron in geological materials by inductively coupled plasma emission spectrometry. *Analytica Chimica Acta.* Vol. 135, pp. 169-172.

Parker, D. R. & Gardner, E. H. (1981). The determination of hotwater-soluble boron in some acid Oregon soils using a modified azomethine-H procedure. *Communications in Soil Science and Plant Analysis.* Vol. 12, pp. 1311-1322.

Parr, A. J. & Loughman, B. C. (1983). Boron and membrane function in plants. In: *Metals and micronutrients: Uptake and utilization by plants,* 87-107, D. A. Robb & W. S. Pierpoint (Eds.), Academic Press, New York, NY, U.S.A.

Patel, P. C. & Patel, K. P. (2003). Effects of zinc and boron application on seed yield of lucerne (*Medicago sativa* L.). *Journal of the Indian Society of Soil Science. Vol.* 51, pp. 320-321.

Patil, G. D.; Patil, M. D.; Patil, N. D. & Adsule, R. N. (1987). Effect of boronated superphosphate, single superphosphate and borax on yield and quality of groundnut. *Journal of Maharashtra Agriculture University.* Vol. 12, pp. 168-170.

Ponnamperuma, F. N.; Clayton, M. T. & Lantin, R. S. (1981). Dilute hydrochloric acid as an extractant for available zinc, copper, and boron in rice soils. *Plant and Soil.* Vol. 61, pp. 297-310.

Porter, S. R, Spindler, S. C. & Widdowson, A. E. (1981). An improved automated colorimetric method for the determination of boron in extracts of soils soil-less peat-based composts plant materials and hydroponic solutions with Azomethine-H. *Communications in Soil Science and Plant Analysis.* Vol. 12, pp. 461-473.

Pougnet, M. A. B. & Orren, M. J. (1986a). Determination of boron by inductively coupled plasma atomic emission spectroscopy: Part 1. Method development. *International Journal of Environmental Analytical Chemistry.* Vol. 24, pp. 253-266.

Pougnet, M. A. B. & Orren, M. J. (1986b). Determination of boron by inductively coupled plasma atomic emission spectroscopy. Part 2. Applications to South African environmental samples. *International Journal of Environmental Analytical Chemistry.* Vol. 24, pp. 267-282.

Rahmatullah, Badr-uz-Zaman & Salim, M. (1999). Plant utilization and release of boron distributed in different fractions in calcareous soils. *Arid Soil Res. Rehab.* Vol. 13, pp. 293-303.

Rammah, A. M. & Khedr, M. S. (1984). Response of alfalfa to iron and boron in some Egyptian soils. *Journal of Plant Nutrition,*Vol. 7, pp. 235-242.

Rashid, A. (1993). Nutritional disorders of rapeseed-mustard and wheat grown in Potohar area. In: *Micronutrient Project Annual Report, 1991-1992*, 25-36, National Agricultural Research Center (NARC), Islamabad, Pakistan.

Rashid, A. 1994. Nutrient indexing surveys and micronutrient requirement of crops. In: *Micronutrient Project Annual Report, 1992-1993,11-19*, National Agricultural Research Center (NARC), Islamabad, Pakistan.

Rashid, A. (1995). Nutrient indexing of cotton and micronutrient requirement of cotton and groundnut. In: *Micronutrient Project Annual Report, 1994-95*, 13-28, National Agricultural Research Center (NARC), Islamabad, Pakistan.

Rashid, A. (2006). *Boron deficiency in soils and crops of Pakistan: Diagnosis and management.* Pakistan Agricultural Research Council (PARC), Islamabad, Pakistan. Viii, 34 pp.

Rashid, A. & Akhtar, M. E. (2006). Soil fertility research and nutrient management in Pakistan. In: *Proceedings of Symposium on Balanced Fertilizer Use: Impact on Crop Production*, 90-113, October 30-31, National Fertilizer Development Centre (NFDC), Islamabad, Pakistan.

Rashid, A. & Qayyum, F. (1991). *Cooperative research programme on micronutrient status of Pakistan soils and their role in crop production: Final Report*, 1983-1990. National Agricultural Research Center (NARC), Islamabad, Pakistan.

Rashid, A. & Rayan, J. (2004). Micronutrient constraints to crop production in soils with Mediterranean type characteristics: A review. *Journal of Plant Nutrition*, Vol. 27, pp. 959-975.

Rashid, A.; Muhammad, S. & Rafique, E. (2005). Rice and wheat genotypic variation in boron use efficiency. *Soil Environ.* Vol. 24, pp. 98-102.

Rashid, A.; Rafique, E. & Bughio, N. (1994). Diagnosing boron deficiency in rapeseed and mustard by plant analysis and soil testing. *Communications in Soil Science and Plant Analysis.* Vol. 25, pp. 2883-2897.

Rashid, A.; Rafique, E. and Bughio, N. (1997). Micronutrient deficiencies in rain-fed calcareous soils of Pakistan. III. Boron nutrition of sorghum. *Communications in Soil Science and Plant Analysis.* Vol. 28, pp. 444-454.

Rashid, A.; Yasin, M.; Ali, M. A.; Ahmad, Z. & Ullah, R. (2009). Boron deficiency in rice in Pakistan: A serious constraint to productivity and grain quality. In: *Salinity and water stress: Improving crop efficiency*, 213-219, M. Ashraf; M. Ozturk and H. R. Athar (Eds.). Springer-Verlag, Berlin-Heidelberg, Germany.

Raven, J. A. (1980). Short- and long-distance transport of boric acid in plants. *New Phytol.* Vol. 84, pp. 231-249.

Raza, M.; Mermut, A. R.; Schoenau, J. J. & Malhi, S. S. (2002). Boron fractionation in some Saskatchewan soils. *Canadian Journal of Soil Science.* Vol. 82, pp. 173-179.

Razzaq, A. & Rafiq, M. (1996). Soil classification and survey. In: *Soil Science*, 405-437, A. Rashid and K. S. Memon (managing authors), E. Bashir & R. Bental (Eds.), National Book of Foundation, Islamabad, Pakistan.

Reid, R. J.; Hayes, J. E.; Post, A.; Stangoulis, J. C. & Graham, R. D. (2004). A critical analysis of the causes of boron toxicity in plants. *Plant Cell Environ.* Vol. 27, pp. 1405–1414.

Reid, R.; & Fitzpatrick, K. (2009). Influence of leaf tolerance mechanisms and rain on boron toxicity in barley and wheat. *Plant Physiology.* Vol. 151, pp. 413–420.

Rerkasem, B. & S. Jamjod. (2004). Boron deficiency in wheat: A review. *Field Crops Research.* Vol. 89, pp. 173-186.

Rerkasem, B.; Bell, R. W.; Lodkaew, S. &. Loneragan, J. F. (1993). Boron deficiency in soybean, peanut and black gram: Symptoms in seeds and differences among soybean cultivars in susceptibility to boron deficiencies. *Plant and Soil.* Vol. 150, pp. 289-294.

Rerkasem, B.; Jamjod, S. & Nirantrayagul, S. (2004). Increasing boron efficiency in many international bread wheat, durum wheat, triticale and barley germplasm will boost production on soils low in boron. *Field Crops Research.* Vol. 86, pp. 175-184.

Reuter, D. J. & Robinson, J. B. (1986). *Plant analysis an interpretation manual.* Inkata Press, Melbourne, Australia.

Russell, E. W. (1973). *Soil condition and plant growth,* 849 pp, 10th ed. Longman Ltd.; London, UK

Ryan, J.; Miyamoto, S. & Stroehlein, J. L. (1977). Relation of solute and sorbed boron to the boron hazard of irrigation water. *Plant and Soil.* Vol. 47, pp. 253-256.

Sah, R. N. & Brown, P. H. (1997). Techniques for boron determination and their application to the analysis of plant and soil samples. *Plant and Soil.* Vol. 193, pp. 15–33.

Sakal, R. & Singh, A. P. (1995). Boron research and agricultural production. In: *Micronutrient research and agricultural production,* 1-64, H. L. S. Tandon (Ed.), Fertilizer Development and Consultation Organization, New Delhi, India. ISBN, 8185116601.

Sakal, R.; Singh, A. P.; Sinha, R. B. & Bhogal, N. S. (1988). Annual progress reports. ICAR All India Coordinated Scheme of Micro-and Secondary Nutrients in Soils and crops of Bihar, *Res. Bull., Department of Soil Science,* RAU, Samastipur, Bihar, India.

Salinas, R.; Cerda, A.; Romero, M. & Caro, M. (1981). Boron tolerance of pea (*Pisum sativum*). *Journal of Plant Nutrition.* Vol. 4, pp. 205-217.

Sarwar, N. & Mubeen, K. (2009). *Nutrient deficiency in rice crop.* DAWN [Online] Available:24 August. http://www.dawn.com/wps/wcm/connect/dawn-content-library/dawn/in-paper-magazine/economic-and-business/nutrient-deficiency-in-rice-crop-489   [15 May 2010].

Schalscha, E. B.; Bingham, F. T.; Galindo, G. G. & Galvan, M. P. (1973). Boron adsorption by volcanic ash soils in Southern Chile. *Soil Sci.* Vol. 116, pp. 70-76.

Scott, H. D.; Beasley, S. D. & Thompson, L. F. (1975). Effect of lime on boron transport and uptake by cotton. *Soil Science Society of America Proceedings.* Vol. 39, pp. 1116-1121.

Shafiq, M.; Ranjha, A. M.; Yaseen, M.; Mehdi, S. M. & Hannan, A. (2008). Comparison of freundlich and Langmuir adsorption equations for boron adsorption on calcareous soils. *Journal of Agricultural Research.* Vol. 46, pp. 141-148.

Shi, L. & Wang, Y. H. (2009). Inheritance of boron efficiency in oilseed rape. *Pedosphere.* Vol. 19, pp. 403-408.

Shiffler, A. K.; V. D. Jolley, J. E. Christopherson, B.L. Webb, D. C. Farrer, & V. A. Haby. (2005). Pressurized hot water and DTPA-sorbitol as viable alternatives for soil boron extraction. I. Boron-treated soil incubation and efficiency of extraction. *Communications in Soil Science and Plant Analysis.* Vol. 36, pp. 2179–2187.

Shorrocks, V. M. (1997). The occurrence and correction of boron deficiency. *Plant and Soil.* Vol. 193, pp. 121-148.

Sillanpaa, M. (1982). Micronutrients and nutrient status of soils, a global study. In: *FAO Soil Bulletin. No. 48,* Rome, Italy.

Sillanpaa, M. & Vlek, P. L. G. (1985). Micronutrients and the agroecology of tropical and mediterranean regions. In: Micronutrients in tropical food crop production, 151-167, P. L. G. Vlek, (Ed.), Martinus Nijhoff and W. Junk Publishers, Dordrecht, Netherlands.

Sims, J. R. & Bingham, F. T. (1967). Retention of boron by layer silicates, sesquioxides and soil materials: I. Layer silicates. *Soil Science Society of America Proceedings.* Vol. 31, pp. 728-732.

Sims, J. R. & Bingham, F. T. (1968a). Retention of boron by layer silicates, sesquioxides and soils materials: II. Sesquioxides. *Soil Science Society of America Proceedings.* Vol. 32, pp. 364-369.

Sims, J. R. & Bingham, F. T. (1968b). Retention of boron by layer silicates, sesquioxides and soil materials: III. Iron- and aluminum-coated layer silicates and soil materials. *Soil Science Society of America Proceedings.* Vol. 32, pp. 369-373.

Singh, A. L. (1994). Micronutrient nutrition and crop productivity in groundnut. In: *Plant productivity under environmental stress,* 67-72, K. Singh & S. S. Purohit (Eds.), Agro Botanical Publishers, Bikaner, India.

Singh, M. V. (2001). Evaluation of micronutrient stocks in different agroecological zones of India. *Fertilizer News.* Vol. 42, pp. 25-42.

Sinha, R. B.; Sakal, R.; Singh, A. P.; Bhogal, N. S. (1991). Response of some field crops to boron application in calcareous soils. *Journal of Indian Society of Soil Science.* Vol. 39, pp. 342-345.

Smithson, J. B. & Heathcote, R. G. (1976). A new recommendation for the application of boronated superphosphate to cotton in northeastern Beune Plateau States. *Samarau Agri Newsletter.* Vol. 18, pp. 59-63.

Sotiropoulos, T. (1997). Boron toxicity of kiwifruit orchards in northern Greece. *Acta Horticulture.* Vol. 44, pp. 243-247.

Soylu, S. & Topal, A. (2004). Yield and yield attributes of durum wheat genotypes as affected by boron application in boron-deficient calcareous soils: an evaluation of major Turkish genotypes for boron efficiency. *Journal of Plant Nutrition,*Vol. 27, pp. 1077-1106.

Spiers, G. A.; Evans, L. J.; Mcgeorge, S. W.; Moak, H. W. & Su, C. (1990). Boron analysis of soil solutions and plant digests using a photodiode-array equipped ICP spectrometer. *Communications in Soil Science and Plant Analysis.* Vol. 21, pp. 1645-1661.

Spouncer, L. R.; Nable, R. O. & Cartwright, B. (1992). A procedure for the determination of soluble boron in soils ranging widely in boron concentrations sodicity and pH. *Communications in Soil Science and Plant Analysis.* Vol. 23, pp. 441-453.

Stangoulis, J. C.; Reid, R. J.; Brown, P. H. & Graham, R. D. (2001). Kinetic analysis of boron transport in Chara. *Planta.* Vol. 213, pp. 142-146.

Tahir, M.; Hussain, F.; Kausar, M. A. & Bhatti, A. S. (1990). Differential uptake and growth response to micronutrients in various rice cultivars. *Pakistan Journal of Agricultural Sciences.* Vol. 27, pp. 367-373.

Takkar, P. N.; Chibba, I. M. & Mehta, S. K. (1989). Twenty years of coordinated research on micronutrient in soils and plants. *Bull. 314. Indian Institute of Soil Science,* Bhopal, India.

Takkar, P.N. & Nayyar, V. K. (1984). Integrated approach to combat micronutrie nt deficiency. In: *Proceedings of FAI Annual Seminar*, PS, Vol. 111, No. 2, pp. 1-16. New Delhi, India.

Tariq, A.; Gill, M. A.; Rahmatullah & Sabir, M. (2004). Mineral nutrition of fruit trees. In: *Proceedings Plant-nutrition management for horticultural crops under water stress conditions*, 28-33. Agriculture Research Institute Sariab, Quetta, Pakistan.

Tisdale, S. L.; Nelson, W. L.; Beaton, J. D. & Havlin. J. L. (1993). *Soil fertility and fertilizers*. 5th ed. Macmillan Publishing Company, New York, NY, U.S. A.

Tsadilas, C. D.; Yassoglou, C. S.; Cosmas, C. S. & Kallianou, C. H. (1994). The availability of soil boron fractions to olive trees and barley and their relationships to soil properties. *Plant Soil* . Vol. 162, pp. 211- 217.

Tsadilas, C. D.; D. Dimoyiannis, & V. Samaras. (1997). Methods of assessing boron availability to kiwifruit plants growing on high boron soils. *Communications in Soil Science and Plant Analysis*. Vol. 28, pp. 973-987.

Turan, M.;. Ketterings, Q. M.; Gunes, A.; Ataoglu, N.; Esring, A.; Bilgili, A. V. & Huang, Y. M. (2010). Boron fertilization of Mediterranean aridisols improves lucerne (*Medicago sativa* L.) yields and quality. *Acta Agriculturae Scandinavica*, Section B - Plant Soil Sci. Vol. 60, pp. 427 - 436.

US Borax Inc. (1996). The boron bonus canola. US Borax Inc., Valencia, CA. No. 280412/1-96.

US Borax Inc. (2002). B in cotton [Online] Available: http://www.borax.com/agriculture/files/cotton.pdf [May 15 2010].

US Borax Inc. (2009). Functions of boron in plant nutrition [Online] Available: http://www.borax.com/agriculture/files/an203.pdf [August 15 2010].

Vasil, I. K. (1987). Physiology and culture of pollen. In: *Pollen cytology and development*, 127-174, K. L. Giles & J. Prakash (Eds.), Orlando, FL, U.S.A.

Vaughan, B. & J. Howe. (1994). Evaluation of boron chelates in extracting soil boron. *Communications in Soil Science and Plant Analysis*. Vol. 25, pp. 1071-1084.

Wang, K., &. Chen, Q. (2003). Effect of Zn, B, and Mo application on growth and quality of alfalfa. *Soils and Fertilizers. Beijing*. Vol. 3, pp. 24-28.

Warington, K. (1923). The effect of boric acid and borax on the broad bean and certain other plants. *Annals of Botany*. Vol. 37, pp. 629-672.

Warncke, D. D. (2005). Ameliorating internal black spot in cranberry bean seed with boron application. *Communications in Soil Science and Plant Analysis*. Vol. 36, pp. 775-781.

Wear, J. I. & Patterson, R. M. (1962). Effect of soil pH and texture on the availability of water-soluble boron in the soil. *Soil Science Society of America Proceedings*. Vol. 26, pp. 344-346.

Welch, R. M. (1995). Micronutrient nutrition of plants. *Critical Reviews in Plant Sciences*. Vol. 14, pp. 49-82.

Welch, R. M.; Allaway, W. H.; House W. A. & Kubota, J. (1991). Geographic distribution of trace element problems. In: *Micronutrients in agriculture*, 31-57, J. J. Mortvedt (Ed.). 2nd edition, Madison, WI, U.S.A.

Wild, A. & Mazaheri, A. (1979). Prediction of the leaching rate of boric acid under field conditions. *Geoderma*. Vol. 22, pp. 127-36.

Wong, M. (2003). Monitor crops closely for signs of low boron. *Farming Ahead*. No. 135. March 2003.

Wooding, F. J. (1985). Interior Alaska crops respond to boron applications. *Agroborealis*. Vol. 17, pp. 47-49.

Woodruff, J. R.; Moore, F. W. & Musen, H. L. (1987). Potassium, boron, nitrogen and lime effects on corn yield and ear leaf nutrient concentrations. *Agronomy Journal*. Vol. 79, pp. 520-524.

Wu, M. (1986). A study on boron deficiency in soybean. *Soybean Science*. Vol. 5, pp. 167-174.

Yang, Y. H. & Gu, H. J. (2004). Effects of boron on aluminum toxicity on seedlings of two soybean cultivars. *Water Air & Soil Pollution*. Vol. 154, pp. 239-248.

Zhang, J. H.; Zhang, D. J. Jiang, Z. L. & Liu, C. X. (1986). A study on boron nutrition and application of boron to groundnut. *Journal of Soil Science. (China)*. Vol. 17, pp. 173-176.

Zhao, D. & Oosterhuis, D. M. (2003). Cotton growth and physiological responses to boron deficiency. *Journal of Plant Nutrition*.Vol. 26, pp. 855-867.

Zia, M. H.; Ahmad R.; Khaliq, I.; Ahmad, A. & Irshad, M. (2006). Micronutrients status and management in orchards soils: applied aspects. *Soil & Environment*. Vol. 25, pp. 6-16.

Zia, M. S. (1993). Fertilizer use efficiency project and soil fertility. *Agricultural Research Project-II (ARP-II) Annual Report*. 1992-93. NARC, Islamabad, Pakistan.

Part 2

General

6

Comparative Analyses of Extracellular Matrix Proteome: An Under-Explored Area in Plant Research

Kanika Narula, Eman Elagamey, Asis Datta,
Niranjan Chakraborty and Subhra Chakraborty*
*National Institute of Plant Genome Research, Aruna Asaf Ali Marg, New Delhi,
India*

1. Introduction

Within their social milieu, cells are petite and deformable, enclosed in a flimsy plasma membrane which swerves from their default spherical shape to more polar shapes due to the local deposition, complex interactions and the remodelling of the extracellular matrix (ECM). Consequently, multicellularity has evolved, albeit independently in plants and animals. Although animals are truly multicellular, plants are supracellular organisms because their immobile cells divide via phragmoplast-based incomplete cytokinesis, which results in the formation of cytoplasmic cell-to-cell channels known as plasmodesmata (Baluska et al., 2003). The ECM in plants, often referred as the cell wall, is integrated into the apoplast—a structurally coherent superstructure extending throughout the plant body. In lieu, plant cells are not fully separated and both the plasma membrane and endoplasmic reticulum traverse cellular borders through plasmodesmata (Baluska et al., 2003; Fincher, G. 2009.). The ECM is a fundamental component of the microenvironment of both animal and plant cells that has been substantially expanded during evolution. Throughout the plant kingdom, the formation and regulation of the ECM architecture has been shown to have the potential to influence many conduits of development, position-dependent differentiation, patterning and totipotent cell niches, besides environmental stress response and pathobiology (Brownlee & Berger, 1995; Degenhardt & Gimmer, 2000; Wilson, 2010). Furthermore, it has been reported that the ECM plays an important morphoregulatory role during somatic embryogenesis and organogenesis in plants, besides its pivotal role in cellular osmo- and volume-regulation (Šamaj et al., 1999; Rose et al., 2004). The plant ECM has biomechanical and morphogenetic functions with the immense ability to turn cells into hydraulic machines which establish a crucial functional difference between cell walls and other cellular surface structures. It encloses the cell hermetically and constrains the hydrostatic pressure evoked by osmotic gradients between the cell and its environment which controls cellular osmo- and volume-regulation (Peters et al., 2000; Cosgrove, D. J. 2005). Plasticity in the ECM allows the cellular uptake of massive amounts of water into

---

*Corresponding author

a central vacuole while rigidity in the ECM determines the conductance of enormous amounts of water and dissolved solutes through vascular bundles. The secretion of an ECM by one cell can also influence the neighbouring cells, conceivably the best exemplified paracrine interaction known in the plant kingdom (for a review, see Brownlee, 2002). Beyond their paramount importance in the generation of form, cell walls are frequently considered 'growth-controlling' (Wolf et al., 2009). Cells devoid of the ECM inevitably lose their polar shape and the loss of cellular polarisation prevents cell-to-cell interactions and communication. The ECM/cell wall is evolutionary and inherently bestowed with information that can be both stored and relayed to cell interior via templating processes. It serves as the first line mediator in cell signalling for perceiving and transmitting extra- and intercellular signals in many cellular pathways. Communication between the cytoplasm and the cell wall is necessary and evident because of events such as cell expansion (Cosgrove, 1997, Schröder, F et al 2009), mechanical stress (Kumar et al., 2006; Telewski, 2006), environmental perturbation (Gail McLean et al., 1997; Thelen, J. and Peck, S. 2007) and pathogen infection (Hammond-Kosack & Jones, 1996) which lead to altered biosynthesis and the modification of wall components and downstream cytoplasmic events. In addition, it can act as a substrate for migration and has also been recognised as a surrogate for providing inputs into cell behaviour (Hall et al., 2002), although the available data is rather scarce for higher plants and critical linker molecules between the cytoskeleton and the ECM are still missing. Thus, the ECM/cell wall primarily serves a dual function, as a cell support system and for signalling during development and stress. The ECM/cell wall must therefore be dynamic as cells divide and elongate, modulating its composition and architecture during its synthesis and after it has been deposited. The wall function is a multi-step, complex process and the underlying mechanisms governing these steps are not fully understood.

Proteome research holds promise of understanding the molecular basis of the ECM function using an unbiased comparative and differential approach. We and others have identified several hundred plant proteins that include both predicted and non-canonical ECM components, presumably associated with a variety of cellular functions; viz. cell wall modification, signal transduction, cellular transport, metabolism, cell defence and rescue, all of which impinge on the complexity of ECM proteins in crop plants (Bhushan et al., 2006; Telewski, 2006). In recent years, reports have also been published focusing on changes in the ECM proteome in varied cellular events (Jones et al., 2004; Irshad et al., 2008, Bhushan et al., 2007, Cheng et al., 2009; Pandey, et al., 2010, Bhushan et al., 2011). The proteins that have been identified reveal the presence of complex regulatory networks that function in this organelle. Currently, we are focusing on disease-responsive ECM proteomes in order to understand the ECM-related pathobiology in plants. Although over the past few years there have been rapid advances in cell wall proteome research, the study of the complexity of ECM proteins remained secondary, irrespective of the fact that they correspond to about 10% of the ECM's mass and are comprised of several hundred different molecules with diverse functions. Moreover, a vast array of post-translational modifications to these proteins adds diversity to the structure and ligand-binding properties of matrix components, leading to their differential activity. Therefore, characterisations of the ECM proteome in plants hold the promise of increasing our understanding about the gene's function.

In this report, we begin by summarising the essential and unique features of the ECM and we discuss recent findings concerning the regulation and biochemistry of it, with specific emphasis on the fundamental role of ECM proteins in development, environmental stress and signalling by analysing the ECM's proteomes. Furthermore, we report here the comparative analysis of ECM proteomes towards crop specificity, organ-based, developmental and environmental adaptations based on our own findings, the available literature and databases focusing on ECM proteins in view of the current understanding and perspectives of the ECM's functions.

1.1 Exploring the sink and link in ECM

Ubiquitously present, the ECM/cell wall is composed of different molecules with diverse functions to meet the specialised requirements of different tissues. It is a dynamic milieu, having homeostatic properties and a reservoir for bioactive molecules, such as carbohydrates and proteins. Long before the determination of comprehensive chemical differences between plant and animal ECMs, Boerhaave proposed in the early 18th century that fermented plant material which is rich in carbohydrate is acidic whereas putrefied animal material which is rich in protein is basic (ammoniacal) (Rose, 2003). It alludes only briefly to the differences and similarities between the ECMs of higher plants and animals. Consequently, proteins are largely responsible for the chemical transformation properties that distinguish plants from animals. The ECM in higher plants and higher animals consists of a mixture of fibrous and amorphous components. In higher animals, a protein-based collagen - elastin, a fibronectin fibre matrix is infiltrated by mucopolysaccharides, peptidoglycans and calcium phosphate, whereas in higher plants, non-protein cellulose and protein-based extensin fibre matrices are infiltrated by a varied assortment of non-nitrogenous hemicelluloses, pectins and lignins and, to a much smaller extent (on a mass basis), by various structural proteins and enzymes (Irshad et al., 2008). Similarities in ECM design may be apparent as it is likely that ancient functional protein domains and carbohydrate backbones have been used in a variety of arrangements and combinations to affect the function of convergent biological structures. On the contrary, as stated by Darwin the "web of wall molecules have a long evolutionary history," and it is therefore relevant that different family members show highly regulated and specific patterns of the expression of ECM components in an evolutionary context. In addition to protein heterogeneity and the presence of various metals as linkers, carbohydrate compositions can vary between cell types and even within one wall of a given cell, suggesting that the cell wall serves as a sink of variability in terms of macromolecules or microelements. On a fresh mass basis, the vegetative growth of all organisms (70-90% water) is predominantly owing to water uptake, but on a dry mass basis the vegetative growth of plants differs markedly from that of animals (Rose, 2003). During differentiation, plant cells increase in size from typically 10^2 mm^3 (volume of a meristematic cell) to up to 10^7 mm^3 (e.g., a xylem vessel). This increase in cellular volume requires the addition of building materials in the form of cell wall polymers and membranes. While new cell wall materials are incorporated, the existing material is deformed and stretched mechanically. The force for this deformation is supplied by the turgor pressure (Geitmann, 2010). The ECM serves as the first line mediator in cell signalling to perceive and transmit extra- and intercellular signals in many cellular pathways. ECM proteins constitute more than just a structural framework but they also play a variety of roles in growth and development, defence against environmental stresses as well as giving structural support.

1.1.1 The ECM protein sink: A dynamic framework for multiple functions

Earlier it was believed that ECM proteins were large and complex, with multiple distinct domains, and were highly conserved among the different taxa (Hall & Cannon, 2002). However, it is not necessary that proteins be large or complex in order to generate strong, stable fibrils and intermediate filament proteins. The conserved domains are now known to be arranged in specific juxtapositions, sometimes controlled by highly regulated alternative splicing (Hynes, 2009), indicating thereby that the specific domains and architectures of ECM proteins contain information of biological importance and evolutionary value. In plants, abundant wall proteins include those rich in hydroxyproline or proline (HRGPs, PRPs), glycine (GRPs) and arabinogalactan (AGPs). Expansins, which relax the linkages of the wall during cell elongation, play a crucial role in development. Peroxidases, methyltransferases, galactosidases, glycanases and proteases have also been identified as the cell wall resident proteins having an N-terminal targeting sequence. Perhaps the protein most expected to be similar to their metazoan counterparts in the plant cell wall is aggrecan, which binds hyaluronan orthologs. At least three classes of hydroxyproline-rich glycoproteins exist in higher plants, namely extensins, arabinogalactans and solanaceae lectins (Hall & Cannon, 2002,). Extensins, which comprise 5-10% of wall proteins, are assumed to play a role in the structure of plant cell walls and may, therefore, be important in controlling growth. Increasing evidence suggests that the level of extension is developmentally regulated. It also accumulates upon wounding and pathogen attack, suggesting its involvement in plant defence (Cassab, 1998). The fact that extensins and collagens are hydroxyproline containing glycoproteins means that they may have common evolutionary precursors (Chen & Verner, 1985). In addition, the primary cell wall includes numerous enzymes, viz. endoglucanases, xyloglucan endotransglycosylases and a number of other glycosyl transferases that alter carbohydrate linkages and modify secreted cell wall components. Tetraspanin - one of the important classes of ECM protein in higher plants but absent from unicellular eukaryotes - is known as the secretory carrier membrane protein, important for synaptic vesicle recycling in stigma-pollen interaction. Other cell wall proteins, some of which are heavily glycosylated, have been proposed as structural cell wall components and have been implicated in mediating multiple aspects of plant development (Irshad et al., 2008). Germin is another ECM protein that signals the onset of growth and determines plant immunity. A chronic theme proverbial to the class of ECM-cytoskeleton linker proteins of plant cells is that these mechano-transducing transmembrane molecules communicate and interact preferentially with the actin cytoskeleton on the cytoplasmic side of the plasma membrane. Generally, the actin cytoskeleton has been optimised during eukaryotic evolution for acting as a structural scaffold for diverse signalling complexes (Baluska et al., 2003). Bruce Kohorn classified putative plant-specific linker molecules in four categories, focusing on the four most appealing candidates: cell wall-associated kinases (WAKs), arabinogalactan proteins (AGPs), pectins and cellulose synthases. Progress made during the last three years has resulted in additional candidates, including formins, plant-specific class VIII myosins, phospholipase D and callose synthases. Unexpectedly, formins represent a new candidate for a putative ECM-cytoskeleton linker in plant cells. Current bioinformatic analyses show that there is one plant-specific group of formins not only abundant in cell wall but also moonlighting in cytosol.

1.1.2 ECM proteins: Cross talk in signalling and stress

ECM senses and physiologically responds to environmental stress via signalling pathways. Signalling events are clearly not linear and induce many different reactions, including stress-related processes that crosstalk with hormone signalling pathways. It is known that cell wall stress provokes a transient depolarised distribution of the cell wall biosynthetic enzyme glucan synthase and its regulatory subunit RHO1, possibly as a mechanism to repair general damage to the wall. Both environmental and patho-stress are thought to cause wall weakening which in turn transduces a signal to the interior of the cell as a homeostatic mechanism to repair the wall. Various kinases mediate the stress-induced synthesis of ECM proteins to combat cell wall interfering factors, such as pathogens, osmotic stress, dehydration and other environmental stresses. Recently, it has been found that wall-associated kinase (WAK) expression was induced when Arabidopsis plants were infected with a pathogen or stimulated by exogenous SA or its analogue INA. WAK1 mRNA induction requires the positive regulator NPR1/NIM1 (Cheng et al., 2009). It provides a direct link between a protein kinase that could mediate signals from the ECM to the events that are precipitated by pathogen infection. It also suggests that while pathogen infection induces protective hangs in cells, these changes can be detrimental if certain cellular components, such as WAK1, are not present in sufficient amounts (Jones et al., 2004). In osmotic, salinity and dehydration stress, the expansion ability of the cell wall decreases. Correlated with this weakening was a substantial decrease in the proportion of crystalline cellulose in the primary cell wall while the amount of insoluble proteins (such as HRGPs) associated with the wall was increased relative to other wall components (Sakurai et al., 1998).

2. Methodology and strategy

We have compared the ECM proteome of six plants, namely Arabidopsis thaliana, Cicer arietinum Medicago sp, Oryza sativa, Zea mays and Brassica napus. The modus operandi in investigating the cell wall proteomes of available crops was the extensive literature and availability of relevant databases (wallprotDB, Phosida, UniPro, ProtAnnDB pep2proandSwissprot) search. The CWPs identified in these works were classified into functional categories. This classification is only tentative, since the biological role of many of the proteins identified has not been established experimentally. Furthermore, we applied a cross-species comparison on the available datasets. When analysing proteomes within the specified group of plants, a logical strategy was used to maximise efficiency and the overall comparative results. Thus, it was imperative to first evaluate the available proteomes, followed by an analysis of organ-specific proteomes of the model plant Arabidopsis. Once the organ-specific differential proteomes of the model plant were analysed, we then tentatively evaluated the developmental proteomics of rice at various leaf stages so as to understand the acquisition of major pathways involved in the development of the cereal. We then moved on to assessing the stress-responsive plant proteomes in order to understand the overlap and specificity amongst different environmental and patho-stress. These comparative studies were customised for specific protein families. For example, when the environmental stress-responsive proteomes were compared, the parallel analysis of the proteomes of different clades of vascular plants were performed, viz. Arabidopsis vs. maize for osmotic stress, and chick pea vs. rice for dehydration. Similarly, in case of patho-stress, Arabidopsis and Brassica proteomes were compared. It is to be noted that protein consensus can be obtained across any combination of proteomes based on the type of extraction procedure.

2.1 Description of tools

An outline of the procedure and an illustration of the data that can be generated with the methodology are shown in Figure 1. Each proteomic study is described through a simplified flowchart showing its different steps, from plant material to protein identification. As illustrated in Figure 1, two types of methods can be used to prepare a CWP fraction. Non-destructive methods leave the cells alive and allow the elution of CWPs from cell walls using different buffered solutions, while destructive methods start with tissue grinding, thus mixing CWPs and intracellular proteins (Boudart et al., 2005; Bayers et al., 2006). The CWP fraction needs to be fractionated in order to allow for the identification of proteins by mass spectrometry (MS). Proteins can be directly submitted to enzymatic digestion with the appropriate proteases, such as trypsin, or to chemical treatment to get peptides of the appropriate mass (usually between 750 and 4000 Da). Alternatively, proteins are separated prior to cleavage into peptides. Since most CWPs are basic glycoproteins which are poorly resolved by bi-dimensional electrophoresis (2D-E), the most efficient means to separate them are either mono-dimensional electrophoresis (1D-E) (Boudart et al., 2005) or else cationic exchange chromatography followed by 1D-E of protein fractions eluted with a salt gradient (Irshad et al., 2008). The identification of proteins can then be done either by peptide sequencing through liquid chromatography (LC) coupled to MS (LC-MS/MS) or by peptide mass mapping using the matrix-assisted laser desorption/ionisation-time of flight (MALDI-TOF) MS followed by *in silico* analyses. At the bioinformatics end, custom ECM protein databases markedly increase the identification of extensively modified peptides. New generations of mass spectrometers will help meet the demand for high-throughput identification and the localisation of biologically significant peptide modifications.

3. Results and discussion

Proteomics has turned out to be an imperative benefactor for studying the acquaintance of plants' ECM structure and functions by allowing the identification of proteins present in this cellular compartment. It is a well known fact that the field of proteomics is evolving from the cataloguing proteins under static conditions to comparative analyses. Defining proteins that change in abundance, form, location or other activities may indicate the presence and functional significance of a protein. Whereas comparative ECM proteome research is quite advanced in animals (Zhu et al., 2007) and yeast (Kim et al., 2007), there is less information as to plants. The identification and cataloguing of plant ECM proteomes in recent years raises the following important questions: What are the essential plant ECM proteins? Do ECM proteins show clade specificity in vascular plants? What are those organ-specific cell wall proteins, if any? Does the cell wall developmental proteomics of one of the clades yield any astonishing or prolific results? How does ECM protein remodelling during environmental and/or patho-stress provide new perspectives? Are some of the ECM proteins unexpected? And, last but not the least, what sort of post-translational modifications have so far been characterised in CWP? Here we analyse and compare the experimental results of the thus far available proteomes so as to elucidate the dynamics of plant ECM /cell wall proteins.

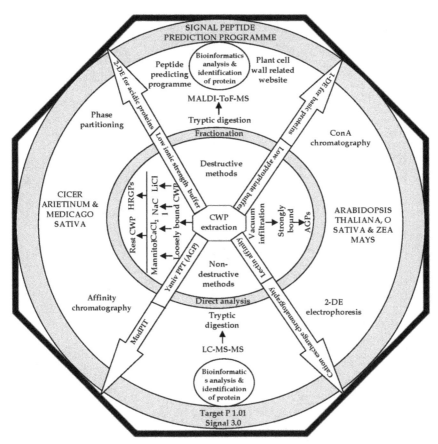

Fig. 1. A flowchart illustrating the overall experimental strategy for the analysis of the ECM proteome.

3.1 Analysis of ECM proteome dynamics in plants: social class vs. diversity

ECM/cell wall design and protein composition has been shown to differ between two major clades, viz. the monocots and dicots of vascular plants. Results have mainly been obtained with the model plants *Arabidopsis thaliana* (Liepman et al., 2010; Basu et al., 2006; Bayer et al., 2006; Borderies et al., 2003; Chivasa et al., 2002; Feiz et al., 2006; Jamet et al. 2008a), *Medicago sativa* (Soares et al., 2007; Watson et al., 2004), and crop plants for, e.g., *Oryza sativa* (Choudhary et al., 2010), *Brassica napus* (Basu et al., 2006) *Zea mays* (Zhu et al., 2006) and *Cicer arietinum* (Bhushan et al., 2006). Around 500 CWPs of *Arabidopsis*, representing about one third of its estimated cell wall proteome, have been described (Liepman et al., 2010) while 219, 143, 102, 58 CWPs were identified in rice, chickpea, maize and Brassica, respectively. Our comparative analysis of different species in relation to their function showed that a high percentage of proteins were found to be unique to each proteome: 87% in *A. thaliana*, 82% in *B. napus*, 84% in *C. arietinum*, 76% in *M. sativa*, 80% in *O. sativa* and 71% in *Z. mays*, with only peroxidase and glycosyl hydrolase being the social class of proteins present ubiquitously in all (Fig. 2).

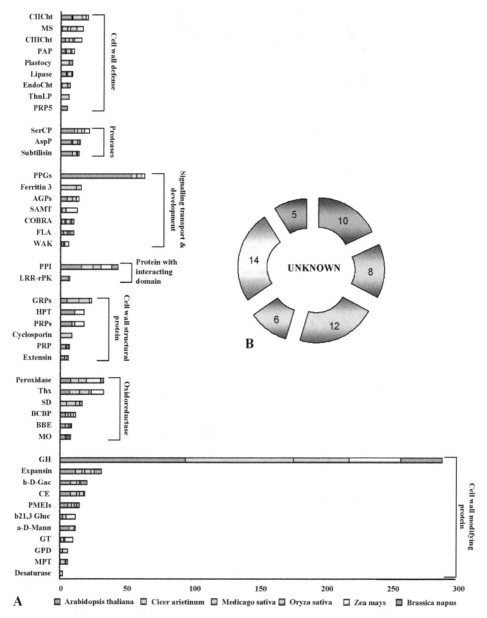

Fig. 2. Cross-species comparison of ECM proteomes. The functional classification of the identified proteins was made according to the biological processes in which they are involved. The length of the bar indicates the number of proteins present in a particular species, such as *Arabidopsis thaliana*, *Cicer arientum*, *Medicago sativa*, *Oryza sativa*, *Zea mays* and *Brassica napus*. The pie chart inset represents the fraction of unknown protein classes in each of these plants.

The available ECM proteome of the six plants compared in Figure 2 varied in molecular weight from 8.9 to 133.8 kDa and had a spread of p*I* values from 5.2 to 10.1. Seventy-six percent of the ECM proteins were basic in nature, concordant with the acidic environment of the wall. Between monocots and dicots, it was found that rice and *Arabidopsis* have a similar number of cell wall-related gene families and members within each family, even though rice has a far greater number of genes than Arabidopsis. This implies that similar numbers of genes are required for wall construction and maintenance, at least among Angiosperms (Yokoyama et al 2004). However, the cell wall proteome data of *Arabidopsis* is better explored than rice and therefore comparison of their proteome may not yield the postulated results as defined by genome analysis. When the maize cell wall proteome was compared with that of *Arabidopsis*, the results revealed an evolutionary divergence as well as tissue specificity, with few conserved proteins (Fig. 2). The protein network of maize (Zhu et al., 2006) revealed the predominance of the inhibitors of hemicellulose-degrading enzymes from monocots, such as endoxylanase inhibitors, and the *Arabidopsis* protein network (Slabas et al., 2004; Peck, 2005) was found to be rich in xyloglucan endoglucanase-inhibiting proteins and glycine rich protein as cell wall remodelling or biosynthetic enzymes. Comparison of the functional classes of cell wall proteins amongst dicot species like *Arabidopsis, Brassica, Medicago* and *Cicer* confirms the dynamic nature of the cell wall, as exemplified by the presence of cellulose synthase and peroxidase in all dicots. However, surprisingly the protein turnover rate of these enzymes are greater in *Medicago*. A more comprehensive investigation of the studied legume proteomes revealed that the proportion of proteins involved in cell wall modification is three times greater in *Medicago* (99 proteins) than in *Cicer* (28 proteins).

This may be due to the fact that ionically bound and soluble ECM proteins can be separated with ease from *Medicago* as compared with *Cicer*. Additional variation of cell wall proteomes in *Cicer* and *Medicago* is provided by the presence of ferritin in the former and Polypolyglutamatesynthase in the latter, illustrating that nature invented vastly different solutions to a common problem, viz. transport and storage. When the studies on the legumes like *Cicer* and *Medicago* were compared to *Arabidopsis* belonging to the Brassicaceae family (Fig. 2), it can be readily observed that the protein machinery of the wall for activating the wall-modifying enzymes is diverse between the two families as well as between the members of the same family, leguminosae. Investigation between *Arabidopsis* and *Brassica* proteomes by MudPIT, using a homology-based search, unambiguously identified 16 proteins which were common to the 52 proteins of *Arabidopsis*. When the cell wall proteomes of *Oryza* (145) and *Zea* (128) were compared, less diversity was observed in Poaceae (Pandey et al., 2010; Zhu et al., 2006) except for the fact that one of the CWP expansins - HPT - is expressed by a moderate amount in maize, whereas in rice PRP is represented by a moderate number (Fig. 2). It may be assumed that the divergence in the resulting proteomes of the vascular plants is due to the presence of the different design of their wall based on their carbohydrate composition. It is known that type I carbohydrates - which typically contains xyloglucan and/or glucomannan and 20–35% pectin - are found in all dicotyledons whereas type II carbohydrate rich in arabinoglycan are only characteristic of the Poaceae family in the monocot, suggesting the occurrence of clade-specific ECM proteins that would bind to their cognitive carbohydrate molecules. Most intriguing are the remaining 10% of ECM proteins that do not have any similarity to the known proteins in other organisms. The challenge is to elucidate their biological role within the cell wall.

3.2 Discerning organ-specific ECM proteomes in *Arabidopsis thaliana*

We further analysed the organ-specific proteome of *Arabidopsis thaliana*, namely root, stem, leaves, etiolated hypocotyl, etiolated seedlings from liquid and cultured media, and protoplast- and leaf-derived cell suspension (Miller & Fry, 1992; Feiz et al., 2006; Minic et al., 2007, Irshad et al., 2008; Minic et al., 2008). Comparative analysis (Fig. 3) revealed that cell wall modifying proteins, structural proteins and proteins involved in signalling and development constitute 58% of the ECM's proteins in mature stems (71) and dark-grown hypocotyls (147) with high and moderate expression. However, it was intriguing to note that most of these CWPs identified by the proteomics study originate from genes whose level of transcripts was low (between 37% and 58%) or below the background (between 18% and 25%) as reported in Minic et al., 2007 indicating thereby the importance of the post-transcriptional regulation of organ-specific ECM proteomes. A further 29 and 54 cell wall modifying proteins were identified in the roots and leaves respectively, in which members of the hydroxyproline-rich glycoprotein family and other major structural proteins were not detected. For a few protein sequences within a particular organ, there also exists a certain degree of heterogeneity in terms of the occasional amino acid substitution as well as their appearance at different molecular weights. The former may be explained due to the origin of these protein species from different genes and the latter by post-translational modifications, such as glycosylation. Expansin, a cell wall modifying component, was the most dominant class in all the major organs, while well-known cell wall enzymes like glycoside hydrolase, pectin methylesterases, peroxidases and glycosyl transferases were represented by several members of the same family (Fig. 3). The analysis of protoplast and suspension-cultured cell derived proteomes in *Arabidopsis* and rice showed the relative abundance of the GH family of ECM proteins. They might be involved in the modification of mixed glycan polymers, only found in monocot cell walls during the regeneration of the cell wall in the protoplast. However, the role of GH family of proteins has not yet been elucidated in *Arabidopsis*. A moderate number of carbohydrate esterases were identified in the ECM proteome of the cell suspension culture, etiolated hypocotyl and leaves while a novel family of HRGP, called LRR-extensin proteins (LRX), has only been found in the case of cell suspension cultures. The only organ in which a few salt-extractable structural proteins were identified is etiolated hypocotyl, possibly because such proteins are not yet completely insolubilised from other organs. Proteins having domains of interaction with proteins or polysaccharides are well-represented in all organs, and especially in rosettes. As expected from the fact that GH represents almost 20% of the identified CWPs (Fig 2 and 3), proteins acting on cell wall polysaccharides are also the category with the highest diversity within each organ. Oxidoreductases are particularly numerous in cell suspension cultures, probably due to the mechanical stress produced by continuous spinning and the oxidative stress that occurs in the liquid media culture. At least 20% of the identified CWPs represent a social class in one organ not found in the others. This may be partially linked to the high redundancy in the number of genes encoding each CWP family, presumably differentially-regulated during organ development (Fig. 3).

3.3 Exploring the variability of the developmental stage specific ECM proteome

A cornerstone of evolution is associated with the diversity of individuals within a population. This diversity is generally understood to arise at the genetic level and leads to characteristics that may be advantageous or disadvantageous within the context of the

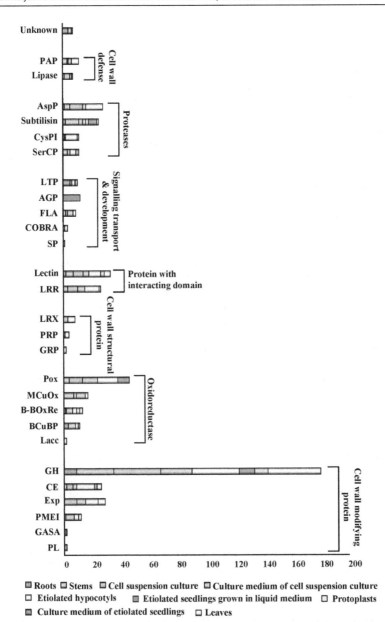

Fig. 3. The organ-specific comparative ECM proteome in *Arabidopsis*. The functional classification of the identified proteins was according to the biological processes in which they are involved. The length of the bars indicates the number of proteins present in a particular organ or culture, such as roots, the stem, the cell suspension culture, the culture medium of the cell suspension culture, etiolated hypocotyls, etiolated seedlings grown in the liquid medium, the culture medium of the etiolated seedlings, protoplast and leaves.

environment (Taraszka et al., 2005). The emerging field of developmental proteomics, in which large mixtures of proteins are characterised in a single experimental sequence, may allow for the assessment of variability or similarity in an individual at the level of the proteome (Hunter et al., 2002). The developmental proteomics of rice is perhaps the least studied, but its importance was realised when the proteome of rice at 5 days and the third and fourth leaf stages were analysed (Jung et al 2008; Chen et al., 2009). When we compared the existing dataset, even though the proteomes were found to be similar, some of the CWPs which were unknown earlier were uniquely present at a particular developmental stage. For example, COBRA and Leucine rich repeat extensins were found only in the third leaf stage while the polysaccharide lyase appeared in the fourth leaf stage (Fig. 4). Although ECM proteins which regulate development and expansion form the major class, very few have been functionally characterised so far. Such a protein, COBRA (COB), anchored to the extracellular surface of the plasma membrane by a glycosyl phosphatidylinositol (GPI) moiety is thought to regulate and link oriented-cell expansion in root cells (Brady, 2007). Another protein, LRX1, a chimeric leucine-rich repeat/extensin is also expressed in root hair cells. The interaction between the cell wall and the LRX1 protein is important for proper root hair development and expansion (Diet et al., 2006). A family of secreted proteins called SCAs (stigma/stylar cysteine-rich adhesion) was identified as a pollen tube adhesion molecule for the wall material of the style found in the lily (Baumberger et al., 2001). One of the ECM protein family Arabinogalactan-proteins (AGPs) belonging to the category of HRGP consists of a rather small and highly glycosylated protein moiety which has been found to play vital role in cell wall development (Gillmor et al., 2005). THESEUS1 (THE1), which is a member of the subfamily of the *Catharanthus roseus* protein kinase1-like receptor kinases also has efficacy in cell wall integrity, sensing and development (Hematy et al., 2007). Thus, the resulting cell wall proteomes were different, showing in another way that the cell wall structure and composition are regulated during development. However, the biological functions of most CWPs involved in development have not yet been experimentally studied.

3.4 *In silico* protein profiling of comparative ECM stress proteomes

The plant cell wall or the extracellular matrix (ECM) is the first compartment that senses stress signals, transmits them to the cell interior and eventually influences the cell fate decision (Ellis et al., 2002), and thus it can be envisaged that ECM proteomes primarily regulate the environmental and patho-stress response in plants. We analysed the cell wall proteomes of *Arabidopsis* and maize in response to osmotic stress (Kachroo et al., 2001; Amaya et al., 1999), and the dehydration responsive ECM proteomes of chickpea and rice (Bhushan et al., 2007; Pandey et al., 2008; Choudhary et al., 2009; Pandey, et al., 2010; Bhushan et al., 2011). Interestingly, a great deal of divergence in the protein classes amongst these organisms was observed (Fig. 5A).

To our surprise, except for peroxidase, serine protease and subtilisin none of the ECM proteins was found to be common in all the organisms under both kinds of the stresses studied. The families of antimicrobial peptides such as thionins, defensins and knottin-like peptides have been found in the dehydration-responsive proteome of chickpea, while it was found that rice DRPs comprised of antimicrobial peptides such as oryzacystatin, thioredoxin and oligopeptidase. The *Cicer* dehydration-responsive protein network showed the exclusive presence of glycine-rich protein, methionine synthase, ferritin, tubby-like protein

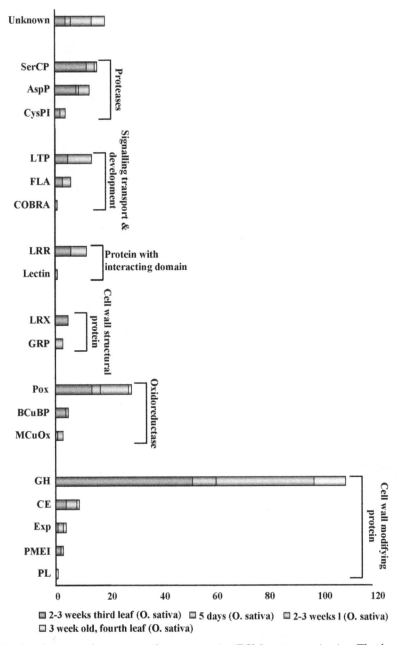

☐ 2-3 weeks third leaf (O. sativa)   ☐ 5 days (O. sativa)   ☐ 2-3 weeks I (O. sativa)
☐ 3 week old, fourth leaf (O. sativa)

Fig. 4. The developmental stage specific comparative ECM proteome in rice. The functional classification of the identified proteins was according to the biological processes in which they are involved. The length of the bars indicates the number of proteins present in a particular leaf stage/day of suspension culture.

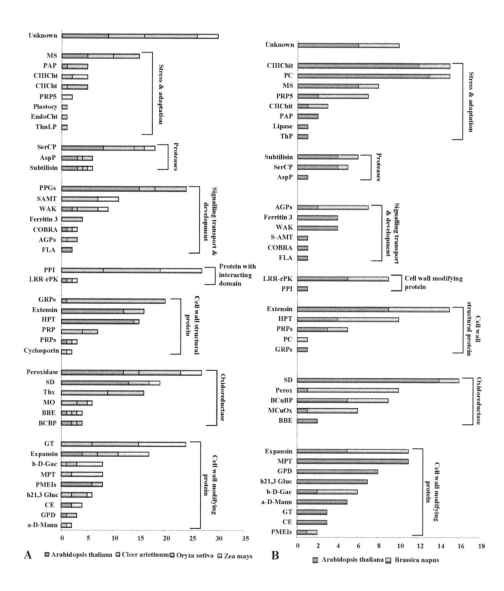

Fig. 5. Comparative stress proteome: A comparison of various functional classes of the extracellular matrix protein in environmental stress (A) and in patho-stress (B). The functional classification of the identified proteins was according to the biological processes in which they are involved. The length of the bars indicates the number of proteins present in a particular stress.

and leucine aminopeptidases. Another important finding was the presence of falacinin-like AGPs during osmotic stress in *Arabidopsis*, but not in other cases. Extensin, hydroxyproline transferase and carbohydrate esterase were predominantly found during the dehydration response but were absent in response to osmotic stress. Interestingly, rice as well as maize cell wall proteomes under both types of abiotic stresses revealed the presence of class III chitinase, plastocyanin, S-adenosylmethionine transferase and cyclosporine, suggesting their clade-specific expression. Our analysis revealed the presence of monocot and dicot peroxidases having specific protein sequences that clearly demonstrate the diversity of the identical CWPs in two divisions of angiosperm. This may be attributed to the evolution of orthologs vs. paralogs.

Moreover, plant cell walls constitute the first stage of defence against invading pathogens. The endogenous wall metabolism might facilitate pathogen infection, either because wall substrates are made more physically accessible to pathogens or because the plant enzymes convert wall polymers into appropriate nutritional substrates for the invading micro-organism. In addition to the crucial role of CWPs in growth and development, these proteins or peptides are also involved in plant defence mechanisms in response to patho-stress. Earlier, a number of ECM proteins have been shown to play a crucial role in plant defence against microbes (Sakurai, 1998), including pathogenesis-related (PR) proteins, chitinases and endo-b-1,3-glucanases, that are known to directly interact with pathogens (Jung et al., 2004; Jones et al., 2006). However, plants also deploy a repertoire of proteins in the wall that act as a surveillance system to allow the early detection of an impending pathogen assault. We analysed the cell wall proteomes of *Arabidopsis* and *Brassica napus* in response to fungal stresses (Ndimba et al., 2003; Floerl et al., 2008), and elicitor-induced ECM proteome of *Zea mays* (Chivasa et al., 2005). The common ECM proteins identified in fungal stress were jacalin-related, LRR-containing proteins, chitinase, thaumatin-like proteins, esterase/lipase thioesterase and the GLIP1 lipase. On contrary the, S-AMT, COBRA, FLA, BBE, CE and GRPs were found to be exclusive in the case of *Arabidopsis-Fusarium* interactions, suggesting that the cell wall is a dynamic milieu and responds differently in response to different pathogen within the divisions or in between the divisions of the angiosperm (Fig. 5B). Likewise, in order to assess the generality of the cell wall proteome of *A. thaliana* (Oh et al., 2005) and *B. napus* (Floerl et al., 2008) under patho-stress were compared. The results indicate that the fungal stress-induced changes in CWPs were diverse in both of the plants except for the oxidoreductases, stress- and adaptation-related proteins and structural proteins. Meanwhile signalling-, transport- and development-related proteins were induced mostly in *A. thaliana*, except for the AGPs which were commonly present in both of the proteomes (Fig. 5B). Thus, *B. napus* may depend exclusively on AGPs-mediated stress signalling responses, whereas diverse signalling pathways operate in *A. thaliana*. Pathogen elicitor-induced changes in maize ECM proteomes revealed the involvement of lipases, esterases and thiols similar to the response of *Arabidopsis* to pathogen invasion. However, how monocot cell wall proteins respond to pathogens still needs further investigation in order to comprehend the monocot-dicot difference in response to pathogen invasion. In addition, a number of extracellular proteins and peptides have been identified that contribute to signalling and the recognition of not only pathogens but also other cell-type responses, such as in pollen–pistil interactions and the phosphate deficiency proteome of *Arabidopsis* (Kachroo et al., 2001; Tran & Plaxton, 2008).

The comparative analysis of clade and organ-specific, developmentally-regulated, stress-responsive plant ECM proteomes revealed the presence of certain proteins that were unexpected, either in their abundance, form, number or else localisation. These unexpected or non-canonical proteins suggest the constant remodelling of cell wall proteomes. The exact function and specificity of these candidates can only be comprehended once they are functionally characterised.

4. Conclusion and perspectives

In this study, cross-species as well as cross-condition comparisons of ECM proteomes in vascular plants illustrates the divergence in protein profiles within only a few social classes. Across species, cell wall modifying proteins (23%) represent the largest category, followed by oxidoreductase (19%) and cell wall structural protein (18.5%). In total, 213 and 110 glycoside hydrolase were found in the organ specific proteome of Arabidopsis and development specific proteomes of rice, respectively. Oxidoreductase constitutes the second largest category in both these cases. Furthermore, the dehydration responsive comparative proteome in legumes, chickpeas and cereals, rice showed both genotypic- and crop-specific adaptation. As expected, the proteins involved in cell-wall remodelling were found to be the most predominant across all conditions. Nonetheless, a large number of proteins were unique or novel to each of the plant species, organs, stages of development and different stresses. It may be thought that the ubiquitously present classes of proteins are the essential proteins for sustenance while the unique classes bring out the condition-specific special function. The differences in terms of protein pattern and protein function appear to encompass both genetic and physiological information. It may be speculated that the differential proteome is shaped by the cellular environment and the ecological niche of the corresponding organism. The divergence may arise due to codon bias, amino acid composition and protein length. However, a much more comprehensive survey of the ECM proteomes in several plants will ultimately draw a more complete picture of the social class vs. protein diversity in this organelle. We are witnessing a significant but inadequate progress in the understanding the ECM proteomes of various crops of agricultural importance. Our understanding of ECM composition, organisation and homeostasis has been greatly enhanced through targeted biochemical and genetic approaches. Unbiased 'discovery' methods, such as proteomics, have only recently gained traction in the field of matrix biology. To date, a key word search using "ECM proteome" retrieves only 43 results in a pubmed search, emphasising the need for in-depth study in the field of the plant ECM proteome. Our future efforts will focus on the development and analysis of comparative ECM proteomes towards an understanding of crop- and genotype-specific adaptation as an important amendment for the determination of protein networks influenced by the internal and external cues associated with the complex cellular biochemical and physiological process that bring about phenome variation.

5. Acknowledgements

This work was supported by a grant from the Department of Biotechnology (DBT), Govt. of India and a grant from National Institute of Plant Genome research. KN and EE are the recipients of pre-doctoral fellowships from the Council of Scientific and Industrial Research and DBT-TWAS, Govt. of India. Authors thank Mr. Jasbeer Singh for illustrations in the manuscript.

6. Abbreviations

Polysaccharide lyase, PL; gibberellin acid-stimulated Arabidopsis (GASA) protein, GASA; pectin methyl esterase inhibitor, PMEI; Expansin, Exp; carbohydrate esterase, CE; glycoside hydrolase, GH; Laccase, lacc; blue copper binding protein, BCuBP; berberine-bridge oxido-reductase, B-BOxRe; multicopper oxidase, MCuOx; Peroxidase, Pox; glycine-rich protein, GRP; proline-rich protein, PRP; leucine-rich repeat extensin, LRX; LRR protein, LRR;signal eptidase, SP; COBRA-like family, COBRA; fasciclin-like arabinogalactan protein, FLA; arabinogalactan protein, AGP; lipid transfer protein, LTP Ser carboxypeptidase, SerCP; Cys protease inhibitor, CysPI; Asp protease, AspP; purple acid phosphatase, PAP; glycosyl transferases, GT; a-D-mannosidases, a-D-Mann; b-D-galactosidases, b-D-Gac; b21,3 Glucanase, b21,3 Gluc; glycerophosphodiesterases, GPD; Mannose-1-phosphateguanyltranferase, MPT; berberine-bridge enzyme, BBE; Superoxide dismutase, SD; Putative cyclosporin, PC; Proline-rich proteins (PRPs), PRPs; Hydroxyl proline transferase, HPT; Putative protease inhibitor, PPI; LRR-receptor protein kinases; LRR-rPK; S-adenosylmethioninetransferaae, S-AMT; Wall- associated kinase, WAK; Arabinogalactan-proteins, AGPs; Thaumatin-like protein, ThP; Class II chitinase, CIIChit; Pathogenesis-related protein 5, PRP5; methionine synthase, MS; Class III chitinase, CIIIChit; threonine-hydroxyproline-rich glycoprotein, THRGP;

7. References

Amaya, I., Botella, M.A., de la Calle, M., Medina, M.I., Heredia, A., Bressan, R.A., Hasegawa, P.M., Quesada, M.A. and Valpuesta,V. (1999). Improved germination under osmotic stress of tobacco plants overexpressing a cell wall peroxidase. *FEBS Lett.*, 457: 80-84.

Baluska, F., Samaj, J., Wojtaszek, P., Volkmann, D. and Menzel, D. (2003) Cytoskeleton-plasma membrane-cell wall continuum in plants. Emerging links revisited. *Plant Physiol.*, 133: 482-491.

Basu, U., Francis, J.L., Whittal, R.M., Stephens, J.L., Wang, Y., Zaiane, O.R., Goebel, R., Muench, D.G., Good, A.G. and Taylor, G.J. (2006). Extracellular proteomes of Arabidopsis thaliana and Brassica napus roots: analysis and comparison by MudPIT and LC-MS/MS. *Plant Soil*, 286: 357-376.

Baumberger, N., Ringli, C. and Keller, B. (2001). The chimeric leucine-rich repeat/extensin cell wall protein LRX1 is required for root hair morphogenesis in *Arabidopsis thaliana*. *Genes Dev.*, 15: 1128-1139.

Bayer, E.M., Bottrill, A.R., Walshaw, J., Vigouroux, M., Naldrett, M.J., Thomas, C.L. and Maule, A.J. (2006). Arabidopsis cell wall proteome defined using multidimensional protein identification technology. *Proteomics*, 6 : 301-11.

Bhushan, D., Jaiswal, D.K., Ray, D., Basu, D., Datta, A., Chakraborty, S. and Chakraborty, N. (2011). Dehydration-responsive reversible and irreversible changes in the extracellular matrix: comparative proteomics of chickpea genotypes with contrasting tolerance. *J. Proteome Res.*, 10: 2027-2046.

Bhushan, D., Pandey, A., Chattopadhyay, A., Choudhary, M.K., Chakraborty, S., Datta, A. and Chakraborty, N. (2006) Extracellular matrix proteome of chickpea (Cicer arietinum L.) illustrates pathway abundance, novel protein functions and evolutionary perspect. *J. Proteome Res.* 5: 1711-1720.

Bhushan, D., Pandey, A., Choudhary, M.K., Datta, A., Chakraborty, S. and Chakraborty, N. (2007) Comparative proteomics analysis of differentially expressed proteins in chickpea extracellular matrix during dehydration stress. *Mol. Cell. Proteomics,* 6: 1868 -1884.

Borderies, G., Jamet, E., Lafitte, C., Rossignol, M., Jauneau, A., Boudart, G., Monsarrat, B., Esquerré-Tugayé, M.-T., Boudet, A. and Pont-Lezica, R. (2003). Proteomics of loosely bound cell wall proteins of *Arabidopsis thaliana* cell suspension cultures: A critical analysis. *Electrophoresis,* 24:3421-3432.

Boudart, G., Jamet, E., Rossignol, M., Lafitte, C., Borderies, G., Jauneau, A., Esquerré-Tugayé, M.-T. and Pont-Lezica, R. (2005). Cell wall proteins in apoplastic fluids of *Arabidopsis thaliana* rosettes: Identification by mass spectrometry and bioinformatics. *Proteomics,* 5: 212-221.

Brady, S.M., Song, S., Dhugga, K.S., Rafalski, J.A. and Benfey, P.N. (2007). Combining expression and comparative evolutionary analysis. The COBRA gene family. *Plant Physiol.,* 143: 172-187.

Brownlee, C. (2002). Role of the extracellular matrix in cell-cell signaling: paracrine paradigms. *Curr. Opin. Plant Biol,* 5: 396-401.

Brownlee, C. and Berger, F. (1995). Extracellular matrix and pattern in plant embryos: on the lookout for developmental information. *Trends Genet.,* 11: 344-348.

Cassab, G.I. (1998). Plant cell wall proteins. *Annu. Rev. Plant Physiol. Plant Mol. Biol.,* 49: 281-309.

Chaudhary, M.K., Basu, D., Datta, A., Chakraborty, N. and Chakraborty, S. (2009) Dehydration-responsive nuclear proteome of rice (*Oryza sativa* L.) illustrates protein network, novel regulators of cellular adaptation and evolutionary perspective. *Mol. Cell. Proteomics,* 8: 1579-1598.

Chen, J. and Varner, J.E. (1985). An extraceliular matrLx protein in plants: characterization of a genomic clone for carrot extension. *EMBO J.,* 4: 2145-2151.

Chen, X.Y., Kim, S.T., Cho, W.K., Rim, Y., Kim, S., Kim, S.W., Kang, K.Y., Park, Z.Y. and Kim, J.Y. (2009) Proteomics of weakly bound cell wall proteins in rice calli. *J. Plant Physiol.,* 166: 675-685.

Cheng, F.-Y., Blackburn, K., Lin, Y.-M., Goshe, M.B. and Wiliamson, J.D. (2009). Absolute protein quantification by LC/MS for global analysis of salicylic acid-induced plant protein secretion responses. *J. Proteome Res.,* 8: 82-93.

Chivasa, S., Ndimba, B., Simon, W., Robertson, D., Yu, X.-L., Knox, J., Bolwell, P. and Slabas, A. (2002). Proteomic analysis of the Arabidopsis thaliana cell wall. *Electrophoresis,* 23: 1754-1765.

Chivasa, S., Simon, W.J., Yu, X.-L., Yalpani, N. and Slabas, A.R. (2005). Pathogen elicitor-induced changes in the maize extracellular matrix proteome. *Protgeomics,* 5: 4894-4904.

Cosgrove, D. J. (2005). Growth of the plant cell wall. *Nat. Rev. Mol. Cell. Biol.,* 6: 850-61.

Cosgrove, D.J. (1997). Assembly and enlargement of the primary cell wall in plants. *Annu. Rev. Cell Dev. Biol.* 13, 171–201.

Degenhardt, B. and Gimmer, H. (2000). Cell wall adaptation to multiple environmental stresses in maize roots. *J. Exp. Bot.,* 51: 595-603.

Diet, A., Link, B., Seifert, G.J., Schellenberg, B., Wagner, U., Pauly, M., Reiter, W.-D. and Ringli, C. (2006) The arabidopsis root hair cell wall formation mutant lrx1 Is suppressed by mutations in the RHM1 gene encoding a UDP-L-Rhamnose synthase. Plant Cell, 18: 1630-1641.

Ellis, C., Karafyllidis, I., Wasternack, C. and Turner, J.G. (2002). The arabidopsis mutant cev1 links cell wall signaling to jasmonate and ethylene responses. *Plant Cell*, 14L 1557-1566.

Elortza, F., Nuhse, T.S., Foster, L.J., Stensballe, A., Peck, S.C. and Jensen, O.N. (2003). Proteomic analysis of glycosylphosphatidylinositolanchored membrane proteins. *Mol. Cell. Proteomics*, 2: 1261-1270.

Feiz, L., Irshad, M., Pont-Lezica, R.F., Canut, H. and Jamet, E. (2006). Evaluation of cell wall preparations for proteomics: a new procedure for purifying cell walls from Arabidopsis hypocotyls. *Plant Methods*, 2: 10.

Fincher, G. (2009). Revolutionary times in our understanding of cell wall biosynthesis and remodeling in the grasses. *Plant Physiol.*, 149: 27-37.

Floerl, S., Druebert, C., Majcherczyk, A., Karlovsky, P., Kües, U. and Polle, A. (2008). Defence reactions in the apoplastic proteome of oilseed rape (Brassica napus var. napus) attenuate Verticillium longisporum growth but not disease symptoms. *BMC Plant Biol.*, 8, 129.

Gail McLean, B., Hempel, F.D. and Zambryski, P.C. (1997). Plant intercellular communication via plasmodesmata. *Plant Cell.*, 9: 1043-1054.

Geitmann, A. (2010). Mechanical modeling and structural analysis of the primary plant cell wall. *Curr. Opin. Plant Biol.*, 13: 693-699.

Gillmor, C.S., Lukowitz, W., Brininstool, G., Sedbrook, J.C., Hamann, T., Poindexter, P., Somerville, C. (2005) Glycosylphosphatidyl inositol-anchored proteins are required for cell wall synthesis and morphogenesis in *Arabidopsis. Plant Cell* 17:1128–1140

Hall, Q. and Cannon, M.C. (2002). The Cell Wall Hydroxyproline-Rich Glycoprotein RSH Is Essential for Normal Embryo Development in Arabidopsis. Plant Cell, 14: 1161-1172.

Hammond-Kosack, K.E. and Jones, J.D.G. (1996). Resistance gene–dependent plant defense responses. *Plant Cell,* 8: 1773-1791.

Hematy, K., Sado, P.E., VanTuinen, A., Rochange, S., Desnos, T., Balzergue, S., Pelletier, S., Renou, J.P. and Hofte, H. (2007). A receptor-like kinase mediates the response of Arabidopsis cells to the inhibition of cellulose synthesis. *Curr. Biol.*, 17: 922-931.

Hunter, T.C., Andon, N.<., Koller, A., Yates, J.R. and Haynes, P.A. (2002). The functional proteomics toolbox: methods and applications. *J. Chromatogr,* B 782: 161-181.

Hynes, R.O. (2009). The extracellular matrix: not just pretty fibrils. *Science,* 326: 1216.

Irshad, M., Canut, H., Borderies, G., Pont-Lezica, R. and Jamet, E. (2008). A new picture of cell wall protein dynamics in elongating cells of Arabidopsis thaliana : confirmed actors and newcomers. *BMC Plant Biol.*, 8: 94.

Jamet, E., Albenne, C., Boudart, G., Irshad, M., Canut, H. and Pont-Lezica, R. (2008a) Recent advances in plant cell wall proteomics. *Proteomics*, 8: 893-908.

Jones, D.A. and Takemoto, D. (2004) Plant innate immunity — direct and indirect recognition of general and specific pathogen-associated molecules. *Curr. Opin. Immunol.* 16: 48-62.

Jones, G., Jonathan, D. and Dangl, J.L. (2006). The plant immune system. *Nat. Biotechnol.* 16: 312-329.

Jung, E.H., Jung, H.W., Lee, S.C., Han, S.W., Heu, S. and Hwang, B.K. (2004) Identification of a novel pathogen-induced gene encoding a leucinerich repeat protein expressed in phloem cells of Capsicum annuum. *Biochim. Biophys. Acta*, 1676: 211-222.

Jung, Y.-H., Jeong, S.-H., Kim, S.H., Singh, R., Lee, J.-E., Cho, Y.-S., Agrawal, G.K., Rakwal, R. and Jwa, N.-S. (2008) Systematic secretome analyses of rice leaf and seed callus suspension-cultured cells: Workflow development and establishment of high-density two-dimensional gel reference maps. *J. Proteome Res.*, 7: 5187-5210.

Kachroo, A., Schopfer, C.R., Nasrallah, M.E. and Nasrallah, J.B. (2001) Allelespecific receptor–ligand interactions in Brassica self-incompatibility. *Science*, 293: 1824-1826.

Kim, I., Yun, H. and Jin, I. (2007)Comparative proteomic analyses of the yeast Saccharomyces cerevisiae KNU5377 strain against menadione-induced oxidative stress. *J. Microbiol. Biotechnol.*, 17: 207-217.

Kumar, S., Maxwell, I.Z., Heisterkamp, A., Polte, T.R., Lele, T.P., Mazur, M.S.E. and Ingber, D.E. (2006). Viscoelastic retraction of single living stress fibers and its impact on cell shape, cytoskeletal organization, and extracellular matrix. *Mech. Biophy. J.*, 90: 3762-3773.

Kwon, H.-K., Yokoyama, R. and Nishitani, K. (2005). A proteomic approach to apoplastic proteins involved in cell wall regeneration in protoplasts of Arabidopsis suspension-cultured cells. *Plant Cell Physiol.*, 46: 843-857.

Liepman, A.H., Wightman, R., Geshi, N., Turner, S.R. and Scheller, H.V. (2010). Arabidopsis – a powerful model system for plant cell wall research. *Plant J.*, 61: 1107-1121.

Mann, M. and Jensen, O.N. (2003). Proteomic analysis of post-translational modifications. *Nat. Biotechnol.*, 21: 255-261.

Miller, J.G. and Fry, S.C. (1992). Production and harvesting of ionically wall-bound extensin from living cell suspension cultures. *Plant Cell, Tissue Organ Culture*, 31: 61-66.

Minic, Z. (2008). Physiological roles of plant glycoside hydrolases. *Planta*, 227: 723- 740.

Minic, Z., Jamet, E., Negroni, L., der Garabedian, P.A., Zivy, M. and Jouanin, L. (2007). A sub-proteome of Arabidopsis thaliana trapped on Concanavalin A is enriched in cell wall glycoside hydrolases. *J. Exp. Bot.*, 58: 2503-2512.

Ndimba, B.K., Chivasa, S., Hamilton, J.M., Simon, W.J., and Slabas, A.R. (2003). Proteomic analysis of changes in the extracellular matrix of Arabidopsis cell suspension culture induced by fungal elicitors *Proteomics*, 3: 1047-1059.

Oh, I.S., Park, A.R., Bae, M.S., Kwon, S.J., Kim, Y.S., Lee, J.E., Kang, N.Y., Lee, S., Cheong, H. and Park, O.K. (2005) Secretome analysis reveals an Arabidopsis lipase involved in defense against Alternaria brassicicola. *Plant Cell*, 17: 2832-2847.

Pandey, A., Rajamani U., Verma, J., Subba, P., Chakraborty, N., Datta, A., Chakraborty, S. and Chakraborty, N., (2010). Identification of extracellular matrix proteins of rice (Oryza sativa L.) involved in dehydration-responsive network: a proteomic approach. *J. Proteome Res.*, 9: 3443-3464.

Pandey, A., Chakraborty, S., Datta, A. and Chakraborty, N. (2008). Proteomics approach to identify dehydration responsive nuclear proteins from chickpea (*Cicer arietinum L.*). *Mol. Cell. Proteomics*, 7: 88-107.

Pandey, A., Rajamani, U., Verma, J., Subba, P., Chakraborty, N., Datta, A., Chakraborty, S. and Chakraborty, N. (2010). Identification of extracellular matrix proteins of rice (*Oryza sativa L.*) involved in dehydration-responsive network: a proteomic approach. *J. Proteome Res.*, 9: 3443-3464.

Peck, S. (2005). Update on proteomics in Arabidopsis. Where do we go from here? *Plant Physiol*, 138: 591-599.

Peters, W.S., Hagemann, W. and Tomos, A.D. (2000) What makes plants different? Principles of extracellular matrix function in 'soft' plant tissues. Comp. Biochem. Physiol. 152: 151-167.

Rose, J.K., Saladié, M. and Catalá, C. (2004). The plot thickens: New perspectives of primary cell wall modification. *Curr. Opin. Plant Biol.*, 7: 296-301.

Rose, J.K.C. (2003). The plant cell wall. *Annu. Plant Rev.*, 8: 190-513.

Sakurai, N. (1998). Dynamic function and regulation of apoplast in the plant body. *J. Plant Res.* 111: 133-148.

Šamaj, J., Bobák, M. and Volkmann, D. (1999). Extracellular matrix surface network of embryogenic units of friable maize callus contains arabinogalactan-proteins recognized by monoclonal antibody JIM4. *Plant Cell Rep.*, 18: 369-374.

Schröder, F., Lisso, J., Lange, P. and Müssig, C. (2009). The extracellular EXO protein mediates cell expansion in Arabidopsis leaves. *BMC Plant Biol.*, 13: 9-20.

Slabas, A.R., Ndimba, B., Simon, W.J. and Chivasa, S. (2004). Proteomic analysis of the Arabidopsis cell wall reveals unexpected proteins with new cellular locations. *Biochem. Soc. Trans.*, 32: 524-528.

Soares, N.C., Francisco, R., Ricardo, C.P., and Jackson, P.A. (2007). Proteomics of ionically bound and soluble extracellular proteins in Medicago truncatula leaves. *Proteomics*, 7: 2070-2082.

Taraszka, T.A., Gao, X., ValentineRena, S.J., Sowell, A., Koeniger, S.L., Miller, D.F., Kaufman, T.C. and Clemmer, D.E. (2005). Proteome Profiling for Assessing Diversity: Analysis of Individual Heads of Drosophila melanogaster Using LC-Ion Mobility-MS J. of Proteome Res. 4: 1238-1247.

Telewski, F.W. (2006). A unified hypothesis of mechanoperception in plants. *Ame. J. Bot.*, 93: 1466-1476.

The Arabidopsis Genome Initiative. (2000). Analysis of the genome sequence of the flowering plant Arabidopsis thaliana. *Nature*, 408: 796-815.

Thelen, J. and Peck, S. (2007). Quantitative proteomics in plants: choices in abundance. *Plant Cell*, 19: 3339-3346.

Tran, H.T. and Plaxton, W.C. (2008) Proteomic analysis of alterations in the secretome of Arabidopsis thaliana suspension cells subjected to nutritional phosphate deficiency. *Proteomics*, 8: 4317-4326.

Watson, B.S., Lei, Z., Dixon, R.A. and Sumner, L.W. (2004). Proteomics of Medicago sativa cell walls. *Phytochemistry*, 65:1709-1720.

Wilson, R. (2010). The extracellular matrix: an underrexplored but important proteome. *Expert Rev. Prtoemics*, 7: 803-806.

Wolf, S., Hematy, K. and Hoft, H. (2009). Growth control and cell wall signaling in plants. *Annu. Rev. Plant Biol.*, 63: 162-181.

Yokoyama, R. and Nishitani, K. (2004). Genomic basis for cell-wall diversity in plants. A comparative approach to gene families in Rice and Arabidopsis. *Plant Cell Physiol.*, 5: 1111-1121.

Zhu, J., Chen, S., Alvarez, S., Asirvatham, V. S. (2006). Cell wall proteome in the maize primary root elongation zone. Extraction and identification of water-soluble and lightly ionically bound proteins. *Plant Physiol.*, 140: 311-325.

Zhu, Q., Zheng, X., Luo, J., Gaut, B.S. and Ge, S. (2007). Multilocus analysis of nucleotide variation of *Oryza sativa* and its wild relatives: severe bottleneck during domestication of rice. *Mol. Biol. Evol.*, 24: 875-888.

7

Leaves Material Decomposition from Leguminous Trees in an Enriched Fallow

José Henrique Cattanio
Federal University of Pará - UFPA
Brazil

1. Introduction

In Amazon human activities such as slashing and burning converted large areas of primary forest to intermittently used agricultural land. Thus, the fallow vegetation plays an important role to maintain or restore soil productivity. In the systems with the soil are poor in nutrients and carbon (C) and farmers do not have much subsidies to buy fertilizer, the efficiency with which nutrient in plant residues is used depends on the amount and quality of the organic matter, the rate at which they are mineralized and thus on the time when they are made available relative to crop requirements. It is important to find contrasting litter quality, and mix this organic material with the possibility to alter the pattern of Nitrogen-release and the efficiency of Nitrogen (N) utilization from the residue by a soil microbial biomass and crop system where other sources of mineral N such as fertilizer are limited or excluded.

Soil organic matter (SOM) represents a major proportion of the organic carbon within the terrestrial biosphere and plays an important role in soil fertility (Powlson et al. 2001). An accumulation of organic matter is not only beneficial to soil functions related to agriculture, favouring growth of biomass, promoting and facilitating carbonation processes, reducing erosion and favouring pedogenesis, and developing organic matter-rich horizons recovering degraded or contaminated soils, but also represents a sequestration of C from atmospheric CO_2 (Macías and Arbestain 2010). In contrast, management practices (e.g. slash and burn system in Amazon region) leading to a decline in SOM content release CO_2, the major greenhouse gas (Powlson et al. 2001). SOM also has a range of other environmental functions such as water retention and the regulation of trace greenhouse gases between land surface and the atmosphere.

Fallow trees affect the soil by their litter deposition in terms of quantity and quality, root activity and changes in microclimate brought about by the leaf canopy. However, the intensification of land use, by small farmers in the tropic, has drastically reduced the fallow period with a decline in soil productivity and environmental quality, resulting in a progressive deterioration of natural resources. Therefore, the soil quality has to be restored in shorter time.

Decline in soil productivity and environmental quality and progressive deterioration of natural resources in the tropics have led to a search for new methods to sustain crop production via more efficient nutrient cycling. In Northeastern of Pará (Brazil) the Amazon region was occupied by an intensive colonization process over the last century until today.

The region was to be utilized, initially by clearing the forests for timber and later by the use of the land for subsistence agriculture, based on slash and burn agriculture. In the context of a bilateral German-Brazilian project ("Secondary Forests and Fallow Vegetation in Eastern Amazon – Function and Management", SHIFT project) slash and mulch system are being recommended to realize fire-free land clearing by cutting and chopping the fallow vegetation and leaving mulch layer on site (Denich et al. 2005). In addition, the fallow vegetation is enriched with fast-growing legume trees to support the mulching effect by increasing biomass production and nitrogen input during the fallow period.

The purpose of this technique is to maintain soil organic matter and assure a slow and continuous release of nutrients, improve moisture retention, reduces excessive soil heating and runoff, reduce soil erosion, and prevent weed seed germination (Denich et al. 2005). Hence, mulching may improve flexibility in planting date to cope with unreliable rain due to conserved soil moisture.

The rate of decomposition and the amount of N-mineralization from organic material determines the short-term benefits of tree residues for plant nutrition (Jensen et al. 1995). If burning is to be abandoned, then the synchronization of nutrient release from organic material and nutrient uptake by plants (Addiscott et al. 1991, Myers et al. 1994), accompanying the competition between plant and microorganisms for nutrients (Cattanio et al. 2008), will be the core problem in applied tropical soil biology research. Yield losses in field trials of the SHIFT project have shown that yield losses in mulch practices are evident as compared to burned treatments (Kato et al 1999). The same authors showed that yield losses were eliminated with fertilizer application, indicating nutrient competition with decomposers and/or an unfavorable nutrient release pattern as compared to crop demand was a problem.

The structure and decomposability of leaf litter varies to a large degree, thus affecting the rate of nutrient cycling and the nutrient availability in soil (Priha and Smolander 1997). Soluble C, which includes metabolic and storage C, is of high quality and is primarily responsible for promoting microbial growth and activity. Large amounts of soluble C but little soluble N and P in decomposing plant residues induce net immobilization (Cattanio et al. 2008). The challenge resides in sustaining crop production while maintaining soil fertility through supply and efficient management of organic residues (Isaac et al. 2000). Biederbeck et al. (1994) suggested that it may be possible to manipulate the timing and quality of litter input through appropriate management of mixed stands to improve the synchrony of nutrient release with crop requirements.

Some works in litter manipulated with mixtures was done by Meentemeyer (1978), Melillo et al. (1982), Anderson et al. (1983), Gallardo and Merino (1993), Vitousek et al. (1994), Hobbie (2000), Lonrez et al (2000). But studies with mixtures in soil litter decomposition were scarce (Franagan and van Cleve 1983). Blair et al. (1990) found in litterbags containing mixed residues that there were significantly greater initial releases of N and lower subsequent N immobilization than predicted, and they suggest that it resulted from differences in the decomposer community originated from the mixtures of varied litter resource quality. In the same way, Handayanto et al. (1997), Kuo and Sainju (1998), Zimmer (2002) and Cattanio et al. (2008) showed that soil N-mineralization rate of prunings could be manipulated by mixing different quality materials.

Different organic materials decompose at contrasting rates because they are decomposed differentially by catabolic enzymes produced by saprophytic organisms (Linkins et al. 1984). Furthermore, decomposition rates are affected by nutrient and lignin content of litter (Moorhead et al. 1996), because the initial lignin-to-N and the lignin + polyphenol-to-N ratios are correlated well with the N-mineralization or N accumulation (Constantinides and Fownes 1994; Janssen 1996; Handayanto et al. 1997). In the other hand, the SOM in the organo-mineral fraction of some soils is relatively protected against mineralization and therefore does not immediately influence crop yields in the shortterm (Mapfumo et al. 2007).

The decomposition of organic matter is the key process in soil-plant N cycle (Barraclough 1997), principally governing the availability of this nutrient to crop growth. The chains of processes are very complex, as ammonia NH_4^+, the initial product of N mineralization, can be consumed by several processes (plant uptake, nitrification, immobilization and volatilization). Heterotrophic bacteria involved in the mineralization-immobilization turnover reactions between inorganic and organic pools of N compete more effectively for NH_4^+ than for nitrate NO_3^- (Jansson 1958; Jenkinson et al. 1985; Schimel et al. 1989).

The efficiency with which N in plant residues is used depends on the rate at which they are mineralized and thus on the time when they are made available relative to crop requirements. The present work aims to determine whether, with contrasting legume litter quality in terms of N mineralization, by mixing this organic material of different quality, it will be possible to alter the pattern of N release and the efficiency of utilization of N from the residue by a soil microbial biomass and catch crop.

2. State of the art

In this way four different legume species (*Acacia mangium* Willd., *A. angustissima* Kuntze, *Sclerolobium paniculatum* Vogel and *Inga edulis* Mart.) each used in enrich the fallow were compared with natural fallow vegetation, which is a mixture of different species, and poor soil without added organic material.

- Within this experiment the following points are essential: a) the impact of enriched legume material in soil N mineralization; b) the use of mineralization with the use of contrasting litter quality; c) the influence of organic material quality on soil microbial biomass in terms of N mineralization, immobilization and consumption.

After identifying the contrasting species, two laboratory decomposition experiments with two different techniques will be used to elaborate the effect that mixing these organic materials of different quality has on the pattern of N release and the efficiency of utilization of N from the residue by a soil microbial biomass.

The first decomposition experiment was made using soil incorporated legume leaf material from the two contrasting species, and their mixture, with ^{15}N at natural abundance and fertilized with enriched ^{15}N-urea fertilizer (conventional isotope dilution technique). In parallel, one experiment with the same species and mixture of legume with previously enriched ^{15}N and fertilized with ^{14}N-urea was carried out (pre-labeling plant material).

- Within this experiment the following points are essential: a) the use of contrasting litter quality may improve N-mineralization in terms of the rate at which they are mineralized; b) the quantification the real amount of N stored in the soil microbial biomass; c) the quantification of N-mineralization and immobilization through the use of labeling techniques. These isotope dilution techniques have the objective to quantify the proportion of N that comes from fertilizer or organic matter and is immobilized by soil microbial biomass.

To assess further whether N recovery by rice could be accurately predicted from relationships between pruning-material quality and N mineralization-immobilization, a greenhouse pot experiment was conducted in which the two isotope techniques were used with the same contrasting materials and their mixture, and ^{15}N uptake by rice was measured.

- Within this experiment the following points are essential: a) whether the use of contrasting litter quality may improve N-mineralization rate and thus the time when they make N available to the crop; b) the quantification of N competition between rice and soil microbial biomass; c) the quantification of N-mineralization and immobilization through the use of two techniques of isotope dilution which allow the quantification of the proportion of N coming from fertilizer or organic matter immobilized by soil microbial biomass and used by rice.

One of the hypothesis of this work is that with the elimination of the burning of biomass and the addition of organic matter as mulch, nitrogen immobilization in mulch by microorganisms will be increased and lead to a decrease in the quality of SOM. To confirm this hypothesis a field experiment whit litterbags from different legume treatment was conducted.

- Within this experiment the following points are essential: a) quantifying mulch decomposition during the field incubation on the litter; b) assessing nitrogen and carbon mineralization from mulch system; c) mulch nutrient retention during the field incubation; d) predict of N mineralization.

In generally this study is intended to answer the following questions:

- Can we regulating N release through the use of mixing residues from legume tree material with different patterns of N mineralization?
- Is N immobilization affected by legume tree material and therefore by mixtures?
- Can we fulfill the crop demands with organic matter fertilization (mulching) using materials from enriched fallows?
- What happens with the use of mulch system in terms of N-mineralization and immobilization?

3. Identify contrasting leguminous decomposition on leaf and wood material incorporated in the soil

3.1 Biochemical characteristics of the plant material

In the Brazilian Amazon soil the decomposition of contrasting amended material with regard to the measure quality characteristics show in all treatments a decrease in total-N content at the end of laboratory incubation period. The initial nutrient content in the organic amendment (wood + leaf material) for different treatments and soils (without added organic matter) is shown in Table 1. In this experiment *I. edulis* and *A. mangium* had an initially higher N content than the other single-legume treatments. But the leaf + wood material added to soil from *S. paniculatum* and *A. mangium* showed an initially higher P concentration. For these two important chemical elements, the mixture of two legumes showed an intermediate concentration in comparison to the single species.

Species	OM	N	C	P	Lignin	Cellulose	Phenol
	mg g^{-1} soil						
A. mangium	**14.15**(0.14)	**0.11**(0.001)	**6.84**(0.07)	**0.004**	**2.61**(0.03)	**5.95**(0.06)	**1.50**(0.02)
m*e[¥]	**14.33**(0.14)	**0.14**(0.003)	**6.90**(0.07)	**0.007**	**3.16**(0.04)	**5.76**(0.05)	**1.03**(0.01)
I. edulis	**14.11**(0.04)	**0.17**(0.001)	**6.77**(0.02)	**0.009**	**3.59**(0.01)	**5.42**(0.01)	**0.52**(0.00)
S. paniculatum	**14.16**(0.11)	**0.11**(0.001)	**6.83**(0.05)	**0.004**	**3.15**(0.03)	**3.72**(0.03)	**0.82**(0.01)
p*a[§]	**14.30**(0.14)	**0.15**(0.002)	**6.97**(0.07)	**0.007**	**3.10**(0.03)	**4.53**(0.04)	**1.26**(0.01)
A. angustissima	**14.26**(0.22)	**0.18**(0.005)	**7.01**(0.11)	**0.008**	**3.00**(0.05)	**5.27**(0.08)	**1.67**(0.03)
Fallow	**9.50**(0.06)	**0.08**(0.001)	**4.68**(0.03)	**0.004**	**2.51**(0.02)	**3.91**(0.03)	**1.05**(0.01)
Soil		**1.27**	**16.73**	**0.001**			

[¥] In this study, m*e represents a mixture of *A. mangium* and *I. edulis* (50:50 w/w).
[§] In this study, p*a represents a mixture of *S. paniculatum* and *A. angustissima* (50:50 w/w).

Table 1. Organic matter added to Amazon sandy soil and the nutrient (selected) content and material quality for different treatments in a laboratory incubation experiment. The number represents Mean (standard deviation), with n = 21.

After 128 days of incubation the significant difference between legume species and mixture were found only for the *S. paniculatum* treatment, which showed the lowest total-N concentration founded in soil. This same species showed a higher initial C-to-N ratio and lignin concentration. In this way, legume species and fallow treatment had a significant positive correlation (r^2 = 0.59, P < 0.01) with the total-N losses and initial C-to-N ratio. The total-N losses[1] decrease in the following order: *S. paniculatum* > Fallow vegetation > *A. mangium* = *A. angustissima* > soil > *I. edulis*[2]. The mixture of *S. paniculatum* and *A. angustissima* leaf material showed a significant inhibitory effect in the total remaining nitrogen, and only 10.8% of total-N was lost. However, the other mixture did not exhibit statistical differences in relation to the single species.

3.2 Nitrogen mineralization and immobilization

Soil-nitrogen mineralization without added organic matter more than doubled the amount of mineral N in the soil founded during the incubation period (Figure 1). This control treatment showed a rapid initial increase after it became steadier. In contrast, the amended soils showed immobilization-consumption of the native soil-N, reversing into a release of N after about a month for the legume-amended soil, particularly if m*e mixture or *S. paniculatum* were present. The fallow amended soil was slower in immobilizing soil-N, but continued to immobilize over the entire incubation period.

---

[1]Losses mean the differences of total N concentration in soil between begin and end of incubation period.
[2]">" symbol represents statistical differences at P < 0.05, and "=" no statistical difference.

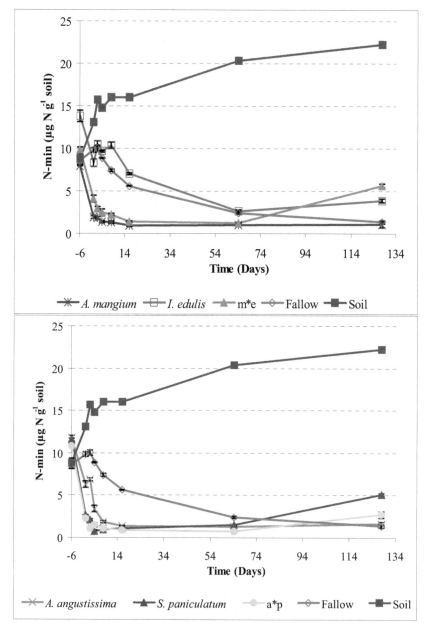

Fig. 1. Soil inorganic N concentration as affected by incorporated wood and leaves from different legumes species in comparison with fallow vegetation (Fallow) and soil with initial organic matter. In all graphics the m*e and p*a symbol correspond to *A. mangium* and *I. edulis* mixture (50:50 w/w), and *A. angustissima* and *S. paniculatum* mixture (50:50 w/w), respectively. Bars represent standard error of the mean.

Total mineral nitrogen in leguminous-amended soil was significantly different with time (P < 0.01) for the two species used in the mixtures, and for the interaction species and time (P < 0.01). The higher differences were found in the beginning of the incubation period and the final inorganic N content in legume-amended soil ranged from 2.0 to 6.1 mg N kg^{-1} soil compared with 1.6 mg N kg^{-1} soil in the fallow-amended soil and 20.7 mg N kg^{-1} soil in the control soil. Thus, at the end of incubation period, mineral nitrogen decreased in the following order: Soil as control > *S. paniculatum* > *I. edulis* > *A. mangium* > *A. angustissima* > natural fallow.

The decomposition patterns of the mixture and total N-mineralization did not reflect the simple mean of the decomposition patterns of single-species organic matter. The m*e mixture showed a higher increase in total N mineral at the end of the experiment, and that of the single species was comparatively lower in the same period. In the same way a*p mixture did not amount to the arithmetic mean of the N-mineralization in of the two single specie

In these sandy soils from Amazon, initial nitrification was significantly higher for soil without added organic matter, *I. edulis* and Fallow treatment. The *I. edulis* treatment presented a significantly higher consumption of NO_3^--N after approximately 4 days of incubation (Figure 2). Only *I. edulis* and the m*e mixture showed a small increase in NO_3^--N concentration at the end of the experiment; the other treatments showed a NO_3^--N consumption and/or denitrification.

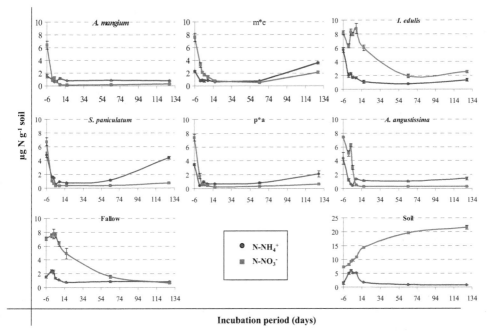

Fig. 2. Concentration of mineral N as influenced by incorporated wood and leaves from different legumes species in comparison with fallow vegetation (Fallow) and soil with initial organic matter (Soil) as control. Bars represent standard error of the mean.

Calculation of net ammonification (a, Figure 3) and net nitrification (n), in order to have a comparative parameters between treatments, was performed by subtracting soil NH_4^+-N in the time x (t_x) from soil NH_4^+-N in the initial time (t_0) and NO_3^--N in the time x (t_x) from soil NO_3^--N in the initial time (t_0), respectively. Apparent microorganism NH_4^+-N-immobilization (i) was calculated by subtracting soil NH_4^+-N microbial biomass in the organic amendment treatment from NH_4^+-N microbial biomass in the control treatment (soil without organic amendment) (Jensen 1997), using the fumigation-extraction method. NO_3^--N consumption (c) was calculated by subtracting soil NO_3^--N in the control treatment from soil NO_3^--N in the organic amendment treatment. This is based on the assumption that the mineralization and losses of indigenous soil N were similar in control and residue-treated soils (Jensen 1997).

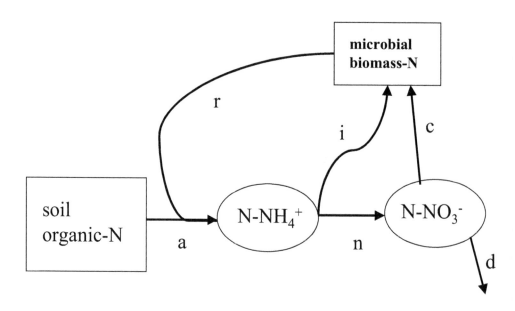

Fig. 3. Flows of soil nitrogen indicated by: (a) apparent N-ammonification, (r) apparent N-remineralization, (n) apparent N-nitrification, (i) apparent NH_4^+-N immobilization, (c) apparent NO_3^--N consumption, (d) denitrification. Adapted from Stark and Schimel (2001).

Apparent ammonification (a) was significantly different ($P < 0.05$) during the incubation period for the majority of treatments (Table 2). Only A. mangium organic material did not show a significant difference in N-ammonification between sampling times during the incubation period. The fallow and I. edulis treatments showed a significantly higher initial N-ammonification. *A. angustissima*, *S. paniculatum* and *A. mangium* presented an intermediate initial N-ammonification; with the mixture of two legume species presenting a significantly lower initial nitrogen ammonification (Table 2).

Treatment	Incubation period (days)[¥]			
	0	16	64	128
A. mangium	1.00[a]	-0.21[a]	-0.15[a]	-0.18[a]
m*e	0.73[b]	-0.16[c]	-0.05[c]	2.76[a]
I. edulis	2.25[a]	-0.95[b]	-1.26[b]	-0.68[b]
S. paniculatum	1.37[b]	-0.87[c]	-0.46[c]	2.79[a]
p*a	0.42[ab]	0.23[b]	0.34[b]	1.63[a]
A. angustissima	1.40[a]	-0.16[b]	-0.25[b]	0.15[b]
Fallow	2.34[a]	-1.64[b]	-1.50[b]	-1.56[b]
LSD$_{\leq 0.05}$	0.67	0.68	0.80	1.16

[Φ] Apparent net N-ammonification (*a*) was calculated by subtracting soil NH_4^+-N in the time *n* (t_n) from soil NH_4^+-N in the initial time (t_0).

[¥] Values within a line that are followed by different letters are significantly different with the Tukey test (P < 0.05).

Least significant difference (LSD) to compare treatments in the same sampling time.

Table 2. Apparent net N-ammonification, a (μg NH_4^+-N g^{-1} soil)[Φ] as a function of time for different treatments.

The m*e mixture, *S. paniculatum* and p*a mixture showed a significantly higher increase in ammonification at the end of incubation period with 2.03, 1.42 and 1.21 μg NH_4^+-N g^{-1} soil, respectively (Table 2). All other treatments showed a decrease in N-ammonification, particularly so for the fallow and *I. edulis* i.e., -3.96 and -2.93 μg NH_4^+-N g^{-1} soil, respectively. The mixture of these two legume species was significantly better mineralizable than the single species. This interaction was not observed with the other mixture, where *S paniculatum* showed a higher increase when mixed with *A. angustissima*.

The treatments with amendment showed a significant difference (P < 0.05) in N nitrification during the incubation period (Table 3). Fallow vegetation, *I. edulis* and *A. angustissima* showed a significantly higher initial nitrification, with m*e, p*a mixture and *S. paniculatum* showing intermediate values, respectively (Table 3). However, a significantly higher nitrification decrease was found in the fallow, *I. edulis* and *A. angustissima*, with variations of 14.4, 10.0 and 9.8 μg N- NO_3^- g^{-1} soil, respectively. *A. mangium* showed a significant lower initial N nitrification. At the end of the incubation period the mineralization decreased in the following order: *S. paniculatum* $\geq$ *A. mangium* $\geq$ m*e = p*a > *I. edulis* $\geq$ *A. angustissima* > Fallow vegetation[3].

---

[3] "$\geq$" and "=" symbols means that the treatments did not show statistical difference (P > 0.05), and ">" symbol showed statistical difference (P < 0.05).

Treatment	Incubation period (days)[¥]			
	0	16	64	128
A. mangium	0.90[a]	-0.76[b]	-0.74[b]	-0.62[b]
m*e	3.29[a]	-2.56[b]	-2.78[b]	-1.18[b]
I. edulis	6.26[a]	-0.28[b]	-4.35[c]	-3.73[c]
S. paniculatum	1.05[a]	-0.68[b]	-0.71[b]	-0.32[b]
p*a	1.83[a]	-1.60[b]	-1.50[b]	-1.18[b]
A. angustissima	5.04[a]	-4.79[b]	-4.77[b]	-4.79[b]
Fallow	7.49[a]	-2.57[b]	-5.91[c]	-6.89[c]
LSD$_{\leq 0.05}$	0.93	1.32	1.28	1.06

[Φ] Net N nitrification (n) was calculated by subtracting soil NO_3^--N in the time n (t_n) from soil NO_3^--N in the initial time (t_0).

[¥] Values within a line that are followed by different letters are significantly different with the Tukey test ($P < 0.05$).

Least significant difference (LSD) to compare treatments in the same sampling time.

Table 3. Net N nitrification, n (μg NO_3^--N g^{-1} soil) [Φ] as a function of time for different treatment.

Differences between the treatments were observed for NH_4^+-N immobilization at the beginning of the experiment ($P < 0.05$) (Table 4). Control treatment[4] decreased NH_4^+-N immobilization from 1.69 μg NH_4^+-N g^{-1} soil in the time 0 to 0.27 μg NH_4^+-N g^{-1} soil at the end of the incubation period. The m*e, p*a mixture and I. edulis treatments presented a significantly higher initial NH_4^+-N immobilization (Table 4), and the treatments showed a lower initial C-to-N ratio in comparison to A. mangium and S. paniculatum (Table 1). A. angustissima also showed a low initial C-to-N ratio (38.9), which did not explain the low initial N immobilization, but this treatment had a higher initial phenol content (Table 1), which contributed to the low initial N-immobilization (Mafongoya et al. 1998).

Only the m*e and p*a mixtures as well as the I. edulis significantly decreased ($P < 0.05$) in net NH_4^+-N immobilization at the end of incubation period. A. mangium showed a smaller initial net NH_4^+-N immobilization and a significantly higher ($P < 0.01$) increase in the net NH_4^+-N immobilization at the end of incubation period. Between 64 and 128 days of the incubation period, NH_4^+-N immobilization was not significantly different ($P > 0.05$) for the m*e, I. edulis, p*a, A. angustissima, and Fallow treatments. However, for this same period, A. mangium and S. paniculatum experienced a stronger increase in NH_4^+-N immobilization, with 3.32 and 1.18 μg NH_4^+-N g^{-1} soil, respectively.

As with the net immobilization, the microbial consumption of NO_3^--N was noticeably different ($P < 0.01$) between treatments at the beginning of the incubation period (Table 5). The strong differences in net consumption at the beginning of the experiment may reflect the differences in organic material quality. The A. mangium, S. paniculatum and p*a mixture treatment showed a significantly higher initial NO_3^--N consumption, and fallow vegetation showed the lowest value at the same time. All treatments showed a significant elevation in

[4]Control treatment mean soil without added organic material.

NO_3^--N consumption during the incubation period. But *I. edulis* showed a significantly lower NO_3^--N consumption ($P < 0.01$) at the beginning of the experiment following by fallow treatment.

Treatment	Incubation period (days)[¥]			
	0	16	64	128
A. mangium	0.76[b]	0.38[b]	1.90[b]	5.22[a]
m*e	6.46[a]	2.61[b]	2.59[b]	2.28[b]
I. edulis	4.15[a]	1.24[c]	3.34[ab]	2.59[b]
S. paniculatum	1.16[ab]	0.31[b]	0.85[ab]	2.34[a]
p*a	5.92[a]	1.02[c]	3.28[b]	3.44[b]
A. angustissima	0.95[ab]	0.19[c]	0.90[b]	1.40[a]
Fallow	1.72[ab]	2.44[a]	0.65[b]	1.33[ab]
LSD$_{\leq 0.05}$	0.96	0.21	0.25	1.26

Φ Apparent microorganism NH_4^+-N immobilization (*i*) was calculated by subtracting soil NH_4^+-N microbial biomass in the organic amendment treatment from NH_4^+-N microbial biomass in the control treatment (soil without organic amendment).
¥ Values within a line that are followed by different letters are significantly different with the Tukey test ($P < 0.05$).
Least significant difference (LSD) to compare treatments in the same sampling time.

Table 4. Apparent net N-microbial immobilization, i (μg NH_4^+-N g^{-1} soil)Φ as a function of incubation period and treatment.

Treatment	Incubation period (days)[¥]			
	0	16	64	128
A. mangium	7.20[d]	14.13[c]	19.33[b]	21.24[a]
m*e	4.81[c]	13.54[b]	18.97[a]	19.41[a]
I. edulis	1.84[c]	8.29[b]	17.57[a]	18.99[a]
S. paniculatum	7.05[d]	13.90[c]	19.14[b]	20.79[a]
p*a	6.27[d]	14.04[c]	19.15[b]	20.87[a]
A. angustissima	3.06[d]	14.02[c]	19.21[b]	21.27[a]
Fallow	0.61[d]	9.35[c]	17.90[b]	20.91[a]
LSD$_{\leq 0.05}$	0.93	1.06	0.25	0.37

Φ Apparent net consumption (*c*) was calculated by subtracting soil NO_3^--N in the control treatment from soil NO_3^--N in the organic amendment treatment in the same incubation period.
¥ Values within a line that are followed by different letters are significantly different with the Tukey test ($P < 0.05$).
Least significant difference (LSD) to compare treatments in the same sampling time.

Table 5. Apparent Net microbial consumption of NO_3^--N, c (μg NO_3^--N g^{-1} soil)Φ as a function of incubation period and treatment.

At the end of the experiment, a significantly lower apparent NO_3^--N consumption was found in *I. edulis* and m*e mixture (P < 0.05), in comparison the other treatment (Table 10). Only for these two treatments, NO_3^--N consumption remained constant after 64 days of incubation. The significantly highest consumption of 20.3 μg NO_3^--N g^{-1} soil was found in the Fallow treatment.

In this study the data indicate that the decline of mineral-N was strongly influenced by immobilization and consumption of mineral N by the microflora. Microbial consumption of NO_3^--N was of a greater magnitude than NH_4^+-N immobilization, thus indicate that the decreases observed in net N-mineralization were due to increasing microbial consumption of N. However, immobilization into soil organic matter (SOM) may be attributed to the apparent net N-mineral loss (Bending et al. 1998).

As was shown by Verchot et al. (2001), the results demonstrate that the patterns of N-mineralization are dependent upon differences between microbial production and consumption. These processes are reliant on organic matter quality. Small changes in mineralization, nitrification, immobilization and consumption may possibly have a large impact on soil N availability for the crop system.

The principal loss of C from SOM is through respiration during decomposition (Woomer et al. 1994). All legume-amended soils showed significantly higher cumulative CO_2 production than fallow-amended soil and the control treatment (soil) (Figure 4). The cumulative CO_2 production was significantly higher for mixtures in comparison with the individual species.

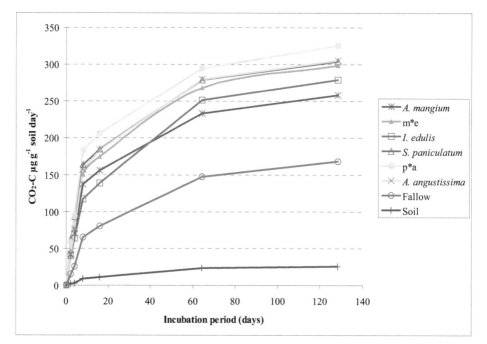

Fig. 4. Accumulative C-CO_2 (μg C g^{-1} soil day^{-1}) production during the incubation period for treatments with different organic matter soil added in comparison with soil without added organic matter as control (Soil).

All treatments showed a strong initial increase in CO_2-C production, which resulted in a high slope of the accumulative CO_2-C curve (Figure 4). It was conform the increase in the initial microbial C, NO_3^- consumption and the high decrease in the initial N mineral.

Soil N mineral decreased with increasing CO_2 production and microbial biomass C. In contrast, N immobilizations by microorganism increased with an increase in CO_2 production due to added organic C, and with a decrease of the N concentrations. This suggests that N dynamics in the legumes amended treatments were highly correlated with organic C dynamics (Figure 5).

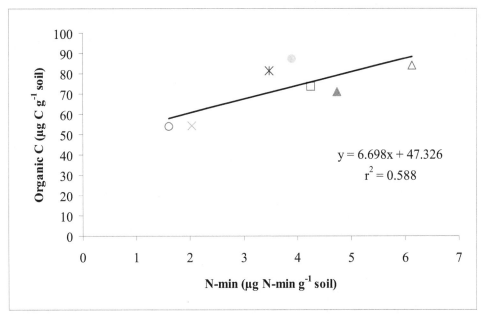

Fig. 5. The relationship between mineral N (N-min) and the extractable organic C resulting from treatment with different legumes, mixture and fallow at the end of incubation period (128 days). The symbols correspond to Fallow (○), *A. angustissima* (×), *A. mangium* (✳), p*a mixture (●), *I. edulis* (□), m*e mixture (▲), *S. paniculatum* (△) treatment.

Since N immobilization is limited by substrate availability in a broad range of ecosystems and soil types, soil organic C concentration would strongly influence N immobilization (Barrett and Burke 2000). The correlation between nitrogen mineralization and carbon mineralization suggests that rapid stabilization of nitrogen is facilitated by an active microbial community and the availability of a readily mineralizable organic substrate. Critical levels of C and plant nutrients, which limit the enzyme activities of microbial decomposition, were found to be important for determining nutrient release (Seneviratne 2000). Soil microbial biomass has been defined as an indicator of soil fertility, long before changes is soil organic matter occur (Powlson and Brookes 1987). Microbial nitrogen NH_4^+ immobilization and NO_3^- consumption appears to be an important variable that needs to be taken into account in studying organic matter decomposition and N mineralization.

The best predictor of N-mineralization was phenol + lignin when all treatments were included in the analyses (Table 6), followed by phenol + lignin-to-N ratios and lignin. Initial N + P-to-phenol ratios were highly correlated with cumulative N-mineralization for leguminous species and mixtures (Table 6).

Object of analyze	Lignin	Phenol + Lignin	Phenol + Lignin / N	Phenol / N	N + P / Phenol
All treatment [Φ]	-0.812*	-0.826*	-0.819*		
Legume and mixture [Ω]	-0.813*	-0.798*	-0.768*	-0.830*	0.950**
Single legumes [δ]	-0.788*	-0.871*	-0.868*	-0.803*	0.916*

[Φ] All treatment means legumes, mixture and fallow treatments.
[Ω] mixture means mixture of two legume species.
[δ] Single legume means legumes species without mixtures treatments. The * is $P \leq 0.05$, ** is $P \leq 0.01$

Table 6. Correlation coefficients relating the cumulative amount of N-mineralization to initial chemical properties in the treatments.

Soil-incorporated plant lignins degrade to polyphenol, which, with the other plant and microbial polyphenol, become the main constituents of recalcitrant N, containing humic polymers (Haynes 1986). Lignin intertwines also with the cell wall, physically protecting cellulose and other cell wall constituents from degradation (Chesson 1997).

Polyphenols include a range of compounds differing in size, solubility, and reactivity. Also, polyphenol can serve as a carbon substrate for decomposers (Mafongoya et al. 1998) but in general they inhibit the growth or function of decomposers and the other organisms (Swift et al. 1981, Zucker 1982). Defense compounds, including phenolics and terpenoids can also influence rates of litter decomposition, by means of direct inhibitor effects on saprophytic organisms (Palm and Sanchez 1991). Condensed tannins, also known as proanthocyanidins, are the polyphenol most noted for their effects on decomposition and nutrient dynamics. This results from their reactions with proteins and nitrogen (Myers et al. 1994; Mafongoya et al. 1998).

This experiment confirms that the resource quality and mixture of contrasting resource quality affect the N-mineralization and -immobilization processes during decomposition. In agreement with Palm (1995), the following factors must be considered when choosing parameters to describe plant quality: a) the processes of decomposition and N release are controlled by different parameters; b) the critical parameters will depend on the time frame of the crop need; and c) the importance of certain parameters change with the type and the mixture of the plant material. The ultimate aim is to identify robust parameters that predict decomposition and nutrient release.

The decomposition patterns and N-mineralization of the mixture were not the arithmetic mean of the decomposition patterns of the component organic material. In this case, there are interactions between components principally in terms of the rate of decomposition and N release, which was demonstrated in N-mineralization (net and total mineralization), microbial biomass C and extractable organic C.

Mafagoya et al. (1998) identified three types of soluble constituents that result in interactions between organic materials: a) compounds that contain available carbon as a substrate, b) compounds that contain readily available N, and c) soluble polyphenols, which can complex with proteins, rendering them resistant to microbial assault. The results showed in this experiment confirm that the carbon availably and soluble polyphenols may be the important parameters that result in interaction between contrasting organic material.

3.3 Decomposition of contrasting leguminous leaf material and gross N dynamics in soil using rice as an indicator plant

A decomposition study of 15N-labeled plant material (*S. paniculatum, I. edulis,* and mixture, p*e) with contrasting litter quality was conducted to assess the rates of mineralization and immobilization, using rice as an indicator plant (Table 7). N-urea fertilizer (3.92 N mg pot^{-1} with N at natural abundance) and ^{15}N-labeled leguminous organic material from *S. paniculatum* and *I. edulis* with 2.02% N and 0.392 atom % [15]N, and 1.93% N and 0.390 atom % [15]N, respectively, were used to find the amount of mineral-N coming from organic matter decomposition, and the extent of competition between microorganisms, soil + organic matter fixation, and rice absorption.

Treatment	Soil	Leaf	N	N-fertilizer	[15]N
	g		mg g^{-1} soil		µg g^{-1} soil
S. paniculatum[15]	50.62(0.10)	70.66(0.002)	1.62(0.001)	0.08(0.002)	6.36(0.012)
p*e[15]	50.62(0.11)	70.48(0.001)	1.64(0.001)	0.08(0.001)	6.40(0.009)
I. edulis[15]	50.52(0.09)	71.00(0.014)	1.69(0.006)	0.08(0.002)	6.61(0.010)
Control	50.68(0.13)			0.08(0.001)	

Φ species name with [14] had leaf with nitrogen at natural abundance and was fertilized with 5.34 atom % [15]N. Species name with [15] had leaf with enriched [15]N (both species at 0.39 % [15]N in leaf material) and was fertilized with fertilizer at natural abundance.

Table 7. Amount of organic matter (OM) mixed with soil in a plastic pot and the total ^{14}N and ^{15}N-excess added with leaf and fertilizer in greenhouse experiment. The numbers represent mean(standard deviation), n = 12.

Total rice biomass and total N in rice was affected by N from the legume leaves in the amended soil during the incubation period (Figure 6). Initial rice dry matter was not statistically different at the time of transplantation. Although the N uptake by rice in the p*e mixture treatment did not differ statistically from that in the *I. edulis* treatment at the end of incubation period, it was significantly higher than the *S. paniculatum* and the control treatments (Figure 6). All treatments showed a significantly (P < 0.01) higher total N content in comparison with the control. But the concentration of nitrogen in seedlings of rice were significantly when are used the interaction between two contrasting leguminous material as a source of nitrogen.

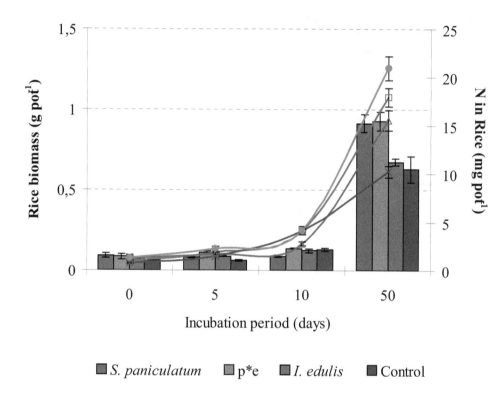

Fig. 6. Rice dry matter (g pot⁻¹, boxes) and total N in plant material (mg N pot⁻¹, lines) growing in soil with leaf-^{14}N legume material and mixture using enriched urea-^{15}N fertilizer. Control was soil that only included urea-^{14}N fertilizer. Bars represent standard error of the mean.

Treatment with the mixture of two legumes species showed a high recovery of N and ^{15}N in comparison to the two legume species (Figure 7). Recovery of ^{15}N from rice was significantly higher (P < 0.001) in the *S. paniculatum* treatment than with the *I. edulis* treatment. In contrast, the recovery of total N was higher for the *I. edulis* than for *S. paniculatum*.

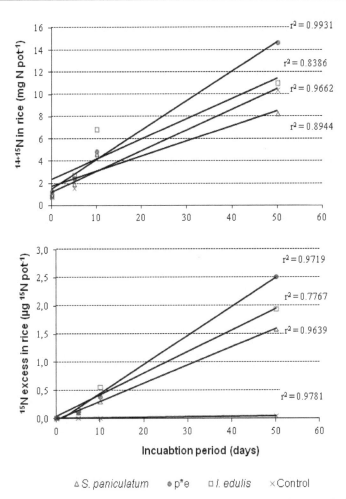

Fig. 7. $^{14}N+^{15}N$ and ^{15}N recovery in rice for soil treatment with ^{15}N-leaf material of two different legumes species and mixture, and ^{14}N-urea fertilizer in comparison with soil without added leaf material and fertilized with ^{14}N-urea, for different incubation periods. The ^{15}N is expressed as the atom % ^{15}N excess abundance above the background (0.3663 atom %).

The N concentration in rice increased until 10 days after transplanting (Table 8) and, during this phase of rice growth, the legume treatments did not differ from the control (P > 0.05). At the end of the incubation period, S. paniculatum showed a significantly (P < 0.01) lower N final concentration in rice (11.6 mg N g^{-1} dry matter) in comparison with the control (17.4 mg N g^{-1} dry matter), I. edulis (18.0 mg N g^{-1} dry matter) and p*e mixture (18.2 mg N g^{-1} dry matter).

The results with leaf enriched with ^{15}N reveal that most of the N absorbed by rice (Table 9), came from the soil, but the interaction between the two leguminous plants was provided to more N for rice after 50 days of incubation (Figure 7).

Treatment / Incubation period (days)	N concentration (mg N g⁻¹ rice dry matter)			
	0	5	10	50
S. paniculatum	11.2	18.0	25.3	11.6
p*e	14.3	19.2	27.8	18.2
I. edulis	14.2	19.6	30.4	18.0
Control	11.6	23.4	31.7	17.4
LSD	3.17NS	1.54*	6.40NS	6.45*

LSD compares different treatment at the same sampling time; NS is not significant; * = $P < 0.05$

Table 8. Nitrogen concentration in rice (mg N g⁻¹ dry matter) for different treatment and control during the incubation time (days). Control was soil without added leaf material and fertilized with N-urea.

Nitrogen microbial immobilization estimated by the fumigation-extraction method showed a contrasting pattern between differently materials (Figure 8 A and B). Generally at the beginning of the experiment, the microbial biomass-N with the small amount of added leaf material behaved exactly opposite to the experiment with the larger amount of leaf material added. The control treatment initially showed a faster decrease in microbial biomass-N.

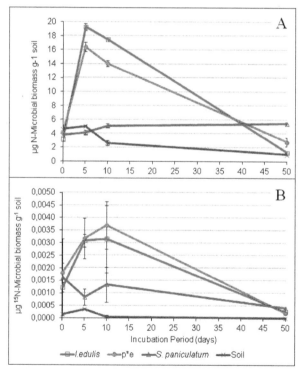

Fig. 8. A) Nitrogen microbial biomass (μg N-microbial biomass g⁻¹ soil) during the greenhouse incubation period (days) for samples with leaf material, and (B) ^{15}N microbial biomass (μg ^{15}N-microbial biomass g⁻¹ soil) samples with ^{15}N enriched leaf material and fertilized with urea-^{14}N. Soil treatment (as control) was soil without added leaf material and fertilized with urea-^{14}N. The bars represent Standard error of the mean

In the *I. edulis* and p*e mixture treatments, a significantly fast initial N microbial immobilization was observed, which increased toward the end of the experiment. However, at the end of the experiments with this species, a decrease in N immobilization was observed. After five days of incubation, the *S. paniculatum* treatment exhibited an approximately constant N immobilization. The immobilization of N by soil microorganisms seems to explain why most of the N absorbed by the rice comes from the ground. The experiment showed that most of the N released from the leaf material, through the process of decomposition, was immobilized by soil microorganisms (Figure 8).

Nitrogen derived from residues (Ndfr; Hood, 2001) of different ^{15}N-legume leaf material was not statistically different between treatments until five days of incubation (Table 9). At ten days, a significantly higher amount of total N in rice came from the *I. edulis* (34.1%) and p*e mixture (31.9%) than from the *S. paniculatum* (25.2%). However, at the end of the incubation period, N in rice originating from the *S. paniculatum* leaf material increased by 11.4% in comparison with a decrease of 3.4% and 2.5% for the p*e mixture and *I. edulis* treatments, respectively, reflecting the slowly biodegradability of *S. paniculatum*.

Treatment	N derived from residue (Ndfr, %)			
	Incubation period (days)			
	0	5	10	50
S. paniculatum	8.62	21.5	25.2	36.2
p*e	8.87	21.7	31.9	28.6
I. edulis	7.66	22.4	34.1	31.6
LSD	1.21^{NS}	0.92^{NS}	2.18**	0.66**
N derived from residue (Ndfr, mg)				
S. paniculatum	0.08	0.42	1.17	2.99
p*e	0.11	0.52	1.55	4.18
I. edulis	0.06	0.60	2.31	3.45
LSD	0.03*	0.07**	0.10**	0.73*
N recovered from residue (Nrfr, %)				
S. paniculatum	0.10	0.51	1.45	3.68
p*e	0.14	0.64	1.91	5.16
I. edulis	0.08	0.75	2.90	4.32
LSD	0.03*	0.03**	0.13**	0.84*

Table 9. Nitrogen derived from residue (NdfR) in % and mg, and N recovered from residue (NrfR, %), according to Hoods (2001), during the greenhouse incubation period in soil treated with 15N-leaf legume and 14N-urea fertilizer.

The amount of N derived from residues increased significantly for all treatments during the incubation period (Table 9). The higher increase in N recovered from the added leaf material was found in the p*e mixture treatment, which showed an increase of 4.07 mg N in comparison with 3.39 and 2.91 mg N for the *I. edulis* and *S. paniculatum* treatments, respectively.

With this higher increase, the total N recovered from added legume leaf material was significantly higher for the p*e mixture than the other treatments (Table 9). However, for the first 10 days of the incubation period, the *I. edulis* treatment showed the higher percentage

(3.65%) of total N recovered from the added leaf material in comparison with the p*e mixture (2.55%) and *S. paniculatum* treatments (1.96%).

Lower rates of ^{15}N-recovery could be due to mineralization-immobilization turnover (Thönnissen et al. 2000). The ^{15}N-release from the legume residue into the soil inorganic pool could be exchanged for ^{14}N in microbial biomass, which could lead to a lower ^{15}N-recovery. On the other hand, lower rates for ^{15}N recovery than for total N may result partly from an overestimation of apparent total N recovery and partly from the importance of soil conditions, in terms of C-quality, during the rapid degradation of ^{15}N-labeled material.

A high (^{14}N+^{15}N)-microbial biomass was observed at the beginning of the greenhouse experiment for the p*e mixture (Table 10) in comparison to the other treatments. All treatments showed an increase in N absorption from microorganisms during the first five days. This increase was very high for the *I. edulis* treatment, which showed an increase of 500% of the initial concentration, followed by the p*e mixture (27.0%), *S. paniculatum* (10.0%) and control (8.8%).

Treatment	(^{14}N+^{15}N)-microbial biomass (µg ^{14}N+^{15}N g^{-1} soil)			
	Incubation period (days)			
	0	5	10	50
S. paniculatum	3.8[b]	4.2[c]	5.1[c]	5.5[a]
p*e	12.9[a]	16.4[b]	14.1[b]	2.8[b]
I. edulis	3.2[b]	19.3[a]	17.5[a]	1.3[c]
Control	4.7[b]	5.1[c]	2.7[d]	1.0[c]
LSD	1.451[***]	0.915[***]	0.310[***]	0.298[***]
	^{15}N-microbial biomass (µg ^{15}N excess g^{-1} soil)			
S. paniculatum	0.01618[a]	0.00084[b]	0.00134[a]	0.00041[a]
p*e	0.00180[b]	0.00316[a]	0.00369[a]	0.00021[ab]
I. edulis	0.00123[b]	0.00308[a]	0.00314[a]	0.00026[ab]
Control	0.00002[c]	0.00004[c]	0.00001[b]	0.00001[b]
LSD	0.0018[***]	0.00070[**]	0.00307[*]	0.00025[*]

LSD compares treatment at the same sampling time; * = P < 0.05; ** = P < 0.01; *** = P < 0.001.

Table 10. (^{14}N+^{15}N)-microbial biomass (µg ^{14}N+^{15}N g^{-1} soil) and ^{15}N-microbial biomass (µg ^{15}N excess g^{-1} soil) in soil amended with ^{14}N-legume leaf material and enriched ^{15}N-urea fertilizer with rice as an indicator plant. Control was soil without added leaf material and fertilized with N-urea. The ^{15}N excess is expressed as the atom % ^{15}N excess abundance above the background (0.3663 atom %).

The high N-microbial biomass level remained high in the *I. edulis* and p*e mixture treatments for the first 10 days of the greenhouse experiment (Table 10). However, at the end of the incubation period, these treatments showed a significantly higher decrease in the microbial biomass-(^{14}N+^{15}N), whereas the *S. paniculatum* treatment displayed a significantly high increase in the concentration of ^{14}N+^{15}N in the microbial biomass.

In all treatments, a significantly higher microbial biomass-^{15}N was observed during the first 10 days of the incubation period in comparison with the control (Table 10). The initially higher microbial biomass-^{15}N concentration for the *S. paniculatum* treatment indicated intensive microbial activity leading to decomposition of recalcitrant leaf material in this treatment. However, after fertilizer had been added, the ^{15}N-excess immobilization in the soil increased to 155% of the initial concentration in the *I. edulis* treatment and 105% in the p*e mixture treatment, whereas the *S. paniculatum* treatment showed a decrease of 92% in the microbial biomass- [15]N concentration. At the end of the experiment, all legume treatments showed a significant decrease in the microbial biomass-^{15}N, and only the *S. paniculatum* showed a significantly higher microbial ^{15}N-concentration in comparison with control (Table 10).

The decrease in soil mineral nitrogen over time was attributable not only to rice plant uptake, but also to considerable microbial immobilization. The use of a mixture of two contrasting litter qualities (called p*e mixture) improved the nitrogen recovered by rice, which was derived from urea-fertilizer by 35.1% and 41.3% of total fertilized N in comparison with *S. paniculatum* and *I. edulis* treatments, respectively. In the same way, the rice recovery of N derived from the legume material in the p*e mixture improved by 73.3% compared with the *S. paniculatum* treatment, but was 0.9% lower than that of the *I. edulis* treatment.

Moreover, the use of a larger amount of leaf material resulted in a higher increase in N immobilization and mineral-N in the soil, depending on the quality of the leaf material. The total cumulative nitrogen mineralization increased 90.9%, 10.3%, and 18.8% for the *S. paniculatum*, p*e mixture and *I. edulis* treatments, respectively. In the same way, the increases in cumulative microbial biomass-N during the incubation period were 109%, 190% and 344%, respectively.

The p*e mixture treatment showed an intermediate cumulative soil microbial- [15]N immobilization, higher cumulative rice biomass and total N, and higher recovery of ^{15}N from urea fertilizer. This indicates that the interaction of two different leguminous species increases the nitrogen absorption by rice through the increase in mineral-N and the decrease of gross microbial-N immobilization.

Decline in soil productivity and environmental quality and progressive deterioration of natural resources in the tropics have led to a search for new methods to sustain crop production via more efficient nutrient cycling. In agricultural ecosystems in the tropics with limited access to fertilizers, plant residues are often used to meet the N requirements of annual food crops (Constantinides and Fownes 1994). The added organic materials are potentially important sources of N, C and P in crop production, especially for resources-poor farmers on tropical agricultural land. In order to successfully manage organic materials, the release and uptake of N by crops must be identified (Hood 2002). But the predictions of net N-mineralization are in many cases unreliable because net N-mineralization is affected by N immobilization and remineralization and losses (Stark and Schimel 2001).

The slash and burn system destroys the above-ground biomass of the fallow vegetation including the litter by burning, which causes loss of nutrients through volatilization and leaching of free nutrients in ash by rainfall. The losses of nitrogen by volatilization and leaching can reach 95-98% (Mackensen et al. 1996). Cerri et al. (1991) observed a reduction of 25% in soil carbon content two years after a plot was cleared, burned and a satisfactorily managed pasture established. However, Kato (1998) showed no reduction in carbon content in the mineral soil in the slash and burn system, and attributed this to the biomass accumulated by the rice crop.

The concept of pools of organic matter that differ in their susceptibilities to microbial decomposition and their longevity in soil has provided a basis for understanding the dynamic nature of soil organic matter and how nutrient availability is influenced by management practices and changes in the soil environment (Stevenson and Cole 1999). Our study showed that the patterns of the added organic carbon need to be taken into consideration. This was supported by the strong correlation between nitrogen dynamics in the contrasting legumes-amended soil and soil microbial biomass and organic carbon, found in this study.

Dissolved organic substances contribute to plant nutrition with nitrogen. Due to their water solubility there is a considerable risk that leaching of these substances will result in enhanced soil degradation. The dynamics of dissolved organic substances is influenced by the quantity and the quality of soil organic matter, the sorption characteristics of the soils and the microbial activity. All these parameters are modified by land use.

When immobilization and mobilization processes of N in soil are managed, it is important to quantify the real amount of N stored in the soil microbial biomass (Joergensen and Mueller 1996). Transient immobilization of soil N in the microbial biomass may contribute to improved conservation of soil N sources (Jensen 1997).

The higher initial concentrations of soil inorganic nitrogen in the high-N treatments would unlikely increase nitrogen immobilization significantly in the absence of added organic matter because nitrogen immobilization is generally limited by available carbon (Recous et al. 1988, Bremer and Kuikman 1997). Thus, fertilizer added as urea-N did not lead to differences in total mineral-N and microbial biomass-N in soil without added organic material during the incubation period (Table 11). Urea increased total microbial biomass-N and decreased total mineral-N, which suggests that fertilizer increased microbial biomass and thus nitrogen consumption by soil microorganism.

On the other hand, the increase in soil organic carbon and nitrogen due to added legume leaf + wood material resulted in a decrease in N-min and microbial biomass in comparison with control treatment (Table 11). Assuming that the fumigation-extraction method did not measure the fungal N-absorption, resulting in a underestimation of N-immobilization, and that the mineralization and losses of indigenous soil nitrogen were similar in the control and residue-treated soil, the real nitrogen immobilization and consumption was 63.2 and 37.7 µg N g^{-1} soil for *S. paniculatum* and *I. edulis* treatment, respectively. This means that the nitrogen in microbial biomass (microbial nitrogen immobilization and consumption) was approximately 9.3 and 1.0 times more than the mineral-N found in the same treatments, respectively.

The withdrawal of the wood material and added nitrogen as fertilizer yielded a strong increase in N-mineralization in comparison with wood material only and control. This suggests that these soils in this Amazon region are very nitrogen and carbon limiting and the microbial competition is very intensive.

S. paniculatum	N-mineral (µg N g⁻¹ soil)		N-microbial biomass (µg N g⁻¹ soil)	
Incubation period (days)	leaf + wood material	leaf material + fertilizer	leaf + wood material	leaf material + fertilizer
0	2.7(0.04)	19.5(1.81)	0.6(0.50)	4.8(1.33)
4	0.5(0.14)	15.5(1.42)	0.8(0.06)	5.2(1.88)
16	0.7(0.11)	13.4(0.28)	0.3(0.09)	6.3(3.38)
64	1.1(0.12)	31.5(1.46)	0.1(0.07)	5.5(0.71)
Total	5.0	79.9	1.8	21.8
I. edulis	Total N-mineral (µg N g⁻¹ soil)		Total N-microbial biomass (µg N g⁻¹ soil)	
Incubation period (days)	leaf + wood material	leaf material + fertilizer	leaf + wood material	leaf material + fertilizer
0	8.3(0.44)	54.5(0.84)	3.6(0.24)	19.3(3.48)
4	9.7(0.08)	51.7(0.80)	1.1(0.14)	18.2(1.82)
16	7.1(0.11)	57.7(1.23)	1.2(0.10)	19.4(1.62)
64	2.7(0.09)	52.0(1.66)	2.6(0.17)	14.2(2.61)
Total	27.8	215.9	8.5	71.1
Control	Total N-mineral (µg N g⁻¹ soil)		Total N-microbial biomass (µg N g⁻¹ soil)	
Incubation period (days)		fertilizer		fertilizer
0	13.1(0.16)	17.8(2.04)	1.3(0.25)	4.8(0.36)
4	14.8(0.26)	12.9(0.95)	4.9(0.25)	1.8(0.41)
16	16.0(0.09)	13.3(0.10)	2.2(0.08)	0.7(0.55)
64	20.4(0.13)	12.7(0.18)	1.3(0.01)	4.6(1.37)
Total	64.3	56.7	9.7	11.9

Φ Nitrogen-mineral was examined with steam distillation procedure.
Ψ N-microbial biomass was measured with fumigation-extraction procedure.

Table 11. Nitrogen-mineral^Φ (µg N g⁻¹ soil) and N-microbial biomass^Ψ (µg N g⁻¹ soil) for incorporated leaf+wood material in comparison with incorporated leaf material+Urea as fertilizer, during the incubation time and total. The numbers represent mean (Standard error).

Microbial immobilization of labelled nitrogen was unaffected by rice plant growth, but was strongly affected by organic matter and nitrogen addition. Immobilization of nitrogen by organic matter decomposers was determined primarily by the amount and accessibility of available nitrogen. Differences in nitrogen immobilization by decomposers of the legume organic matter were greatest between N treatments, but were also affected by mixture of two contrasting legume materials

Rice growth and nitrogen accumulation closely reflected the differences in chemical composition and mineralization between the residues and their mixture. Approximately the same amount of nitrogen was added in all legume treatments, and yet the amount of nitrogen accumulated differed with the legume quality. The use of a mixture of two contrasting litter qualities improved the rice recovery of nitrogen derived from urea-fertilizer and the rice recovery of nitrogen derived from legume material.

This study showed that the quality and the quantity of organic carbon presented an important factor affecting soil nitrogen mineralization and immobilization. Changes in soil carbon substrates influenced the dynamics of soil inorganic nitrogen because of the importance of labile carbon in the microbial immobilization and consumption of nitrogen. Compton and Boone (2002) showed that the light fraction of soil organic matter incorporated more ^{15}N than the heavy fraction per unit of carbon, which indicated that not simply the amount but the composition of organic matter controls its function as a site for N incorporation.

Soil microbial biomass immobilizes a higher amount of the residue N mulched or incorporated into the soil and this needs to be taken into consideration. On the other hand, soil microbial biomass immobilization is a labile repository of nitrogen, their turnover and remineralization may conserve this N in the system and release this N for later plant use.

Legume-enriched mulch material had different patterns of mulch decomposition and nutrient release. As was pointed out earlier by Constantinides and Fownes (1994), Fox et al. (1990), the contents of N, lignin and polyphenol are the principal chemical factors controlling degradability of plant material. The high correlation with the ratio of N + P-to-phenol and N-mineralization implies that the plant nutrients which limit microbial action govern decomposition and nutrient release. This is likely associated with the formation of stable polymers with many forms of N binding N released to the soil by the incorporated organic matter.

4. Conclusion

The use of fast-growing legumes for fallow enrichment, as for example *A. mangium*, does not necessarily translate directly in a CO_2 sequestration, because the fast decomposition rate. Species, contrasting in lignin and polyphenol concentration with higher N- and P-content must be used for enrichment fallow. The *I. edulis* and *S. paniculatum*, two Amazon species, showed greater promise as enrichment candidates, because of their high organic C input in combination with low losses of OM.

Since immobilization of N is generally determined by the amount of decomposable carbon present in the soil rather than by the amount of inorganic N, the addition of compost showing a wide C-to-N ratio accelerates N-immobilization through increased microbial activity. Contrasting leguminous species had different patterns of net N-mineralization and immobilization in comparison with the single species. The use of two contrasting leguminous species increased the nitrogen absorption by rice through the increase of mineral N and decrease of microbial N-immobilization.

Soil microbial biomass immobilizes a higher amount of the residue N mulched or incorporated into the soil and this needs to be taken into consideration. On the other hand, soil microbial biomass immobilization is a labile repository of nitrogen, their turnover and remineralization may conserve N in the system and release this N for later plant use. Managing soil biological processes is a key aspect of sustainable development. The researchers must better understand soil organisms, their functions and their interactions with the chemical and physical environment. Many aspects of soil biology and ecology are worthy of research in view of their fundamental scientific interest and their role in ecosystem functioning.

The high correlation with the ratio of N + P-to-phenol and N-mineralization implies that the plant nutrients which limit microbial action govern decomposition and nutrient release. This is likely associated with the formation of stable polymers with many forms of N binding N released to the soil by the incorporated organic matter.

The slash-and-mulch systems with thick mulch mats need to be improved for the synchronization of nutrient release from organic material and nutrient uptake by crop systems. The use of contrasting plant material in terms of litter quality, C reduction in the vegetation with the selective removal of wood, and soil incorporation of fallow residues need to be further tested. On the other hand, the increase of agriculture in the Amazon region cannot be done by the increase of deforestation and the scope of an increase in fertilizer use is limited. Thus, the intensification and improvement of currently managed land would have to be attempted. Fallow-mulch system is a considerable challenge and calls for more research. The agricultural policy in the Amazon region could promote organic agriculture with incentives to production and facilitating the commercialization of Amazon organic agricultural products.

5. References

Addiscott, T. M., Whitmore, A. P. and Powlson, D. S. (1991). Farming fertilizers and the Nitrate Problem. Wallingford, UK, CAB International.

Anderson, J. M., Proctor, J. and Vallack, H. W. (1983). "Ecological Studies in Four Contrasting Lowland Rain Forests in Gunung Mulu National Park, Sarawak: III. Decomposition Processes and Nutrient Losses from Leaf Litter." Journal of Ecology 71: 503-527.

Barraclough, D. (1997). "The direct or MIT route for nitrogen immobilization: a 15N mirror image study with Leucine and Glycine." Soil Biology and Biochemistry 29(1): 101-108.

Barrett, J. E. and Burke, I. C. (2000). "Potential nitrogen immobilization in grassland soils across a soil organic matter gradient." Soil Biology and Biochemistry 32: 1707-1716.

Bending, G. D., Turner, M. K. and Burns, I. G. (1998). "Fate of nitrogen from crop residues as affected by biochemical quality and the microbial biomass." Soil Biology and Biochemistry 30(14): 2055-2065.

Biederbeck, V. O., Janzen, H. H., Campbell, C. A. and Zentner, R. P. (1994). "Labile soil organic matter as influenced by cropping practices in an arid environment." Soil Biology and Biochemistry 26(12): 1647-1656.

Blair, J. M., Parmelee, R. W. and Beare, M. H. (1990). "Decay Rates, Nitrogen Fluxes, and Decomposer Communities of Single- and Mixed-Species Foliar Litter." Ecology 71(5): 1976–1985.

Bremer, E. and Kuikman, P. (1997). "Influence of competition for nitrogen in soil on net mineralization of nitrogen." Plant and Soil 190(1): 119-126.

Cattanio, J.H., Kuehne, R., Vlek, P.L.G. (2008). "Organic material decomposition and nutrient dynamics in a mulch system enriched with leguminous trees in the Amazon." Revista Brasileira Ciencia do Solo 32: 1073-1086.

Cerri, C. C., Volkoff, B. and Andreaux, F. (1991). "Nature and behaviour of organic matter in soils under natural forest, and after deforestation, burning and cultivation, near Manaus." Forest Ecology and Management 38: 247-257.

Chesson, A. (1997). Plant degradation by ruminants: parallels with litter decomposition in soils. Drive by Nature: Plant Litter Quality and Decomposition. G. Cadisch and K.E. Giller (eds.). Wallingford, UK, CAB International: 47-66.

Compton, J. E. and Boone, R. D. (2002). "Soil nitrogen transformations and the role of light raction organic matter in forest soils." Soil Biology and Biochemistry 34(7): 933-943.

Constantinides, M. and Fownes, J. H. (1994). "Nitrogen mineralization from leaves and litter of tropical plants: Relationship to nitrogen, lignin and soluble polyphenol concentrations." Soil Biology and Biochemistry 26(1): 49-55.

Denich, M., Vlek, P.L.G., Abreu-Sa, T. D., Vielhauer, K., Luecke, W. (2005). "A concept for the development of fire-free fallow management in the Eastern Amazon, Brazil." Agriculture, Ecosystems and Environment 110(1-2): 43–58.

Fox, R., Myers, R. and Vallis, J. K. (1990). "The nitrogen mineralization rate of legume residues in soil as influenced by polyphenol, lignin, and nitrogen contents." Plant and Soil 129: 251-259.

Franagan, P. W. and van Cleve, K. (1983). "Nutrient cycling in relation to decomposition and organic-matter quality in taiga ecosystems." Canadian Journal of Forest Research 13: 795-817.

Gallardo, A. and Merino, J. (1993). "Leaf Decomposition in Two Mediterranean Ecosystems of Southwest Spain: Influence of Substrate Quality." Ecology 74(1): 152–161.

Handayanto, E., Cadisch, G. and Giller, K. E. (1997). Regulating N Mineralization from Plant Residues by Manipulation of Quality. Driven by Nature: Plant Quality and Decomposition. G. Cadisch and K.E. Giller (eds.), Wallingford, UK. CAB International: 175-185.

Haynes, R. J. (1986). The decomposition process: Mineralization, immobilization, humus formation and degradation. Mineral Nitrogen in the Plant-Soil System. R. J. Haynes (ed.). Orlando, FL., Academic Press: 52-176.

Hobbie, S. E. (2000). "Interactions between litter lignin and soil nitrogen availability during leaf litter decomposition in a Hawaiian Montane Forest." Ecosystems 3(4): 484-494.

Hood, R. (2001). "Evaluation of a new approach to the nitrogen-15 isotope dilution technique, to estimate crop N uptake from organic residues in the field." Biology and Fertility of Soils 34: 156-161.

Isaac, L., Wood, C. W. and Shannon, D. (2000). "Decomposition and nitrogen release of prunings from hedgerow species assessed fro alley dropping in Haiti." Agronomy Journal 92: 501-511.

Janssen, B. H. (1996). "Nitrogen mineralization in relation to C:N ratio and decomposability of organic materials." Plant and Soil 191(1): 39-45.

Jansson, S. L. (1958). "Tracer studies on nitrogen transformations in soil with special attention to mineralization-immobilization relationships." Annals of the Royal Agricultural College of Sweden 24: 101-361.

Jenkinson, D. S., Fox, R. H. and Rayner, J. H. (1985). "Interaction between fertilizer nitrogen and soil nitrogen: The so-called priming effect." Journal of Soil Science 36: 425-444.

Jensen, B. K., Jensen, E. S. and Magid, J. (1995). "Decomposition of 15N-labelled rye grass in soils from a long-term field experiment with different manuring strategies." Nitrogen Leaching in Ecological Agriculture: 221-228.

Jensen, E. S. (1997). "Nitrogen immobilization and mineralization during initial decomposition of 15N-labelled pea and barley residues." Biology and Fertility of Soils. 24: 39-44.

Joergensen, R. G. and Mueller T. (1996). "The fumigation-extraction method to estimate soil microbial biomass: calibration of the kEN value." Soil Biology and Biochemistry 28(1): 33-37.

Kato, M. S. A., Kato, O. R., Denich, M. and Vlek, P. L. G. (1999). "Fire-free alternatives to slash-and-burn for shifting cultivation in the eastern Amazon region: the role of fertilizers." Field Crops Research 62(2-3): 225-237.

Kato, O. R. (1988). Fire-free Land Preparation as an Alternative to Slash-and-burn Agriculture in the Bragantina Region, Eastern Amazon: Crop Performance and Nitrogen Dynamics. Dissertation submitted for the degree of Doctor of Agricultural Sciences of the Facultay of Agricultural Sciences. Goettingen, George-August-University: 132.

Kuo, S. and Sainju, U. M. (1998). "Nitrogen mineralization and availability of mixed leguminous and non-leguminous cover crop residues in soil." Biology and Fertility of Soils 26: 346-353.

Linkins, A. E., Melillo, J. M. and Sinsabaugh, R. L. (1984). Factors affecting cellulase activity in terrestrial and aquatic ecosystems. Current Perspectives in Microbial Ecology. M.J. Klug and C.A. Reddy. Washington, American Society for Microbiology: 572-579.

Lonrez, K., Preston, C. M., Raspe, S., Morrison, I. K. and Feger, K. H. (2000). "Litter decomposition an humus characteristics in Canadian and German spruce ecosystems: information from tannin analysis and 13C CPMAS NMR." Soil Biology and Biochemistry 32: 779-792.

Mackensen, J., Hölscher, D., Klinge, R. and Fölster, H. (1996). "Nutrient transfer to the atmosphere by burning of debris in eastern Amazonia." Forest Ecology and Management 86: 121-128.

Macías, F. and M. C. Arbestain (2010). "Soil carbon sequestration in a changing global environment." Mitig Adapt Strateg Glob Change 15: 511–529.

Mafongoya, P. L., Giller, K. E. and Palm, C.A. (1998). "Decomposition and nitrogen release patterns of tree prunings and litter." Agroforestry Systems 38(1-3): 77-97.

Mapfumo, P., Mtambanengwe, F., and Vanlauwe, B. (2007). "Organic matter quality and management effects on enrichment of soil organic matter fractions in contrasting soils in Zimbabwe." Plant and Soil 296(1-2): 137–150

Meentemeyer, V. (1978). "Macroclimate the lignin control of litter decomposition rates." Ecology 59(3): 465–472.

Melillo, J.M., Aber, J.D. and Muratore, J.F. (1982). "Nitrogen and lignin control of hardwood leaf litter decomposition dynamics." Ecology 63(3): 621-626.

Moorhead, D. L., Sinsabaugh, A. E., Linkins, A. E. and Reynolds, J. F. (1996). "Decomposition processes: modelling approaches and applications." The Science of Total Environment 183: 137-149.

Myers, R. J. K., Palm C. A., Cuevas, E., Gunatilleke, I. U. N. and Brossard, M. (1994). The synchronization of nutrient mineralization and plant nutrient demand. Biological

Management of Tropical Soil Fertility. P.L. Woomer and M.J. Swift (eds.). Chichester, UK, Wiley-Sayce Publication.: 81-116.

Palm, C. A. (1995). "Contribution of agroforestry trees to nutrient requirements of intercropped plants." Agroforestry Systems 30: 105-124.

Palm, C. A. and Sanchez, P. A. (1991). "Nitrogen release from the leaves of some tropical legumes as affected by their lignin and polyphenolic contents." Soil Biology and Biochemistry 23(1): 83-88.

Powlson, D. S. and Brookes, P.C. (1987). "Measurement of soil microbial biomass provides an early indication of changes in total soil organic matter due to straw incorporation." Soil Biology and Biochemistry 19: 159-164.

Powlson, D. S., Hirsch, P. R. and Brookes, P. C. (2001). "The role of soil microorganisms in soil organic matter conservation in the tropics." Nutrient Cycling in Agroecosystems 61(1-2): 41-51.

Priha, O. and Smolander, A. (1997). "Microbial biomass and activity in soil and litter under Pinus sylvestris, Picea abies and Betula pendula at originally similar field afforestation sites." Biology and Fertility of Soils 24(1): 45-51.

Recous, S., Fresneau, C., Faurie, G. and Mary, B. (1988). "The fate of labeled 15N urea and ammonium nitrate applied to a winter wheat crop: Nitrogen transformation in soil." Plant and Soil 112: 205-214.

Schimel, J. P., Jackson, L. E. and Firestone, M. K. (1989). "Spatial and temporal effects on plant-microbial competition for inorganic nitrogen in a California annual grassland." Soil Biology and Biochemistry 21: 1059-1066.

Seneviratne, G. (2000). "Litter quality and nitrogen release in tropical agriculture: a synthesis." Biology and Fertility of Soils. 31: 60-64.

Stark, J. M. and Schimel, J. (2001). "Errors in `Overestimation of gross N transformation rates in grassland soils." Soil Biology and Biochemistry 33(10): 1433-1435.

Stevenson, F. J. and Cole, M. A. (1999). Cycles of Soil : Carbon, Nitrogen, Phosphorus, Sulfur, Micronutrients. New York, NY, Wiley. Pp 427.

Swift, M. J., Russell-Smith, A. and Perfect, T. J. (1981). "Decomposition and Mineral-Nutrient Dynamics of Plant Litter in a Regenerating Bush-Fallow in Sub-Humid Tropical Nigeria." Journal of Ecology 69(3): 981-995.

Thönnissen, C., Midmorea, D. J., Ladha, J. K., Olk, D. C. and Schmidhalter, U. (2000). "Legume decomposition and nitrogen release when applied as green manures to tropical vegetable production systems." Agronomy Journal 92: 253-260.

Verchot, L. V., Holmes, Z., Mulon, L., Groffman, P. M. and Lovett, G. M. (2001). "Gross vs net rates of N mineralization and nitrification as indicators of functional differences between forest types." Soil Biology and Biochemistry 33(14): 1889-1901.

Vitousek, P. M., Turner, D. R., Parton, W. J. and Sanford, R. L. (1994). "Litter Decomposition on the Mauna Loa Environmental Matrix, Hawai: Patterns, Mechanisms, and Models." Ecology 75(2): 418–429.

Woomer, P. L., Martin, A., Albrecht, A., Resck, D. V. S. and Scharpenseel, H. M. (1994). The importance and management of soil organic matter in the tropics. P. L. Woomer and M. J. Swift (eds.). The Biological Management of Tropical Soil Fertility. West Sussex, UK, John Wiley & Sons: 47-80.

Zimmer, M. (2002). "Is decomposition of woodland leaf litter influenced by its species richness?" Soil Biology and Biochemistry 34(2): 277-284.

Zucker, W. V. (1982). "How aphids choose leaves: the roles of phenolics in host selection by a Galling Aphid." Ecology 63(4): 972-981.

8

Legume Crops, Importance and Use of Bacterial Inoculation to Increase Production

María A. Morel[1*], Victoria Braña[1*] and Susana Castro-Sowinski[1, 2]
[1]Molecular Microbiology, Biological Sciences Institute Clemente Estable
[2]Biochemistry and Molecular Biology, Faculty of Science, Montevideo,
Uruguay

1. Introduction

Legumes are flowering plants that produce seedpods. They have colonized several ecosystems (from rain forests and arctic/alpine regions to deserts; Schrire et al., 2005), and have been found in most of the archaeological record of plants. Early in 37 B.C. Varro said "Legumes should be planted in light soils, not so much for their own crop as for the good they do to subsequent crops" (Graham & Vance, 2003), recognizing the importance of multiple cropping and intercropping production.

Leguminosae or Fabaceae is the third most populous family of flowering plants (behind Asteraceae and Orchidaceae) with 670 to 750 genera and 18,000 to 19,000 species. Legumes include important grain, pasture and agro-forestry species. They are harvested as crops for human and animal consumption as well as used as pulp for paper production, fuel-woods, timber, oil production, sources of chemicals and medicines, and are also cultivated as ornamental, used as living fences and firebreaks among others (Lewis et al., 2005).

The legumes provide many benefits to the soil so they are usually utilized as cover crop, intercropped with cereals and other staple foods. They do produce substantial amounts of organic nitrogen (see below, Improving legume yield by inoculation with rhizobia), increase soil organic matter, improve soil porosity and structure, recycle nutrients, decrease soil pH, reduce soil compaction, diversify microorganisms and mitigate disease problems (U.S Department of Agriculture [USDA], 1998). In rotation with cereals, legumes provide a source of slow-release nitrogen that contributes to sustainable cropping systems. The improvement in the production of these crops will therefore contribute substantially to better human nutrition and soil health (Popelka et al., 2004).

Based on total harvested area and production, cereals are the most important crops, and they are followed by legumes (Fig.1). Close up to 180 million Ha (12-15% of the Earth's arable surface) are worldwide used to produce grain and forage legumes. These numbers point the central importance of world legumes production. In addition, the promise of low-cost production of legume biomass, mainly soybean, for bioenergy purpose focus attention of investors in the improvement of legume production, and deserves an entirely section for discussion.

---

[*] M. Morel and V. Braña contributed equally to this chapter

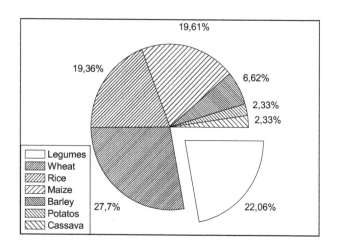

Fig. 1. Net worldwide harvested area for several crops, in the year 2009. Data obtained from Food and Agriculture Organization database [FAO] (http://faostat.fao.org/site/567/default.aspx#ancor) and Graham & Vance (2003).

1.1 Forage and grain legumes

Forage legumes play an important role in dairy and meat production being sources of protein, fibre and energy. They are usually richer in protein, calcium, and phosphorus than other non-legume forages, such as grass. They include alfalfa (*Medicago sativa*), clover (*Trifolium* spp.), birdsfoot trefoil (*Lotus corniculatus*) and vetch (*Vicia* spp.) among others. Alfalfa is one of the most important forage crops. In 2006, the worldwide production was around 436 million tons. U.S. is the largest alfalfa producer, with 15 million Ha planted in 2010. Canada, Argentina (primarily grazed), Southern Europe, Australia, South Africa, and the Middle East have also considerable production (FAO, 2011).

Grain legumes also called pulses, which according to FAO are crops harvested exclusively for the dry seeds, play an important role in the nutrition of many people due to their high protein content in seeds. They represent a major source of protein in many developing countries, especially among the poorest population, and are rich in essential amino acids such as lysine, supplementing thus the nutritional value of cereal and tuber diets (Graham & Vance, 2003). The world pulse production has almost increased by half during the period of 1980 – 2004, overtaking the 60 million tons in 2005 (FAO, 2005). According to FAO Statistical Yearbook 2010, in the year 2008, Canada, China and United States were the main exporters of pulses (28%, 12% and 11% of total exports, respectively). Interestingly, India, the world's 12th largest economy and the third largest in Asia behind Japan and China, is the main importer, responsible of 21% of global trade in of pulses (2.5-3.5 million tons). India produces (15-18 million tons; the world's largest producer), imports and consumes (18.5-20.5 million tons) a wide range of pulses. Thus, considering pulse relevance in the world´s largest economies such as U.S., China and India, incomes and a raising world population, it is obvious the interest of farmers and investors for improving pulse production.

1.2 Soybean – The new legume-star

The soybean (U.S.) (*Glycine max*), also called soya bean (U.K.), is an annual summer legume native of South-eastern Asia, which is used as human food (Liu, 1999) and livestock feed as well as for several industrial purposes (Ali, 2010). According to the newest available information, this legume is one of the main crops cultivated for oil extraction (35.9 million tons oil and 57% global oilseed production), preceded only by the oil of palm (FAO, 2011). Interestingly, over half of the world's 2007 soybean crop (58.6%) was genetically modified (GM), achieving 77% in the year 2009. These GM-soybeans possess a gene that confers herbicide resistance. The nations that produce almost exclusively GM-soybean are U.S. (85%) and Argentina (98%), tending to 100%. The global production and utilization of soybean have increased by ten during the last century (Qiu & Chang, 2010). In 2009, world's soybean cultivated area and production were 99.5 million Ha and 223.2 million tons (FAO, 2011), respectively. U.S. is the world's leader soybean producer and exporter, responsible of 41% global production, followed by Brazil (26%), Argentina (14%), China (7%) and India (4%) (FAO, 2011).

In U.S. the soybean farm gate value raised more than double, ranging from 12.6 billion USD (in 2001) to 29.6 billion USD (in 2009). The price of soybean has increased more than 80% because of soybean-oil's use in soy-biodiesel and as feed for fish farming. Biodiesel is in demand and soybean represents about 25% total worldwide global biodiesel raw material (Pahl, 2008). The net energy balance when the soybean-oil is used for fuel has improved since soybean is a legume, it fixes nitrogen and does not require nitrogen fertilizer (see below) (Kinney & Clemente, 2010).

2. Improving legume yield by inoculation with rhizobia

Leguminous plants are relevant economic and cultural important crops because their exceptional diversity, manifested in variety of vegetable forms that adapted to a wide range of ecological conditions, the high protein content of some grains, their use as pastures, increased world production and commodities. In this scenario, many farm investors, industries and researchers have focussed attention in the development of biological and eco-friendly technologies for legume growth improvement and establishment. The ability of many legumes to form associations with bacteria that fix atmospheric nitrogen (the symbiotic association that improve growth) is thus a big matter of ecological and economic interest (Zahran, 2009).

2.1 Biological vs chemical nitrogen fertilization

Microorganisms are essential to the Earth's nitrogen cycle and to the Biological Nitrogen Fixation (BNF) process in leguminous plants, playing a very important role in terms of plant production in agriculture. Nitrogen fixing microorganisms could be used in live formulations (biofertilizer) that when applied to seed, root or soil colonize the rhizosphere, or the interior of the plant, and promote growth by increasing the nitrogen supply to the host plant and building up soil health. The evaluation, in terms of economic and ecological costs, between chemical- and biological-nitrogen fertilizers support that BNF represents an economic, sustainable and environmentally friendly resource to guarantee the nitrogen requirement of an agro-ecosystem.

Chemical-fertilizer demand has historically been influenced by changing and often interrelated factors such as increasing populations and economic growth, agricultural production, prices, and government policies. In 2007, the production of chemical nitrogen fertilizers was 130 million tons which is likely to increase further in the coming years (FAO, 2011). Their production requires a great consumption of fossil fuels (1-2 % global fossil fuel) and is subjected of constant variations in prices (Vieira et al., 2010). Although their direct contribution to energy consumption seems minimal, it is unnecessary and unsustainable. On average, U.S. farmers apply 30-40 % more chemical nitrogen than is needed for optimal crop yield, thus wasting most of the applied chemical nitrogen. Given the rising cost of chemical nitrogen fertilizers, nitrogen fixation cover crops offer significant economic benefits. In 2006, the price of nitrogen fertilizers in U.S. raised to 521 USD per ton (Huang, 2007), estimating an over cost of 7 to 10 billion USD annually compared with FBN. For instance, the modest use of alfalfa in rotation with corn by U.S. farmers saved 200 to 300 million USD (Graham & Vance, 2003).

In addition to the inconvenience of increasing prices, chemical nitrogen fertilization is associated with environmental problems because watershed contamination by nitrogen leaching, volatilization and denitrification. These problems could be avoided offering to farmers low-cost biofertilizer technologies. These are ecologically sound and their application could help to minimize the global warming as well as to reduce the fertilizer input in farming practices (Herridge et al., 2008a).

2.2 The biological nitrogen fixation (BNF)

BNF benefits not only the legumes themselves but also any intercropped or succeeding crop, reducing or removing the need for nitrogen fertilization. In soils with low mineral nitrogen content, nitrogen fixing microorganisms provide ammonium into the legume biomass, allowing faster growing than their plant competitors. In contrast, if nitrogen is abundant, nitrogen fixing microorganisms tend to be competitively excluded by non-fixing species because the nitrogen fixation process is bio-energetically costly (Houlton et al., 2008). It means that there is a range of physiological and ecological situations that tend to constrain BNF in legume systems, mainly by the nitrogen demand of the plant and by the C:N stoichiometry of the ecosystem. In fact, the hypothesis of a feedback control between legume demand and BNF in a particular ecosystem has been now supported by evidence from both experimental and theoretical models (Soussana & Tallec, 2010).

There is the potential to increase BNF by the use of well adapted and efficient nitrogen fixing microorganisms and/or genetic modified plant species to ensure legume crop at high levels of productivity. Farmers are familiar with the application of commercially available microorganisms (inoculants) that have been especially selected for their ability to effectively nodulate plants and to fix nitrogen from the atmosphere. These kind of microbial inoculants, also known as soil inoculants, are agricultural amendments that use microorganisms known as rhizobia to promote legume growth. These bacteria form symbiotic relationships with the target leguminous plant, and both parts benefit. The legume supplies energy and photosynthates to rhizobia, and rhizobia provide the legume with nitrogen, mainly in the form of ammonium (Howard & Rees, 1996). The symbiosis is initiated through the legume root infection by the rhizobia and formation of root nodules where BNF occurs through the action of a bacterial enzyme, called "Nitrogenase" (Masson-Boivin et al., 2009).

2.3 Rhizobia: The master microbe

The current taxonomy of rhizobia consists of several genera in the subclass Alpha- and Beta-Proteobacteria. *Rhizobium, Mesorhizobium, Ensifer* (formerly *Sinorhizobium*), *Azorhizobium, Methylobacterium, Bradyrhizobium, Phyllobacterium, Devosia* and *Ochrobactrum* are genera that belong to rhizobial Alpha-Proteobacteria. In rhizobial Beta-Proteobacteria the following genera have been described: *Burkholderia, Herbaspirillum* and *Cupriavidus* (NZ Rhizobia, 2011). It is important to clarify that this classification is based on taxonomically important strains that may not necessarily be important reference strains for legume growth improvement. Rhizobial strains commonly used in inoculants have good field performance and stability of symbiotic properties in culture, but are not necessarily well documented or used in taxonomy or molecular biology studies (Lindström et al., 2010). The legume-rhizobia association is specific (each rhizobial strain establishes a symbiosis with only a limited set of host plants and *vice versa*). Thus, there is a restricted number of inoculants that fit with a leguminous plant, and farmers must know which inoculant must be applied according plants and characteristics of soil (Mabrouk & Belhadj, 2010). In other words "Be sure to buy the right inoculant for the legume the farmer intends to plant". Such information must be given by the manufacturer and should be clearly specified in the label. Plants mutually compatible with the same species of rhizobia were listed in earlier years in so-called "cross-inoculation groups" (Table 1). This concept was used in rhizobial taxonomy, but is it unreliable as taxonomic marker because of aberrant cross-infection among plant groups.

Rhizobia	Legume Cross-inoculation group
Ensifer meliloti	Alfalfa Group: alfalfa (*Medicago sativa*), sweet clover (*Melilotus* spp.) (yellow and white), fenugreek (*Trigonella* spp.)
Rhizobium leguminosarum bv *trifolii*	Clover Group (Clover I, II, III and IV): clovers (*Trifolium* spp.)
Bradyrhizobium japonicum	Soybean Group: soybean (*Glycine max*)
Bradyrhizobium spp.	Cowpea Group: pigeon pea (*Cajanus cajan*); peanut (*Arachis hypogaea*); cowpea, mungbean, black gram, rice bean (*Vigna* spp.); lima bean (*Phaseolus lunatus*); *Acacia mearnsii*; *A. mangium*; *Albizia* spp.; *Enterlobium* spp., *Desmodium* spp., *Stylosanthes* spp., Kacang bogor (*Voandzeia subterranea*), *Centrosema* sp., winged bean (*Psophocarpus tetragonolobus*), hyacinth bean (*Lablab purpureus*), siratro (*Macroptilium atropurpureum*), guar bean (*Cyamopsis tetragonoloba*), calopo (*Calopogonium mucunoides*), puero (*Pueraria phaseoloides*)
Rhizobium leguminosarum bv *viciae*	Pea Group: peas (*Pisum* spp.), lentil (*Lens culinaris*), vetches (*Vicia* spp.), faba bean (*Vicia faba*)
Rhizobium leguminosarum bv *phaseoli*	Bean Group: beans (*Phaseolus vulgaris*), scarita runner bean (*Phaseolus coccineus*)
Mesorhizobium loti	Chickpea Group: chickpea (*Cicer* spp.), Birdsfoot trefoil (*Lotus corniculatus* L.)
Rhizobium lupini	Group Lupines
Rhizobium spp	Crownvetch

Table 1. Cross-inoculation group and *Rhizobium*-legume association

The occurrence of a wide diversity of microorganisms in a particular soil increases the opportunity for a legume host to find compatible rhizobia. The principle of specific legume-rhizobia association is commonly used for the isolation of well adapted and efficient rhizobial strains (Castro-Sowinski et al., 2002; Florentino et al., 2010). Usually trap-plants are used to catch the rhizobial strain with highest performance and the strain is used for the design of new inoculants. Details about inoculation technology for legumes can be read in Herridge (2008b).

2.4 Formulation and low-cost are crucial aspect of producing inoculants

Formulation is the industrial "art" of converting a promising laboratory-proven microorganism into a commercial field product. But, the development of successful inoculants involves more than the selection of the most efficient rhizobial strain, it involves the choice of a carrier (powder, granule, and liquid), packaging and marketing, avoiding of microbial contaminations. Inoculant preparations for agricultural use constitute a stressful environment because bacterial cells may have to be stored for long periods, and should survive desiccation and transportation conditions. Some aspects related to inoculant preparation, production and application are described by Hungria et al. (2005).

The formulation should maintain or enhance activity in field. In order to survive in nutrient-poor ecosystems, bacteria use different strategies, among them, the use of polyhydroxyalkanoates (PHA) as intracellular carbon storage compounds. Cells with higher PHA content can survive longer than those with lower amounts, and PHA degraded elements can be used rapidly for numerous metabolic needs. Accumulation of PHA can provide the cell with the ability to endure a variety of harmful physical and chemical stresses (Castro-Sowinski et al., 2010; Kadouri et al., 2005).

A good formulation contains microorganisms (active ingredient) in an active metabolic state, immersed in a suitable carrier together with additives that are responsible for the microbial cells stabilization and protection during storage and transportation. Most of the research done in the improvement of inoculant quality is based on improving carrier properties, by adding elements that can prolong survival, such as nutrients, or other synthetic products (López et al., 1998). Most commercial inoculants are in powder (finely ground peat mixed with the nitrogen-fixing bacteria), ready for mixing with the seed. Granular formulations are designed to be placed in the seed rut at planting. Liquid inoculants and other non-peat-based inoculants are also being used. Liquid inoculants simplify the production of the inoculant and the application to seeds or field. However, bacterial survival in the inoculant and on inoculated seeds is not as good as when using peat as a carrier, because bacteria lack carrier protection (Tittabutr et al., 2007). Peat provides bacterial protection and prevents drying and death, compared to the inoculants that do not contain peat. However, alternative substrates to peat can be used as carriers: compost cork, perlite, volcanic pumice, alginate beads and coal, among many others, also gave good results in terms of supporting bacterial growth and long survival, as well survival on seeds (Albareda et al., 2008; Ben Rebah et al., 2007).

Another important consideration in formulation is the cost-effectiveness that must be low enough to allow sufficient incoming compared to chemical fertilization. In U.S. and Canada, a seed inoculant is sell for 5.00 and 2.50 USD per Ha, respectively, while granular inoculants range from 15.00 to 18.00 (US) per Ha (Xavier et al., 2004). But, inoculants need only a modest increase in yield to offset the cost. A good inoculant will usually provide at least a 70- to 140-Kg per Ha return on yield.

2.5 The input of BNF in legume yield

The annual input of fixed nitrogen was calculated to be 2.95 Mton for the pulses and 18.5 Mton for the oilseed legumes, being the soybean the dominant crop legume (50% global crop legume area and 68% global production). In addition to the annual legume nitrogen fixation inputs of 12-15 Mton (pasture and fodder legumes), there is an input by nitrogen fixation in rice (5 Mton), sugar cane (0.5 Mton), non-legume crop lands (<4 Mton) and extensive savannas (<14 Mton). Thus, the total overall estimated in agricultural systems is of 50–70 Mton biologically fixed nitrogen (Herridge et al., 2008a). These numbers show that the process of BNF is an economically attractive and eco-friendly alternative to reduce the external nitrogen (chemical fertilizers) input, which improves the quality and quantity of crop resources.

A successful BNF is capable of improving agricultural productivity while minimizing soil loss and ameliorating adverse edaphic conditions. Conditions such as drought, salinity, unfavorable soil pH, nutrient deficiency, mineral toxicity, high temperature, insufficient or excessive soil moisture, inadequate photosynthesis, and plant diseases conspire against a successful symbiotic process. Many inoculant manufactures worldwide have developed formulations with high symbiotic efficiency under stress conditions. However, the actual view of plant growth promoting preparations focuses their investigations in the design and development of new-formulations supplemented with plant and/or microbe exudates. These exudates contain molecules involved in the microbe-plant interaction: flavonoids, sugars, acids, amino acids, amines and other low molecular weight compounds that promote plant growth (Skorupska et al., 2010; Garg & Geetanjali, 2009). Macchiavelli & Brelles-Mariño (2004) showed increased plant nodulation treating *Medicago truncatula* roots and seeds with Nod Factors prior to inoculation. Lipo-chito-oligosaccharides (LCOs), or Nod Factors (NFs), are bio-signals produced by the rhizobia which act as bacteria-to-plant communication molecule that mediates recognition and nodule organogenesis (Masson-Boivin et al., 2009). The inclusion of NFs in formulations might have technological applications since presoaking seeds with submicromolar concentrations of this oligo-saccharide before sowing leaded to increased nodulation under field conditions. In fact, a soybean inoculant based on NFs technology was introduced on the market many years ago (Zhang & Smith, 2002). Currently, many companies like Rizobacter (www.rizobacter.com.ar) and Nitragin (www.nitragin.com.ar) are marketing formulations with bio-signals that improve the symbiotic relationship, activate mechanisms to resist abiotic stress conditions, and induce defensive response.

3. The use of microbial consortium in legume agronomic production

The new fashion in agriculture is the use of microbial consortiums of plant-growth promoting bacteria (PGPB, which includes rhizobia). PGPB are exogenous bacteria introduced into agricultural ecosystems that act positively upon plant development (Castro-Sowinski et al., 2007). It is possible to increase agricultural productivity and, eliminate or decrease the use of chemical fertilizers and pesticides (Adesemoye et al., 2009a; Vessey, 2003) even in marginal soils (Gamalero et al., 2009) when the formulation contains different PGPB.

3.1 Getting more from legumes

Current studies indicate that we are still detecting new bacteria and fungi with diverse growth-promoting characteristics, and that the combination of different PGPB into a single-formulation increases plant yield, compared with single-inoculation. On the other hand, efforts have been done manipulating PGPB to produce master inoculants by the introduction of foreign DNA that provides new abilities (GMM, Genetic Modified Microorganisms). Globally, it was expected a big explosion in this area of research, the use of recombinant DNA-technological tools for the production of inoculants (Barea et al., 2005; Valdenegro et al., 2001). However, the use of GMM is in discussion and needs clear regulatory policies, controls and suitable legislation (Fedoroff et al., 2010).

Some cooperative microbial activities can be exploited for developing new sustainable, environmentally-friendly, agro-technological practices (Barea et al., 2005). In this regard, the plant co-inoculation with rhizobia and other PGPB received considerable attention for legume growth promotion (Cassán et al., 2009; Bai et al., 2002a; 2002b; Zhang et al., 1996). Results from many studies concerning the effect of co-inoculation on legume growth are summarized in Table 2. Several genera of bacteria have been identified as *"helpers"* of the rhizobia-legume symbiotic process (Beattie, 2006). Examples are bacteria of the genus *Azospirillum* (Cassán et al., 2009; Itzigsohn et al., 1993), *Azotobacter* (Qureshi et al., 2009; Yasari et al., 2008), *Bacillus* (Bullied et al., 2002), *Pseudomonas* (Barea et al., 2005; Fox et al., 2011), *Serratia* (Bai et al., 2002b; Lucas-Garcia et al., 2004a; Zhang et al., 1996), *Thiobacillus* (Anandham et al., 2007), and *Delftia* (Morel et al., 2011), among many other. The stimulation of the legume–rhizobia symbiosis by non-rhizobial-PGPB implicates different processes such as production of phytohormones (usually indole-acetic acid; IAA) that stimulates root growth; qualitative change of flavonoids pattern secreted for the plant; solubilization of non-available nutrients (mainly re-fixation of exogenously applied phosphorus), among others (Medeot et al., 2010). In this section, we summarize the knowledge about bacteria that promote the symbiotic relationship between legumes and rhizobia (from now, the symbiotic enhancer), and the mechanisms involved in this phenomenon. The effect of other microorganisms, such as micorrhizal fungi is not discussed.

Probably the most studied bacterial consortium is the rhizobia-azospirilla one. Azospirilla species are being used as seed inoculants under field conditions for more than a decade (Dobbelaere et al., 2001; Puente et al., 2009). The positive effect of *Azospirillum* in the nodulation and nitrogen fixation by rhizobia on several forage legumes was early reported (Yahalom et al., 1987). Since then, many works have been done and mostly are summarized in Bashan et al. (2004). It proven that the combined inoculation with rhizobia and azospirilla increases the shoot length and weight, root hairs number, root diameter, the main- and total-root nodule number and the percentage of infected root hairs, thus resulting in increased legume yields (Cassán et al., 2009). Worldwide, salinity is one of the most important abiotic stresses that limit crop growth and productivity. It was shown that the rhizobia-azospirilla co-inoculation significantly reduces the negative effects of abiotic stresses (such as caused by irrigation with saline water) on root development and nodulation (Dardanelli et al., 2008).

Under stress conditions, such as drought, salinity, S-deficient or heavy metal (HM)-contaminated soils, several associations between plants and beneficial bacteria showed a defensive response and an increased yield (Anandham et al., 2007; Dary et al., 2010; Fuentes-Ramírez & Caballero-Mellado, 2005; Han & Lee, 2005). However, the physiological mechanism involved in stress mitigation is still unknown (Figueiredo et al., 2008; Furina & Bonartseva, 2007).

3.2 Enhancing the legume – Rhizobia symbiosis by co-inoculation: Modes of action

Many evidences have been accumulated showing that co-inoculation with beneficial microorganisms, having different mechanisms of plant-growth promotion, have additive or synergistic effect on plant growth and crop yield (Table 2). Diverse mechanisms are implicates in the co-inoculation benefits and some of them have been discussed in Barea at al. (2005).

Legume	Bacterial system	Increase (%) compared to single rhizobial inoculation	Experiments done in	Proposed mechanism of action	Reference
Soybean (*Glycine max*)	*B. japonicum - Serratia* spp.	50 in NN; 30 in SDW; 32 in RDW	Greenhouse	Production of LCO- analogue	Bai et al., 2002a Bai et al., 2002b
		40 in NN under sub-optimal temperature	Laboratory	Unknown	Zhang et al., 1996
	B. japonicum - B. cereus	10 in SDW	Field	Unknown	Bullied et al., 2002
	B. japonicum - S. proteamaculans /B. subtilis	12 in SDW; 10 in P-uptake	Greenhouse (saline stress)	Limited Na-uptake	Han & Lee, 2005
	B. japonicum - A. brasilense	47 in NN	Laboratory	Production of IAA, GA3 and Zeatin	Cassán et al., 2009
	B. japonicum - A. brasilense	16-40 in RDW; 200-700 in total RL	Laboratory	Unknown	Molla et al., 2001a
		30 in NN	Greenhouse	Production of plant hormones	Molla et al., 2001b
	E. fredii - Chryseobacterium balustinum	56 and 44 in SDW; 100 and 200 in RDW; 155 and 286 in NN under non-saline and saline conditions respectively	Laboratory (saline stress)	Unknown	Estevez et al., 2009
	B. japonicum - P. putida	40 in SDW; 80 in NN; 45 in RDW	Laboratory	P-solubilization and production of siderophores	Rosas et al., 2006

Legume	Bacterial system	Increase (%) compared to single rhizobial inoculation	Experiments done in	Proposed mechanism of action	Reference
Common bean (*Phaseolus vulgaris*)	*R. tropici/etli - A. brasilense*	18-35 and 20-70 in RDW; 29 and 28 in SDW under non saline and saline conditions, respectively	Hydroponic (saline stress)	Production of flavonoid-like compounds	Dardanelli et al., 2008
	R. etli - C. balustinum	35 in SDW; 35 in NN under non-saline conditions; and 39 in SDW; 63 in RDW under saline conditions	Laboratory (saline stress)	Unknown	Estevez et al., 2009
	R. tropici - Paenibacillus polymyxa	50 in NN; 40 in N uptake in non-drought stress	Greenhouse (drought stress)	Unknown	Figuereido et al., 2008
	Rhizobium spp. - *A. brasilense /B. subtilis/P. putida*	30 in NN; 20 in SDW; 30-45 in RDW	Greenhouse (two levels of P-fertilization)	IAA production or 1-aminocycloprop ane-1-carboxylate (ACC) deaminase activity	Remans et al., 2007
	Rhizobium spp. - *A. brasilense*	70 in NN	Hydroponic	IAA production	Remans et al., 2008a
		30 total yield	Field	IAA production	Remans et al., 2008b
	Rhizobium spp. - *P. fluorescens /A. lipoferum*	25 in NN; 13 in SDW; 74 in seed yield	Field	P-solubilization; auxin and siderophores production	Yadegari et al., 2010

Legume	Bacterial system	Increase (%) compared to single rhizobial inoculation	Experiments done in	Proposed mechanism of action	Reference
Chickpea (*Cicer arietinum*)	*Rhizobium spp - Pseudomonas/ Bacillus* spp.	20 in SDW; 30-120 in RDW	Greenhouse	Production of flavonoid-like	Parmar & Dadarwal, 1999
	Mesorhizobium sp. *Cicer - Pseudomonas* spp.	70 in NN; 30 in SDW, 30 in N-uptake	Laboratory	Unknown	Goel et al., 2002
		1,2-1,86 in NN; 1,3-2,11 NFW; 1-2,93 in PDW	Laboratory	IAA production	Malik & Sindhu., 2011
	Rhizobium - B. subtilis/ megaterium	18 in SDW; 16-30 in RDW; 14 in total biomass yield in field	Laboratory and Field	N-fixation by *B. subtilis* or/and P-solubilization by *B. megaterium*	Elkoca et al., 2008
	M. ciceri - Azotobacter chroococcum	15 in NN; 25 in P-soil availability	Field (two levels of N-fertilization)	Unknown	Qureshi et al., 2009
	M. ciceri - Pseudomonas sp/ *Bacillus* sp.	20 in PDW; 30 in NN; 100 in P-uptake	Field	P-solubilization by PGPB	Wani et al., 2007
Peanut (*Arachis hypogaea*)	*Thiobacillus* sp. - *Rhizobium* sp.	50 in PDW; 80 in NN	Greenhouse (S-deficiency) and Field	S-oxidation	Anandham et al., 2007
Clover (*Trifolium repens*)	*R. leguminosarum bv.trifolii - P. fluorescens*	20 in SDW; 100 in NN	Laboratory	Production of B-group vitamins	Marek-Kozaczuk & Skorupska, 2001
	R. leguminosarum bv. trifolii - Delftia sp.	50 in SDW and 80 in nodulation rate	Laboratory	IAA production	Morel et al., 2011
Altramuz (*Lupinus luteus*)	*Bradyrhizobium* sp. - *Pseudomonas* sp./ *Ochrobactrum cytisi*	66 in SDW and 20-40, 25, and 30-50 decrease in Cd, Cu and Zn - accumulation in roots, respectively	Field (Heavy metal contaminated soil)	Phyto-stabilization: Biosorption of heavy metals by bacterial biomass	Dary et al., 2010
Alfalfa (*Medicago sativa*)	*S. meliloti - Delftia* sp.	10 in SDW; 30 in nodulation rate	Laboratory	IAA production	Morel et al., 2011

Legume	Bacterial system	Increase (%) compared to single rhizobial inoculation	Experiments done in	Proposed mechanism of action	Reference
Galega (*Galega orientalis*)	*R. galegae* bv. *orientalis* - *Pseudomonas* spp.	70 in SDW; 60 in RDW; 30 in NN; 44 in N-uptake	Greenhouse	Production of IAA and/or cellulase by *Pseudomonas* spp.	Egamberdieva et al., 2010
Vetch (*Vicia sativa*)	*R. leguminosarum* bv. *viciae* - *A. brasilense*	30 in SDW	Greenhouse	IAA production and increased root secretion of flavonoids	Star et al., 2011
		nod gene induction and decreased in indoles content	Hydroponic		
Pea (*Pisum sativum* L. cv. Capella)	*R. leguminosarum* bv *viceae* - *P. fluorescens*	1,3 in Pea DW; 0,5-0,69 in plants with disease	Greenhouse (*Fusarium oxysporum* infected soils)	Antifungal activity by production of siderophores	Kumar et al., 2001
	R. leguminosarum-*B. thuringeinsis*	84 times in NN; 15 in SDW; 15 in RDW	Laboratory and greenhouse	Unknown	Mishra et al., 2009
Lentin (*Lens culinaris* L.)	*R. leguminosarum*-*B. thuringeinsis*	73 in NN; 5 in SDW; 10-30 in RDW	Laboratory and greenhouse	Unknown	Mishra et al., 2009
Pigeon pea (Cajanus cajan)	*Rhizobium* sp.-*Bacillus* spp.	50 in PFW; 300 in NN	Greenhouse (sterile soil)	Cross-utilization of siderophores produced by *Bacillus* sp. and *Rhizobium*	Rajendran et al., 2008
	Rhizobium sp. - *P.putida/ P. fluorescens/ B. cereus*	73 in NN; 30 in grain yield	Greenhouse	Unknown	Tilak et al., 2006
Mung bean (*Vigna radiata* L.)	*B. japonicum* - *P. putida*	20 in total biomass; 48 in NN	Greenhouse	Reduced ethylene production	Shaharoona et al., 2006

Table 2. Ten years of studies on legume co-inoculation (2001-2011). Increase in legume symbiotic parameters and yield by co-inoculation compared to single-inoculation with rhizobia. Abbreviations are as follows: RDW: root dry weight; SDW: shoot dry weight; RL: root length; NN: nodule number; NFW: Nodule fresh weight; PDW: plant dry weight; PFW: plant fresh weight.

Probably, the most reported mechanism that explains the improved rhizobia-legume association by other PGPB is the production of plant-hormones (phytohormones), such as gibberellic acid (GA3) or auxin-type phytohormones (mainly indole-3-acetic acid; IAA; Beattie, 2006). That is the case for *Pseudomonas* (Egamberdieva et al. 2010; Malik & Sindhu, 2011) and *Azospirillum* (Cassán et al., 2009; Dobbelaere et al., 2001; Okon, 1994; Perrig et al., 2007). For information about IAA production and effects, we recommend Baca & Elmerich (2007) and Spaepen et al. (2007). However, the main mechanism involved in improved rhizobia-legume association is still under investigation (Dobbelaere & Okon, 2007). It might be possible that multiple mechanisms, rather than only one are acting. This is known as the "Additive Hypothesis" (Bashan et al., 2004; Bashan & de-Bashan, 2010).

Many other signal molecules or analogues involved in plant-rhizobia communication, different than phytohormones but produced by the non-rhizobial co-inoculant strain, have been implicated in the rhizobia-plant association (Lucas-Garcia et al., 2004b; Mañero et al., 2003). Some direct evidence suggests that the presence of *Pseudomonas* spp. (Parmar & Dadarwal, 1999) and *Azospirillum* spp. cells (Burdman et al., 1996; Dardanelli et al., 2008, Volpin et al., 1996) induce the synthesis of flavonoids by roots of chickpea, common bean and alfalfa, in experiment of co-inoculation with rhizobia. Interestingly, it is not strictly necessary the presence of the bacteria, the application of bacteria-free exudates of symbiotic enhancers to the root exert similar effect that during bacterial-co-inoculation (Molla et al., 2001b). For example, the application of NFs analogues produced by *Serratia proteamaculans* 1-102 promotes soybean-bradyrhizobia nodulation and soybean growth (Bai et al., 2002b). The list of metabolites produced by symbiotic enhancers might become bigger: vitamins that may supplement the nutritional requirement of rhizobia (Marek-Kozaczuk & Skorupska, 2001); hydrolytic enzymes that assist during rhizobial penetration in the root hair, or attack phytopathogenic fungi (Egamberdieva et al., 2010; Sindhu & Dadarwal, 2001; Sindhu et al., 2002); or P-solubilizing acids that increase phosphorus availability (Elkoca et al., 2008). However, in most cases the mechanism underlying the plant growth promotion by co-inoculation is unknown (Bullied et al., 2002; Goel et al., 2002; Lucas García et al., 2004a, 2004b; Vessey & Buss, 2002).

3.3 Increasing crop yield by co-inoculation

On average, an increase of 4-5% in crop yield has an important impact in agricultural production. The data obtained in different growth-systems (gnotobiotic laboratory conditions, hydroponics, greenhouse and field) shows that co-inoculation produces a major increase in legume yield compared with single inoculation (Table 2), overwhelming the agronomic expectations.

Inoculation and co-inoculation experiments must be done in field to provide a realistic assessment of the performance of a living-formulation in practical farming conditions. Table 2 shows examples of legume co-inoculation in field experiments. An increase of 74% in seed yield was detected when *Phaseolus vulgaris* was co-inoculated with *P. fluorescens* or *A. brasilense* compared with single-inoculation with *Rhizobium* spp. (Yadegari et al., 2010). As well, 14% total biomass chickpea yield was detected during co-inoculation with P-solubilizing *Bacillus* isolates compared with single-inoculation with *Rhizobium* sp (Elkoca et al., 2008). Vast areas of agricultural land are not appropriated for cropping because the soil has P-deficiency and the co-inoculation of legumes with rhizobia and P-solubilizing bacteria might supply nitrogen and phosphorus to these poor lands. The examples above provided show a huge increase in yield during co-inoculation in field experiments, pointing the economically relevance of co-inoculation practices in countries with high pulse crop production.

Bacteria	Target crop	Formulation	Yield increase (%)[a]	Reference
Rhizobia - *B. subtilis*	Soybean; peanuts; dry beans	Co-inoculant[b]	4-6	www.histicknt.com
Rhizobia - *A. brasilense*	Soybean; peanut; pea; vetch	Co-inoculant	8-30	www.intxmicrobials.com
Rhizobia – *A. brasilense* - *P. fluorescens*	soybean	Co-inoculant	5-10	www.inoculantespalaversich.com
Rhizobia – *A. brasilense*	soybean	Co-inoculant	10	www.nitrosoil.com.ar
Rhizobia - *B. megaterium* - *Saccharomyces cerevisiae*	All legumes[c]	Co-inoculant	Undeclared	www.iabiotec.com
B. megaterium	All crops[d]	Inoculant[e]	10	www.rajshreesugars.com
			Undeclared	www.manidharmabiotech.com
P-solubilizing bacteria (genus undeclared)	All crops	Inoculant	10-15	www.gsfclimited.com
Frateuria aurantia	All crops	Inoculant	10-20	www.manidharmabiotech.com
P-solubilizing bacteria (genus undeclared)	All crops	Inoculant	20-30	www.varshabioscience.com
Delftia acidovorans	Canola (*B.napus*)	Inoculant	Undeclared	Banerjee & Yesmin, 2004 www.brettyoung.ca

[a] – compared to single inoculation
[b] –the formulation contains both rhizobia and non-rhizobial PGPB in the same package
[c] – recommended for all kind of legumes
[d] – recommended for many crops, including legumes
[e] – the formulation does not contain rhizobia, but it can be used with rhizobial-formulation

Table 3. Some available commercial formulations (containing two PGPB) for legume crops. Note: mycorrhiza and bio-control bacteria are not included in this list.

Chickpea is the most largely produced pulse crop in India accounting for 40% of total pulse crops production, being the leading chickpea producing country in the world. India annually produces around 6 Million tons of chickpea and contributes of approximately 70% in the total world production. On the other hand, Brazil is the world leader in dry bean production (3.3 Million ton), followed by India (3.0 Millon ton) and China (1.9 Millon ton). All these countries belong to "the BRICs". In economics, BRIC is a grouping acronym that refers to Brazil, Russia, India and China, which are considered to be at a similar stage of newly advanced economic development. The BRIC thesis, by Goldman Sachs, recognizes that Brazil, Russia, India and China have changed their political systems to embrace global capitalism, and predicts that China and India, respectively, will become the dominant global suppliers of manufactured goods and services, while Brazil and Russia will become similarly dominant as suppliers of raw materials. In this scenario, of countries with growing world economies and important production and consumption of pulses, the development of new formulations based in bacterial consortiums are being encouraged. However, a major constraint for exploiting living-formulation technologies has been that most farmers are not aware of the technology and its benefits.

3.4 New formulations: The use of bacterial consortium

Some bacterial symbiotic enhancers are promising microorganisms that would be used for the design of new formulations. These formulations could contain different bacteria in one pack, ready for direct placing in the seed at planting. However, some manufacturers also produce formulations that do contain non-rhizobial PGPB, but that can be mixed with rhizobial-formulation at the moment of planting. Information on both kinds of formulations is provided in Table 3.

Despite the great progress and the increasing interest in mixed formulations for legumes inoculation, there are few commercial products with different bacteria. Most of these products are based on *Bacillus* spp. *Azospirillum*-based inoculants are also abundant in the market, but most of them are available for non-legumes crops (Figueiredo et al., 2010). Most commercially available biofertilizers are biopesticides and biofunguicides, but they are not described in this chapter.

4. Concluding remarks

The doubling time world's current growth is 54 years and we can expect the world's population to become 12 billion by 2054. This demographic growth has to be accompanied by an increase in food production. Thus, the humanity has to face a new challenge, by doing a good use of soils (Fedoroff et al., 2010; Godfray et al., 2010) and developing new technologies (Pretty, 2008), mainly based in eco-friendly microorganisms that control pest and improve plant growth. In such scenario, the use of biofertilizers, rhizobia or consortium of plant-beneficial microbes (rhizobia and symbiotic enhancers) in formulations provides a potential solution. The data showed in this chapter support that the design of new formulations with cooperative microbes might contribute to the growth improvement of legumes. The co-inoculation has a positive effect in growth stimulation of legume crops; however, we believe it is necessary to continue studying this subject.

5. Acknowledgments

The authors thank Prof. Yaacov Okon for his valuable suggestions during writing and PEDECIBA for partial financial support. The work of M. Morel and V. Braña was supported by ANII (Agencia Nacional de Investigación e Innovación).

6. References

Adesemoye, A.; Torbert, H. & Kloepper, J. (2009). Plant growth-promoting rhizobacteria allow reduced application rates of chemical fertilizers. *Microbial Ecology*, Vol. 58, pp. 921-929, ISSN 0095-3628

Albareda, M.; Rodríguez-Navarro, DN.; Camacho, M. & Temprano, FJ. (2008). Alternatives to peat as a carrier for rhizobia inoculants: Solid and liquid formulations. *Soil Biology & Biochemistry*, Vol. 40, pp. 2771–2779, ISSN 0038-0717

Ali, N. (2010). Soybean Processing and Utilization, In: *The soybean: botany, production and uses*, Singh, G., (Ed.) 345-362, CAB International, ISBN 9781845936440, UK.

Anandham, R.; Sridar, R.; Nalayini, P.; Poonguzhali, S.; Madhaiyan, M. & Tongmin, S. (2007). Potential for plant growth promotion in groundnut (*Arachis hypogaea* L.) cv. ALR-2 by co-inoculation of sulfur-oxidizing bacteria and *Rhizobium*. *Microbiological Research*, Vol. 162, pp. 139-153, ISSN 0944-5013

Baca, B. & Elmerich, C. (2007). Microbial production of plant hormones. In: *Associative and endophytic nitrogen-fixing bacteria and cyanobacterial associations*. Elmerich, C. & Newton, E. (Eds.), 113-143, Springer, ISBN 978-1-4020-3541-8,Dordrecht, The Netherlands

Bai, Y.; Pan, B.; Charles, T. & Smith, L. (2002a). Co-inoculation dose and root zone temperature for plant growth promoting rhizobacteria on soybean [*Glycine max* (L.) Merr] grown in soil-less media. *Soil Biology & Biochemistry*, Vol. 34, pp. 1953-1957, ISSN 0338-0717

Bai, Y.; Souleimanov, A. & Smith, D. (2002b). An inducible activator produced by a *Serratia proteamaculans* strain and its soybean growth-promoting activity under greenhouse conditions. *Journal of Experimental Botany*, Vol. 373, pp. 1495-1502, ISSN 0022-0957

Banerjee, M.R & Yesmin, L. (2004). BioBoost: a new sulfur-oxidizing bacterial inoculant for canola, *4th International Crop Science Congress*, ISBN 1-920842-21-7, Brisbane, Australia, September 2004

Barea, J.; Pozo, M.; Azcón, R. & Azcón-Aguilar, C. (2005). Microbial co-operation in the rhizosphere. Journal of Experimental Botany, Vol. 56, No. 417, pp 1761-1778, ISSN 0022-0957

Bashan, Y. & de-Bashan, L. (2010). How the plant growth-promoting bacterium *Azospirillum* promotes plant growth – A critical assessment. In: *Advances in Agronomy*, Sparks, D. (Ed.), Vol.108, 77-136, ISBN 978-0-12-374361-9, NewarK, USA

Bashan, Y.; Holguin, G. & de-Bashan, L. (2004). *Azospirillum*-plant relationships: physiological molecular, agricultural, and environmental advances (1997-2003). *Canadial Journal of Microbiology*, Vol. 50, pp. 521-577, ISSN 0008-4166

Beattie, G. (2006). Plant-associated bacteria: survey, molecular phylogeny, genomics and recent advances, In: *Plant-associated bacteria*, Gnanamanickam, S. (Ed.), 1-56, Springer ISBN 978-1-4020-4537-0, Netherlands

Ben Rebah, F.; Prévost, D.; Yezza, A. & Tyagi, RD. (2007) Agro-industrial waste materials and wastewater sludge for rhizobial inoculant production: A review. *Bioresource Technology*, Vol. 98, pp. 3535-3546, ISSN 09608524

Bullied, W.; Buss, T. & Vessey, J. (2002). *Bacillus cereus* UW85 inoculation effects on growth, nodulation, and N accumulation in grain legumes: Fields studies. *Canadian Journal of Plant Science*, Vol. 82, pp. 291-298, ISSN 0008-4220

Burdman, S.; Volpin, H.; Kapulnik, Y. & Okon, Y. (1996). Promotion of *nod* gene inducers and nodulation in common bean (*Phaseolus vulgaris*) root inoculated with *Azospirillum brasilense* Cd. *Applied Environmental Microbiology*, Vol.62, pp. 3030-3033, ISSN 0099-2240

Cassán, F.; Perrig, D.; Sgroy, V.; Masciarelli, O., Penna, C. & Luna, V. (2009). *Azospirillum brasilense* Az39 and *Bradyrhizobium japonicum* E109, inoculated singly or in combination, promote seed germination and early seedling growth in corn (*Zea mays* L.) and soybean (*Glycine max* L.). *European Journal of Soil Biology*, Vol. 45, pp. 28-35, ISSN 1164-5563

Castro-Sowinski, S.; Burdman, S.; Matan, O. & Okon, Y. (2010). Natural functions on bacterial polyhydroxyalkanoates, In: *Plastics from bacteria. Natural Functions and applications*. Chen, GQ. (Ed.), 39-61, Springer, ISBN 1862-5576, New York.

Castro-Sowinski, S.; Carrera, I.; Catalan, AI.; Coll, J. & Martinez-Drets, G. (2002). Occurrence, diversity and effectiveness of mild-acid tolerant alfalfa nodulating rhizobia in Uruguay. *Symbiosis*, Vol.32, pp. 105-118, ISSN 0334-5114

Castro-Sowinski, S.; Herschkovitz, Y.; Okon, Y. & Jurkevitch, E. (2007). Effects of inoculation with plant growth-promoting rhizobacteria on resident rhizosphere microorganisms. *FEMS Microbiology Letters*, Vol.276, No.1, pp. 1–11, ISSN 0378-1097

Dardanelli, M.; Fernández de Córdoba, F.; Espuny, M.; Rodriguez Carvajal, M.; Soria Díaz, M.; Gil Serrano, A.; Okon, Y. & Megías, M. (2008). Effect of *Azospirillum brasilense* coinoculated with *Rhizobium* on *Phaseolus vulgaris* flavonoids and Nod factor production under salt stress. *Soil Biology and Biochemestry*, Vol.40, pp. 2713-2721, ISSN 0038-0717

Dary, M.; Chamber-Pérez, M.; Palomares, A. & Pajuelo, E. (2010). "In situ" phytostabilisation of heavy metal polluted soils using *Lupinus luteus* inoculated with metal resistant plant-growth promoting rhizobacteria. *Journal of Hazardous Materials*, Vol. 177, pp. 323-330, ISSN 0304-3894

Dobbelaere, S. & Okon, Y. (2007). The plant growth-promoting effect and plant responses. In: *Associative and endophytic nitrogen-fixing bacteria and cyanobacterial associations*, Elmerich, C. & Newton, E. (Eds), 145-170, Springer, ISBN 978-1-4020-3541-8, Dordrecht, The Netherlands

Dobbelaere, S.; Croonenborghs, A.; Thys, A.; Ptacek, D.; Vanderleyden, J.; *et al.* (2001). Responses of agronomically important crops to inoculation with *Azospirillum*. *Australian Journal of Plant Physiology*, Vol. 28, pp. 871-879, ISSN 0310-7841

Egamberdieva, D.; Berg, G.; Lindström, K. & Räsänen, L.A. (2010). Co-inoculationof *Pseudomonas* spp. with *Rhizobium* improvesgrowth and symbiotic performance of fodder galega (*Galegaorientalis* Lam.) *European Journal of Soil Biology*, Vol.46, pp. 269-272, ISSN 1164-5563

Elkoca, E.; Kantar, F. & Sahin, F. (2008). Influence of nitrogen fixing and phosphorus solubilizing bacteria on the nodulation, plant growth, and yield of chickpea. *Journal of Plant Nutrition,* Vol.31, pp. 157-171, ISSN 0190-4167

Estevez, J.; Dardanelli, M.S.; Megías, M. & Rodriguez-Navarro, D.N. (2009). Symbiotic performance of common bean and soybean co-inoculated with rhizobia and *Chryseobacterium balustinum* Aur9 under moderate saline conditions. *Symbiosis,* Vol.49, pp. 29-36, ISSN 0334-5114

FAO. (2005). Pulses: Past trends and future prospects, access on August 30, available from: <http://www.fao.org/es/esc/en/15/97/highlight_98.html>

FAO. (2010). FAO Statistical Yearbook, access on August 30, available from: <http://www.fao.org/economic/ess/ess-publications/ess-yearbook/ess-yearbook2010/en/>

FAO STAT. (2011). Food and Agriculture Organization Statistical Database, access on August 30, available from : <http://faostat.fao.org>

Fedoroff, N.; Battisti, D.; Beachy, R.; Cooper, P.; Fischhoff, D.; Hodges, C.; Knauf, V.; Lobell, D.; Mazur, B.; Molden, D.; Reynolds, M.; Ronald, P.; Rosegrant, M.; Sanchez, P.; Vonshak, A. & Zhu, J. (2010). Radically rethinking agriculture for the 21st century. *Science,* Vol. 327, No. 5967, pp. 833-834, ISSN 0036-8075

Figueiredo, M.; Burity, H.; Martínez, C. & Chanway, C. (2008). Alleviation of drought stress in the common bean (*Phaseolus vulgaris* L.) by co-inoculation with *Paenibacillus polymyxa* and *Rhizobium tropici. Applied Soil Ecology,* Vol. 40, pp. 182-188, ISSN 0929-1393

Florentino, LA., Martins de Sousa, P., Silva, IS., Barroso Silva, K. & de Souza Moreira, FM. (2010). Diversity and efficiency of *Bradyrhizobium* strains isolated from soil samples collected from around *Sesbania virgata* roots using cowpea as trap species. *Revista Brasileira de Ciencia do Solo,* Vol. 34, pp. 1113-1123, ISSN 0100-0683.

Fox, S.L; O'Hara, G. & Bräu, L.(2011). Enhanced nodulation and symbiotic effectiveness of *Medicago truncatula* when co-inoculated with *Pseudomonas fluorescens* WSM3457 and *Ensifer (Sinorhizobium) medicae* WSM419. *Plant and Soil,* Vol. 348, No.1-2, pp. 245-254, ISSN 0032-079X

Fuentes-Ramírez, L. & Caballero-Mellado, J. (2005). Bacterial biofertilizers. In: *PGPR: Biocontrol and biofertilization,* Siddiqui, Z. (Ed.), 143-172, Springer, ISBN , Dordrecht, Netherlands

Furina, E. & Bonartseva, G. (2007). The effect of combined and separate inoculation of alfalfa plants with *Azospirillum lipoferum* and *Sinorhizobium meliloti* on denitrification and nitrogen-fixing activities. *Applied Biochemistry and Microbiology,* Vol. 43, No. 3, pp. 286-291, ISSN 0003-6838

Gamalero, E.; Berta, G. & Glick, B. (2009). The use of microorganisms to facilitate the growth of plants in saline soils, In: *Microbial strategies for crop improvement,* Khan, S.; Zaidi, A. & Musarrat, J. (Eds.), 1-22, Springer, ISBN 978-3-642-01978-4, New York.

Garg, N. & Geetanjali, G. 2009. Symbiotic nitrogen fixation in legume nodules: process and signaling: a review, In: *Sustainable Agriculture.* Lichtfouse, E.; Navarette, M.; Véronique, S. & Alberola, C. (Eds.), 519-531, Springer, ISBN 978-90-481-2666-8, Netherlands.

Godfray, H.; Beddington, J.; Crute, I.; Haddad, L.; Lawrence, D.; Muir, J.; Pretty, J.; Robinson, S.; Thomas, S. & Toulmin, C. (2010). Food security: the challenge of feeding 9 billion people. *Science*, Vol.327, No.5967, pp. 812-818, ISSN 0036-8075

Goel, A.; Sindhu, S. & Dadarwal K. (2002). Stimulation of nodulation and plant growth of chickpea (*Cicer arietinum* L.) by *Pseudomonas*. *Biology and Fertility of Soils*, Vol.36, pp. 391-396, ISNN 0178-2762

Graham, PH. & Vance, CP. (2003), Legumes: Importance and Constraints to Greater Use. *Plant Physiology*, Vol. 131, No. 3, pp. 872–877, ISSN: 0032-0889

Han, H. & Lee, K. (2005). Physiological responses of soybean-Inoculation of *Bradyrhizobium japonicum* with PGPR in saline soil conditions. *Research Journal of Agriculture and Biological Sciences*, Vol.1, No.3, pp. 216-221, ISSN 1816-1561

Herridge, D. (2008b). Inoculation technology for legumes. In: *Nitrogen-fixing leguminous symbioses*, Dilworth, M. J.; James, EK; Sprent, JI. & Newton, WE . (Eds.), 77-115, Springer, ISBN 9781402035456, Netherlands

Herridge, DF., Peoples, MB. & Boddey, RM. (2008a). Global inputs of biological nitrogen fixation in agricultural systems. *Plant and Soil*, Vol. 311, pp. 1–18, ISSN 0032079X.

Houlton, BZ.; Wang, YP.; Vitousek, PM. & Field, CB. (2008). A unifying framework for dinitrogen fixation in the terrestrial biosphere. *Nature*, Vol. 454, No. 3, pp. 27–334, ISSN 0028-0836

Howard, JB., & Rees, DC. (1996). Structural Basis of Biological Nitrogen Fixation. *Chemical Review*, Vol. 96, pp. 2965–2982, ISSN 0009-2665

Hungria, M.; Loureiro, M.; Mendes, I.; Campo, R. & Graham, P. (2005). Inoculant preparation, production and application, In: *Nitrogen fixation in agriculture, forestry, ecology, and the environment*, Werner, D. & Newton, W. (Eds.), 223-253, Springer, ISBN 10 1-4020-3542-X , Netherlands.

Itzigsohn, R.; Kapulnik, Y.; Okon ,Y. & Dovrat, A. (1993). Physiological and morphological aspects of interactions between *Rhizobium meliloti* and alfalfa (*Medicago sativa*) in association with *Azospirillum brasilense*. *Canadian Journal of Microbiology*, Vol.39, pp. 610-615, ISSN 0008-4166

Kadouri, D.; Jurkevitch, E.; Okon, Y. & Castro-Sowinski, S. (2005). Ecological and agricultural significance of bacterial polyhydroxyalkanoates. *Critical Review in Microbiology*, Vol. 31, No.2, pp. 55-67, ISSN 1040-841X

Kinney, A., & Clemente, TE. (2010). Soybeans, In: *Energy Crops*, Halford, N., & Karp, A., (Eds.), (148-164), Royal Society of Chemistry Publishing, ISBN: 978-1-84973-204-8, UK

Kumar, B.; Berggren, I. & Martensson, A. (2001). Potential for improving pea production by co-inoculation with fluorescent *Pseudomonas* and *Rhizobium*. *Plant and Soil*, Vol.229, pp. 25-34, ISSN 0032-079X

Lewis, G., Schrire, B., MacKinder, B. & Lock, M. (2005). *Legumes of the world*. Royal Botanical Gardens, Kew Publishing, ISBN 1 900 34780 6, UK

Lindström, K.; Murwira, M.; Willems, A. & Altier, N. (2010). The biodiversity of beneficial microbe-host mutualism: the case of rhizobia. *Research in Microbiology*, Vol. 161, pp. 453-463, ISSN 0923-2508

Liu, K. (1999). Chemistry and nutritional value of Soybean components, In: *Soybeans: Chemistry, Technology and Utilization*, Liu, K., pp (25-94), Aspen Publisher, ISBN 0-8342-1299-4, New York.

López, N.I.; Ruiz, J.A. & Méndez, B.S. (1998). Survival of poly-3-hydroxybutyrate-producing bacteria in soil microcosms. *World Journal of Microbiology and Biotechnology*, Vol.14, pp. 681–684, ISSN 0959-3993

Lucas-Garcia, J.; Probanza, A.; Ramos, B.; Barriuso, J. & Gutierrez-Mañero, F. (2004a). Effects of inoculation with plant growth rhizobacteria (PGPRs) and *Sinorhizobium fredii* on biological nitrogen fixation, nodulation and growth of *Glycine max* cv. *osumi. Plant and Soil*, Vol. 267, pp. 143-153, ISSN 0032-079X

Lucas-Garcia, J.; Probanza, A.; Ramos, B.; Colon-Flores, J. & Gutierrez-Mañero, F. (2004b). Effects of plant growth promoting rhizobateria (PGPRs) on the biological nitrogen fixation, nodulation and growth of *Lupinus albus* I. cv. *multolupa. Engineering in Life Sciences*, Vol. 4, pp. 71–77, ISSN 1618-2863

Mabrouk, Y., & Belhadj, O. (2010). The potential use of *Rhizobium*-legume symbiosis for enhancing plant growth and management of plant diseases, In: *Microbes for Legume Improvement*, Khan, MS., Zaidi, A., & Musarrat, J. (Eds.), 495-514, Springer-Verlag, ISBN 978-3-211-99752-9, New York

Macchiavelli, RE. & Brelles-Mariño, G. (2004). Nod factor-treated *Medicago truncatula* roots and seeds show an increased number of nodules when inoculated with a limiting population of *Sinorhizobium meliloti. Journal of Experimental Botany*, Vol. 55, pp. 2635–2640, ISSN 0022-0957

Malik, D. & Sindhu, S. (2011). Production of indole acetic acid by *Pseudomonas* sp.: effect of coinoculation with *Mesorhizobium* sp. *Cicer* on nodulation and plant growth of chickpea (*Cicer arietinum*). *Physiology and Molecular Biology of Plants*, Vol.17, No.1, pp. 25-32, ISSN 0971-5894

Mañero, F.; Probanza, A.; Ramos, B.; Flores, J. & Garcıa-Lucas, J. (2003). Effects of culture filtrates of rhizobacteria isolated from wild lupin on germination, growth, and biological nitrogen fixation of lupin seedlings. *Journal of Plant Nutrition*, Vol.26, pp. 1101–1115, ISSN 0190-4167

Marek-Kozaczuk, M. & Skorupska, A. (2001). Production of B-group vitamins by plant-growth promoting *Pseudomonas fluorescens* strain 267 and the importance of vitamins in the colonization and nodulation of red clover. *Biology and Fertility of Soils*, Vol.33, pp. 146-151, ISSN 0178-2762

Masson-Boivin, C., Giraud, E., Perret, X., & Batut, J. (2009). Establishing nitrogen-fixing symbiosis with legumes: how many *Rhizobium* recipes? *Trends in Microbiology*, Vol. 17, pp. 458-466, ISSN 0966-842X.

Medeot, DB.; Paulicci, NS.; Albornoz, AI.; Fumero, MV.; Bueno, MA.; Garcia, MB.; Woelke, MR.; Okon, Y. & Dardanelli, MS. (2010). Plant growth promoting rhizobacteria improving the legume-rhizobia symbiosis, In: *Microbes for Legume Improvement*, Khan, MS., Zaidi, A., & Musarrat, J. (Eds.), 473-494, Springer-Verlag, ISBN 978-3-211-99752-9, New York

Mishra, P.; Mishra, S. & Selvakumar, G. (2009). Coinoculation of *Bacillus thuringeinsis*-KR1 with *Rhizobium leguminosarum* enhances plant growth and nodulation of pea (*Pisum sativum* L.) and lentin (*Lens culinaris* L.). *World Journal of Microbiology & Biotechnology*, Vol.25, pp. 753-761, ISSN 0959-3993

Molla, A.; Shamsuddin, Z. & Saud, H. (2001b). Mechanism of root growth and promotion of nodulation in vegetable soybean by *Azospirillum brasilense. Communication in Soil Science and Plant Analalysis*, Vol. 32, pp. 2177-2187, ISSN 0010-3624

Molla, A.; Shamsuddin, Z.; Halimi, M.; Morziah, M. & Puteh, A. (2001a). Potential enhancement of root growth and nodulation of soybean co-inoculated with *Azospirillum and Bradyrhizobium* in laboratory systems. *Soil Biology and Biochemesty*, Vol.33, pp. 457-463, ISSN 0038-0717

Morel, MA; Ubalde, M.; Braña, V. & Castro-Sowinski, S. (2011). *Delftia* sp. JD2: a potential Cr(VI)-reducing agent with plant growth-promoting activity. *Archives of Microbiology*, Vol.193, No.1, pp. 63-68, ISSN 0302-8933

NZ Rhizobia. (2011) New Zealand Rhizobia, access on September 14, available from: <http://www.rhizobia.co.nz>

Okon, Y (Ed). (1994). *Azospirillum /*plant associations, CRC-Press, ISBN 0849349257, Boca Raton, FL, USA

Osburn, RM.; Dénarié, JC.; Maillet, F.; Penna, C.; Díaz-Zorita, M.; Kosanke, JW. & Smith, RS. (2004). New patented growth promoter technology to enhance early season soybean development and grain yield. In: *Proceedings of the 19th North American Nitrogen Fixation Conference*, Bozeman, Montana, June 2004

Pahl, G. (2008). Biodiesel 101, In: *Biodiesel: growing a new energy economy*, Pahl, G., (Ed.), 33-53, Chelsea Green Publishing, ISBN 978-1-933392-96-7, U.S.

Parmar, N. & Dadarwal, K. (1999). Stimulation of nitrogen fixation and induction of flavonoid-like compounds by rhizobacteria. *Journal of Appied Microbiology*, Vol.86, pp. 36-44, ISSN 1365-2672

Perrig, D.; Boiero, ML.; Masciarelli, OA.; Penna, C.; Ruiz, OA; Cassán FD. & Luna, MV. (2007) Plant-growth-promoting compounds produced by two agronomically important strains of *Azospirillum brasilense*, and implications for inoculant formulation. *Applied Microbiolog and Biotechnolog*, Vol.75, No.5, pp. 955, ISSN 0175-7598

Popelka, C., Terryn, N. & Higgins, T. (2004). Gene technology for grain legumes: can it contribute to the food challenge in developing countries?. *Plant Science*, Vol. 167, No. 2, pp 195-206, ISSN 01689452.

Pretty, J. (2008). Agricultural sustainability: concepts, principles and evidence. *Philosophical Transations of The Royal Society Biological Science*, Vol.363, pp. 447-465, ISSN 1471-2970

Puente, M.; Garcia, J. & Alejandro, P. (2009). Effect of the bacterial concentration of *Azospirillum brasilense* in the inoculum and its plant growth regulator compounds on crop yield of corn (*Zea mays* L.) in the field. *World Journal of Agricultural Sciences*, Vol. 5, No. 5, pp: 604-608, ISSN 1817-3047

Qiu, L.J. & Chang, R.Z.. (2010). The origin and history of soybean. In: *The soybean: botany, production and uses*, Singh, G., (Ed.) 345-362, CAB International, ISBN 9781845936440, UK.

Qureshi, M.; Ahmad, M.; Naveed, M.; Iqbal, A.; Akhtar, N. & Niazi, K. (2009). Co-inoculation with *Mesorhizobium ciceri* and *Azotobacter chroococcum* for improving growth, nodulation and yield of chickpea (*Cicer arietinum* L.). *Soil & Environment*, Vol. 28, No. 2, pp. 124-129, ISSN 2074-9546

Rajendran, G.; Sing, F.; Desai, A.J & Archana, G. (2008). Enhanced growth and nodulation of pigeon pea by co-inoculation of *Bacillus* strains with *Rhizobium* spp. *Bioresource Technology*, Vol.99, pp. 4544–4550, ISSN 0960-8524

Remans, R.; Beebe, S.; Blair, M.; Manrique, G.; Tovar, E.; Rao, I.; Croonenborghs, A.; Torres-Gutierrez R.; El-Howeity, M.; Michiels, J. & Vanderleyden, J. (2008a). Physiological and genetic analysis of root responsiveness to auxin-producing plant growth-promoting bacteria in common bean (*Phaseolus vulgaris* L.). *Plant and Soil*, Vol.302, pp. 149-161, ISSN 0032-079X

Remans, R.; Ramaekers, L.; Schelkens, S.; Hernandez, G.; Garcia, A.; Reyes, J.; Mendez, N.; Toscano, V.; Mulling, M.; Galvez, L. & Vanderleyden, J. (2008b). Effect of *Rhizobium-Azospirillum* coinoculation on nitrogen fixation and yield of two contrasting *Phaseolus vulgaris* L. genotypes cultivated across different environments in Cuba. *Plant and Soil, Vol.* 312, pp. 25-37, ISSN 0032-079X

Remans, R.; Croonenborghs, A.; Torres Gutierrez, R.; Michiels, J. & Vanderleyden, J. (2007). Effects of plant growth-promoting rhizobacteria on nodulation of *Phaseolus vulgaris* L. are dependent on plant nutrition. *European Journal of Plant Pathology*, Vol.119, pp. 341-351, ISSN 0929-1873

Rosas, S.; Andrés, J.; Rovera, M. & Correa, N. (2006). Phosphate-solubilizing *Pseudomonas putida* can influence the rhizobia-legume symbiosis. *Soil Biology and Biochemestry*, Vol.38, pp. 3502-3505, ISNN 0038-0717.

Schrire, BD., Lewis, GP., & Lavin, M. (2005). Biogeography of the Leguminosae, In: *Legumes of the world*, Lewis, G., Schrire, G., Mackinder, B., & Lock, M., (Eds.), 21–54, Royal Botanic Gardens, ISBN 8773043044, Kew, UK

Shaharoona, B.; Arshad, M. & Zahir, Z. (2006). Effect of plant growth promoting rhizobacteria containing ACC-deaminase on maize (*Zea mays* L.) growth under axenic conditions and on nodulation in mung bean (*Vigna radiate* L.). *Letter in Applied Microbiology*, Vol.42, pp. 155-159, ISSN 1472-765X

Sindhu, S. & Dadarwal, K. (2001). Chitinolytic and cellulolytic *Pseudomonas* sp. Antagonistic to fungal pathogens enhances nodulation by *Mesorhizobium* sp. *ciceri* in chickpea. *Microbiological Research*, Vol.156, No.4, pp. 353-358, ISSN 0944-5013

Sindhu, S.; Suneja, S.; Goel, A.; Parmar, N. & Dadarwal K. (2002). Plant growth promoting effects of *Pseudomonas* sp. on coinoculation with *Mesorhizobium* sp. *ciceri* strain under sterile and "wilt sick" soil conditions. *Applied Soil Ecology*, Vol.19, No1, pp.57-64, ISSN 0929-1393

Sindhu, S.S.; Dua, S.; Verma, M.K. & Khandelwal, A. (2010). Growth promotion of legumes by inoculation of rhizosphere bacteria, In: *Microbes for Legume Improvement*, Khan, MS., Zaidi, A., & Musarrat, J. (Eds.), 195-235, Springer-Verlag, ISBN 978-3-211-99752-9, New York

Skorupska, A.; Wielbo, J.; Kidaj, D. & Marek-Kozaczuk, M. (2010). Enhancing *Rhizobium*-legume symbiosis using signaling factors, In: *Microbes for Legume Improvement*, Khan, MS., Zaidi, A., & Musarrat, J. (Eds.), 27-54, Springer-Verlag, ISBN 978-3-211-99752-9, New York

Soussana, J-F. & Tallec, T. (2010). Can we understand and predict the regulation of biological N_2 fixation in grassland ecosystems?. *Nutrient Cycling in Agroecosystems*, Vol. 88, pp. 197-213, ISSN 1385-1314

Spaepen, S.; Vanderleyden, J. & Remans, R. (2007). Indole-3-acetic acid in microbial and microorganism-plant signaling. *FEMS Microbiology Review*, Vol.31, pp. 425–448, ISSN 0168-6445

Star, L.; Matan, O.; Dardanelli, M.S.; Kapulnik, Y.; Burdman, S. & Okon, Y. (2011). The *Vicia sativa* spp. nigra-*Rhizobium leguminosarum* bv. *viciae* symbiotic interaction is improved by *Azospirillum brasilense*. *Plant and Soil*, DOI 10.1007/s11104-010-0713-7 , ISSN 0032-079X

Tilak, K.; Ranganayaki, N. & Manoharachari, C. (2006). Synergistic effects of plant-growth promoting rhizobacteria and *Rhizobium* on nodulation and nitrogen fixation by pigeonpea (*Cajanus cajan*). *European Journal of Soil Science*, Vol.57, pp. 67-71, ISSN 1351-0754

Tittabutr, P.; Payakapong, W.; Teaumroong, N.; Singleton, PW. & Boonkerd, N. (2007). Growth, survival and field performance of Bradyrhizobial liquid inoculant formulations with polymeric additives. *ScienceAsia*, Vol. 33, pp. 069-077, ISSN 1513-1874.

USDA. (1998). Soil Quality – Agronomy Technical Note, access on 30 August 2011, available from: <http://soils.usda.gov/sqi/management/files/sq_atn_6.pdf>

Valdenegro, M.; Barea, J. & Azcón, R. (2001). Influence of arbuscular-mycorrhizal fungi, *Rhizobium meliloti* strains and PGPR inoculation on the growth of *Medicago arborea* used as model legume for re-vegetation and biological reactivation in a semi-arid Mediterranean area. *Plant Growth Regulation*, Vol. 34, pp. 233-240, ISSN 0167-6903

Vessey, J. & Buss, T. (2002). *Bacillus cereus* UW85 inoculation effects on growth, nodulation, and N accumulation in grain legumes: controlles-environment studies. *Canadian Journal of Plant Science*, Vol. 82, No. 2, pp. 282-290, ISSN 0008-4220

Vessey, J.k. (2003). Plant growth promoting rhizobacteria as bio-fertilizers. *Plant and Soil*, Vol. 255, pp. 571-586, ISSN 0032-079X

Vieira, RF., Mendes, IC., Reis-Junior, FB., & Hungria, M. (2010). Symbiotic Nitrogen Fixation in Tropical Food Grain Legumes: Current Status, In: *Microbes for Legume Improvement*, Khan, MS., Zaidi, A., & Musarrat, J., (Eds.), 427-472, Springer-Verlag, ISBN 978-3-211-99752-9, New York.

Volpin, H.; Burdman, S.; Castro-Sowinski, S.; Kapulnik, Y. & Okon, Y. (1996). Inoculation with *Azospirillum* increased exudation of rhizobial *nod*-gene inducers by alfalfa roots. *Molecular plant Microbe Interaction*, Vol.9, pp. 388-394, ISSN 0894-0282

Wani, P.; Khan, M.S. & Zaidi, A. (2007). Synergisitc effects of the inoculation with nitrogen-fixing and phosphate-solubilizing rhizobacteria on the performance on field-grown chickpea. *Journal of Plant Nutrition and Soil Science*, Vol.170, pp. 283-287, ISSN 1522-2624

Huang, W. (2007). Impact of Rising Natural Gas Prices on U.S. Ammonia Supply, access on August 30 2011, available from:
www.ers.usda.gov/publications/WRS0702/wrs0702.pdf

Xavier, I..; Holloway, G. & Leggett, M. (2004). Development of Rhizobial Inoculant Formulations, In: *Plant Management Network*, 14.01.2003, Available from http://www.plantmanagementnetwork.org/pub/cm/review/2004/develop/, Crop Manage. ISSN 1543-7833

Yadegari, M.; Rahmani, HA.; Noormohammadi, G. & Ayneband, A. (2010). Plant growth promoting rhizobacteria increase growth, yield and nitrogen fixation in Phaseolus vulgaris. *Journal of Plant Nutrition*, Vol.33, pp. 733-1743, ISNN 0190-4167

Yahalom, E.; Okon, Y. & Dovrat, A. (1987). *Azospirillum* effects on susceptibility to *Rhizobium* nodulation and on nitrogen fixation of several forage legumes. *Canadial Journal of Microbiology*, Vol. 33, No. 6, pp. :510-514, ISSN 0008-4166

Zhang, F.; Dashti, N.; Hynes, R. & Smith, D. (1996). Plant growth promoting rhizobacteria and soybean [*Glycine max* (L.) Merr.] nodulation and nitrogen fixation at suboptimal root zone temperatures. *Annals of Botany*, Vol. 77, pp. 453-459, ISSN 0305-7364

Yasari, E.; Esmaeili-Azadgoleh, A.M.; Pirdashti, H. & Mozafari, S. (2008). *Azotobacter* and *Azospirillum* inoculants as biofertilizers in Canola (*Brassica napus* L.) cultivation. *Asian Journal of Plant Sciences, Vol. 7*, No.5, pp.490-494, ISSN 1682-3974

Zahran, H.H. (2009). Enhancement of Rhizobia-Legumes Symbioses and Nitrogen Fixation for Crops Productivity Improvement. In: *Microbial Strategies for Crop Improvement*, Khan, M.S.; Zaidi, A.; Musarrat, J. (Eds), 227-254, ISBN 9783642019784, Springer, Netherlands

Zhang, F. & Smith, D. 2002. Interorganismal signaling in suboptimum environments: the legume-rhizobia symbiosis, In: *Advances in Agronomy*, Vol. 76, Spark, D. (Ed.), Elsevier, 125-161, ISBN 0-12-000794-0, U.S.

9

The Shade Avoidance Syndrome Under the Sugarcane Crop

Jocelyne Ascencio and Jose Vicente Lazo
Universidad Central de Venezuela, Facultad de Agronomía, Maracay
Venezuela

1. Introduction

Sugarcane is grown mainly for sugar and ethanol production, belongs to the Poaceae family, genus *Saccharum* native to Southest Asia and India and cultivated intensively in tropical and subtropical areas throughout the world, and it plays a significant role in the world economy and the area cultivated yields observed to have progressively increased to remarkable levels in the last 10 years (Azevedo et al., 2011). Commercial sugarcane, mainly the interspecific hybrids of *S. officinarum* and *S. spontaneum* would greatly benefit from biotechnological improvements due to the long duration (10-15 years) required to breed elite cultivars, more importantly there is an ongoing need to provide durable and disease and pest resistance in combination with superior agronomic performance (Lakshman et al., 2005).

There is an increasing pressure worldwide to enhance the productivity of sugarcane cultivation in order to sustain profitable sugar industries (Hanlon et al., 2000), for example, improvement of industrial processes along with strong sugarcane breeding programs in Brazil, brought technologies that currently support a cropland of 7 million hectares of sugarcane with an average yield of 75 tons/ha (Matsuoka et al., 2009). Besides, biomass has gained prominence in the last years as new technologies for energy production from crushed sugarcane stalks developed a sugarcane industry that currently is one the most efficient systems for the conversion of photosynthate into different forms of energy, for example, the production of ethanol as a liquid fuel.

The crop is vegetatively propagated by stalk cuttings, having one to three buds, known as seed pieces or setts, is a perennial crop regrowing from these vegetative buds after the crop has been harvested giving subsequent regrowth or crop cycles known as rattoning. The germinating bud develops its own root system, and several shoots arise by heavy tillering which produces a canopy of leaves during closing-in stages of crop growth; the term "closed crop" defines a community of plants, of uniform height, which extends indefinitely in the horizontal plane. Within a "closed crop" canopy, we might expect the leaves in any particular horizon to experience a uniform environment, and we might further expect the only significant source of environmental variation to be found in the vertical plane (Charles-Edwards, 1981), thus for the sugarcane crop the production of stalks, to quickly achieving a closed canopy, is important as a means of increasing competition with the weeds growing underneath and for crop protection against adverse conditions.

Sugarcane uses the C4 pathway of photosynthesis where CO_2 is efficiently captured, in the mesophyll cells giving a four-carbon organic acid, oxaloacetate which is the first product of CO_2 fixation, and recycled inside the leaves because of the compartmentalized arrangement of leaf tissues into bundle sheath and mesophyll cells (Hatch & Slack, 1966). This photosynthetic specialization of cell types allows leaves to fix CO_2 at higher rates and at lower stomatal conductance; however other C4 species dominate the list of the world's worst weeds which in many cases, like for the sugarcane crop, are among the most noxious plants to the crop. Failure to control weeds during early stages of crop growth can reduce yields appreciably. As a C4 plant sugarcane grows better at high irradiances and temperatures and is also resistant to some environmental conditions very common in the field, especially in the tropics, such as water deficits. Because of these attributes improved cultivars with increased resistance to stressful conditions, adequate management of water and other resources have been developed.

Light interception is an important component of the environment within a crop canopy; in sugarcane solar radiation is intercepted by the extended leaves but canopy architecture can modify photosynthetic performance. Thus commercial sugarcane varieties may have erect or planophille leaves but in a closed canopy most of the light is intercepted by the top fully expanded leaves. Erect leaved varieties, appear to be more efficient capturing light than those with more planophille or droopy leaves, specially at high plant densities. In this context strategy for weed control in order to improve farm management must include the knowledge of the dynamic and biology of plants growing underneath the canopy.

Canopy shade is an important part of weed-crop interference, thus the effect of radiation quality (wavelength) and quantity (irradiance, photon flux) on the diversity of plant species is a serious constraint to production and crop management. The sugarcane crop canopy closes at about three months after planting/sprouting, and the population of plants under the canopy changes in number and diversity depending on the ability of some species to escape or avoid shade by a series of developmental changes at the individual and population levels. Under field conditions, for each sugarcane ratoon cycle (regrowth), recognition of the species diversity as well as their strategies for shade avoidance, before and after canopy closure, is relevant to agricultural applications and plant biology and as research on the shade avoidance syndrome has been mostly restricted to individual plants and under controlled conditions, the objective of the present study is to provide information about what happens under a sugarcane canopy under field conditions, as related to spectral shifts within the canopy and the changes in species diversity.

The shade avoidance syndrome (SAS) in plant neighborhoods such as those underneath a crop canopy, is associated with both the quality (wavelength) and quantity (energy) of light and the decrease in the red/far red ratio (R/FR), as the light environment becomes enriched in far red radiation that is reflected by the leaves of all plants, including the crop itself. A reduced R/FR is the proximity signal that is perceived by the plants alerting that they are being shaded, and in fact it is perceived in early developmental stages, as shown by Ballare & Casal, 2000; Ballare et al., 1991; Smith et al., 1990. Small changes in amounts of red relative to far red light have been shown to alter the equilibrium of different phytochrome forms appreciably, which plays a major role in plant development. Perception of light quality and quantity at the stem level may elicit morphological adaptations in the plants growing beneath the canopy, that may result in shade tolerance or avoidance, and under field conditions, slight differences in height, degree of tillering, earlier flowering and increase in the shoot/root ratio among different species, might imply a greater potential to survive escaping shade; thus recognition of plant species before and after canopy closure as well as the changes in the light profiles under the canopy, are important for weed detection in cultivated crops as a means to ascertain which species avoid shade.

Shade avoidance responses are mediated by multiple forms of phytochromes; despite of initial attempts to adscribe the SAS to the action of a single member of the phytochrome family (Franklin & Whitelam, 2005). In this context and according to Schmitt (1997), the shade avoidance response has undergone adaptive evolution as quantitative genetic variation in R:FR ratio sensitivity has been detected in wild populations. The "Shade Avoidance Syndrome" has been described by Morgan et al (2002); Smith (1982); Smith & Whitelam (1997) as an accelerated extension growth (as seen by an increase in shoot and petiole elongation), reduced branching (increase in apical dominance), earlier flowering (i.e., rate of flowering markedly accelerated) and increase in the shoot to root ratios, changes in plant architecture and leaf shape, among other responses not easily seen under field conditions, however, we have used the term "Agronomic Shade Avoidance Syndrome" to include species that persist and compete successfully with the crop after canopy closure, but not by means of growth responses normally associated to the SAS under field conditions, such as morphological changes in leaf shape, stem elongation or plant architecture. The persistence of such species after canopy closure, may be associated to the seed bank, sunflecks in gaps within the canopy and the production of underground organs (that become well established and almost impossible to control by shade) as well as climbing strategies such as in tie-vines.

From another point of view, the shade avoidance syndrome is not restricted to terrestrial ecosystems but has also been studied in connection to other stress responses, such as submergence (Pierik et al., 2005). In complete submergence, well-adapted plants may overcome the effects by adopting an avoidance strategy to induce growth responses, phenotypically similar to those described when plants are shaded by proximate neighbors.

In this chapter, we will analyze light profiles within sugarcane canopies and how changing light conditions in a closed-in canopy, may affect the development and diversity of plant species, as seen in the field at the individual level. Then, based on the results of experimental trials and under controlled conditions, some of the strategies to escape or avoid shade and to capture light more efficiently, will be discussed for different species known to be noxious to the sugarcane crop.

2. Research methods

2.1 Field experiments

Field experiments were conducted in two different sites where sugarcane is a mayor crop in Venezuela. The first experiment was established with sugarcane plants (*Saccharum* spp hybrid var. PR-692176 (first raton crop) in a 625 m² plot inside a commercial regrowth of a sugarcane field in Chivacoa, Yaracuy State and the second experiment, in 2500 m² experimental area in the Agricultural Experimental Station at the College of Agriculture in the Central University of Venezuela in Maracay, Aragua State, using droopy and erect leaf commercial varieties planted from cane (initial or plant cane crop).

2.1.1 First field experiment

The objectives of the first trial were: 1) to acknowledge the species that were present before and after the sugarcane canopy closure; 2) to determine the effect of canopy shade on the developmental responses that could be associated to the SAS under field conditions and 3) to measure the amount of light in terms of photon flux density of photosynthetically active radiation (PAR), at different points within the crop and weed canopies.

Standard management practices included hand planting of 50 cm long stalks or seed pieces placed at rates of 24000 stalks/ha in the bottom of furrows spaced 1,5 m and covering the sugarcane seed pieces with soil. Soil was regularly irrigated every two weeks, or as needed, and fertilized before planting with ammonium phosphate and with a second application of Urea+KCl at 45 days after planting. Chemical weed control in the drive areas of the experimental plot was not performed for the experiments discussed in this chapter. For species recognition, ten 0,5 m² fixed wooden squares were randomly distributed in the field and all plant species (except for those of the crop) inside the squares were collected at 60 and 90 days after germination (sprouting of the buds from the seed pieces or stalks), to acknowledge for the presence and number of different species, and for plant height and leaf size determinations. Flowering and any other visual symptoms associated to the shade avoidance response were as well registered.

In order to compare the plants growing under a "canopy in a non-shaded condition", in some previously selected rows in the same plot the leaf arrangement of sugarcane plants, was artificially changed by loosely bounding the leaves in an upright position along the stem, simulating an plant biotype with erect leaves.

Light quantity was measured as Photosynthetically Active Radiation (PAR as $\mu mol \ m^{-2} \ s^{-1}$) using a quantum-radiometer-photometer LiCor 185B by positioning the quantum sensor at different heights above and below the sugarcane canopy and above the population of weeds growing in shaded and non-shaded sites beneath the crop, as shown in Fig. 1. Instant measurements, at five different points in each position (sites) within the canopy, were performed between 12M and 1PM at 60 days after germination of buds from the seed pieces or sugarcane stalks, for the two leaf arrangements described in the preceding paragraph.

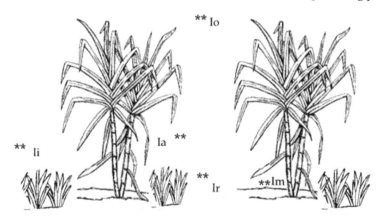

Fig. 1. Sugarcane plants and weed underneath the canopy showing quantum sensor positions (lo, li, la, lm) to register PAR values shown in Table 1.

2.1.2 Second field experiment

For the second field experiment, with the objective of recognizing weed species before and after canopy closure for two commercial sugarcane varieties with contrasting growth habits, and to register radiation profiles within the canopy, a plant crop (first cycle) was established in 2500 m² experimental area in the Agricultural Experimental Station at the College of Agriculture in the Central University of Venezuela in Maracay, Aragua State. The experimental area was divided into four   625 m²   plots where droopy and erect leaf commercial varieties, C 266-70, which is a fairly typical variety, with planophile or droopy leaves and RB 85-5035 with more erect leaves, were planted from initial plant canes. Crop management practices were as described for the first experiment except for the fact that 10 randomly located fixed wooden 1m² squares, were used to collect the plants growing under the crop canopy and that weed control, in some previously selected drive areas between rows, was manually performed.

Spectral profiles within sugarcane canopy: energy distribution of visible and near infrared radiation above, within, and on the soil below a canopy a sugarcane plants with droopy or erect leaf commercial varieties were registered and light spectra (Spectral Irradiance, W m⁻² should be $W\ m^{-2}\ nm^{-1}$ and Quantum Intensity, $\mu mol\ m^{-2}\ s^{-1}\ nm^{-1}$) were measured in the field plot during the growing season, at the beginning of canopy closure (three months from planting date), with a spectroradiometer ASD FieldSpec*pro* VNIR 350-1050 nm, using hyperspectral analysis, approximately at 20 cm above the crop, at 30 cm above and below the weed canopy and at 30 cm above the soil, in shaded and non shaded sites.

2.2 Greenhouse experiments

Three of the most abundant species known to be noxious weeds to the sugarcane crop and present in the experimental fields (*Rottboellia cochinchinensis, Leptochloa filiformis* and *Cyperus rotundus*), were selected for the study of growth responses associated to shade quality (wavelength) and quantity (Photon flux density, PFD 400 - 700 nm). Seeds of *Rottboellia* and *Leptochloa* and corms from *Cyperus* were sowed in pots, containing soil, shaded by cabinets covered with red, blue and green cellophane paper and under low PFD neutral shade while another group was left uncovered. Cabinets were directly exposed to daylight. Effects were compared separately for each species when plants in any of the groups showed visual symptoms of deterioration.

3. Research results

3.1 First field experiment

Under the sugarcane canopy in the **first field trial**, at 60 days after sprouting when canopy closure was not complete, a stratified pattern in plant height for the different species was observed. Plants species identified under the sugarcane canopy before canopy closure were (Lara & Ascencio, unpublished): *Ruellia tuberosa* L., *Trianthema portulacastrum* L., *Amaranthus dubius* Mart, *Eclipta alba* (L.) Hassk, *Tridax procumbens* L., *Lagascea mollis* Cav., *Heliotropium ternatum* Valhl and *H. indicum* L., *Cyperus rotundus* L., *Commelina difusa* Burm, *Ipomoea indica* (Burm.) Merr and *I. batatas* (L.) Lam, *Momordica charantia* L., *Cucumis dipsaceus* Ehremb. ex Spash, *Ceratosanthes palmata* (L.) Urb, *Euphorbia hirta* L. and *E. hypericifolia* L., *Croton lobatus* L. *Phylathus niruri* L., *Desmanthus virgatus* (L.) Willd, *Spigelia anthelmia* L., *Leptochloa filiformis*

(Lam.) Beauv, *Rottboellia cochinchinensis* (Lour) W. Clayton, *Panicum fasciculatum* Sw., *Echinochloa colona* (L.) Link, *Eleusine indica* (L.) Gaerth, *Cynodon dactylon* (L.) Pers, *Dactyloctenium aegyptium* (L.) Wild, *Portulaca oleracea* L., *Capraria biflora* L., *Physalis angulata* L., *Corchorus orinocensis* Kunth, *Priva lappulaceae* (L.) Pers, *Kallstroemia maxima* (L.) Hook & Arn.

It is important to note that 23% of the species listed above belong to the Poaceae (as sugarcane) and that the rest are distributed in 18 families, thus 8 species known to be noxious for the sugarcane crop were selected for this study, where, except for *Trianthema postulacastrum*, *Heliotropium ternatum* and *Cyperus rotundus* the other five (*Leptochloa filiformis*, *Rottboellia cochinchinensis*, *Panicum fasciculatum*, *Panicum maximum* and *Cynodon dactylon*) belong to the Poaceae.

When the first evaluation was made in the field 30 days after sprouting a reduction in the number of plants for the species *Cyperus rotundus* and *Leptochloa filiformis* was observed at first sight under the shade of other plants (either the crop or other plants in the neighbourhood), but not for other species in the site, a first indication that shade was affecting plant performance, plant loss or even causing plant death. Differences in plant height (stem or internode elongation) for the species growing below the sugarcane canopy were observed after 60 days in the following order: *Rottboellia cochinchinensis* > *Panicum maximum* > *Panicum fasciculatum* > *Heliotropium ternatum* > *Trianthema portulacastrum* > *Cyperus rotundus*. Thus, these species except for *Cyperus rotundus*, *Trianthema portulacastrum* and *Heliotropium ternatum*, escaped canopy shade by an increase in plant height. Maximal plant height approaching that of the crop after full canopy closure (90 days) was observed for *Rottboellia exaltata* and *Panicum maximum*, which effectively escaped shade, competing successfully with the crop for light (Fig 2).

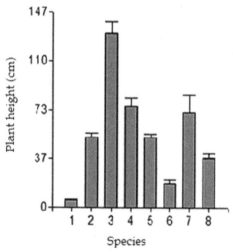

Species: 1. *Cyperus rotundus* L., 2. *Leptochloa filiformis* (Lam.) Beauv., 3. *Rottboellia cochinchinensis* (Lour.) W. Clayton, 4. *Panicum fasciculatum* Sw., 5. *Heliotropium ternatum* Vahl., 6. *Trianthema portulacastrum* L., 7. *Panicum maximum* Jacq., 8. *Cynodon dactylon* (L.) Pers. (Lara & Ascencio, unpublished).

Fig. 2. Plant height for species growing under a sugarcane (*Saccharum spp* hybrid) canopy with droopy leaves (*Saccharum spp* hybrid var PR692176) after 90 days of crope emergence.

Two growth strategies associated to the SAS as seen in the field were observed: increased internodes length and decreased leaf size. The species that showed higher sensitivity towards canopy shade were *Cyperus rotundus* and *Trianthema portulacastrum*, as plants eventually died in this condition apart from *Leptochloa filiformis* in which changes in leaf morphology, such as broader but shorter leaves were observed, as well as an early flowering of the individuals in order to produce seeds, which is also a means of escaping shade for population survival; however, the plants eventually died under the shade. Changes in leaf shape were also observed for plants escaping canopy shade through stem elongation, as in the case of *Rottboellia cocinchinensis* and *Panicum fasciculatum* were shorter leaves were observed as compared to those growing before canopy closure.

The effect of weed canopy shade on the development of plants in the neighborhood was observed when sugarcane leaves, in some selected rows, where loosely bound around the stem simulating an extreme erect biotype. As seen from Figure 3, at beginning of the crop cycle (60 days after sprouting) plant height for the different species were: *Rottboellia cochinchinensis* > *Leptochloa filiformis* > *Panicum fasciculatum* > *Heliotropium ternatum* > *Panicum maximum* > *Trianthema portulacastrum* > *Cynodon dactylon* > *Cyperus rotundus*, and after 90 days at full canopy closure (Fig.4), a steeper gradient for plant heights was observed as follows: *Rottboellia cochinchinensis* > *Panicum maximum* > *Panicum fasciculatum* > *Heliotropium ternatum* > *Leptochloa filiformis* > *Trianthema portulacastrum* > *Cyperus rotundus*. Plants from *Cynodon dactylon* were absent from the stand, unable to tolerate shade.

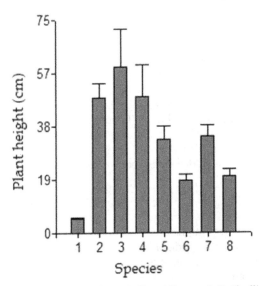

Species: 1. *Cyperus rotundus* L., 2. *Leptochloa filiformis* (Lam.) Beauv., 3. *Rottboellia cochinchinensis* (Lour.) W. Clayton, 4. Panicum fasciculatum Sw., 5. Heliotropium ternatum Vahl., 6. *Trianthema portulacastrum* L., 7. *Panicum maximum* Jacq. y, 8. *Cynodon dactylon* (L.) Pers. (Lara & Ascencio, unpublished)

Fig. 3. Plant height for species growing under a sugarcan (Saccharum spp hybrid var PR692176) canopy for 60 days after emergence with their leaves loosely bound to the stem simulating and erect leaf arrangement.

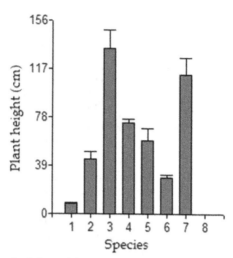

Species: 1. Cyperus rotundus L., 2. Leptochloa filiformis (Lam.) Beauv., 3. Rottboellia cochinchinensis (Lour.) W. Clayton, 4. Panicum fascilatum Sw., 5. Heliotropium ternatum Vahl., 6. Trianthema portulacastrum L., 7. Panicum maximum Jacq. y 8. Cynodon dactylon (L.) Pers. (Lara & Ascencio, unplublished)

Fig. 4. Plant height for species growing under sugarcane (Saccharum spp hybrid var PR692176) canopy for 90 days after emergence with their leaves loosely bound to the stem simulating and erect leaf arrangement.

Light quantity, an important component of shade, was measured in shaded and non-shaded positions within the sugarcane canopy (see Fig. 1), and for two leaf arrangements along the stem: droopy (planophile leaves) and erect (leaves bound to the stem to an erect position). PAR was measured in full sunlight above the crop at different crop heights (Io), above the canopy of weeds at 30 cm above soil level (which is in average the height of the population of weeds underneath the crop), under shaded (Ia) and non shaded (Ii) positions, at soil level at sites not shaded by plants (Ir) and also under the canopy of weeds growing beneath the canopy of sugarcane leaves (Im). Values for (Ii) are more likely to be sunflecks reaching gaps inside the canopy, while Ia are PAR values above the canopy of weeds but underneath the sugarcane leaves.

As seen from the results shown in Table 1, maximum PAR values were observed above the crop (Io = 1310) and above the canopy of weeds exposed to light in the erect leaf arrangement (1150) growing below the crop canopy at 30 cm above soil level (Ii) . PAR was 50% higher for (Ii) than under the shade of leaves (Ia), with low extinction values from Io for droopy and erect leaf arrangement, equal to -9.3 and -12.2% respectively. Another characteristic of the light environment underneath a canopy of leaves, is that direct sunlight may arrive as high intensity light beams known as sunflecks. In Table 1, values for (Ii) are more likely to be sunflecks in drive areas between rows. For droopy leaf arrangement and underneath the crop canopy (Ia), extinction values from Io were -38.3% above the weed canopy, and -70% at soil level, while for the erect leaf arrangement, higher values equal to -65.0 and -82.0% were observed for the extinction from Io. There is no clear explanation for this difference except for the fact that for the erect or vertically inclined leaf arrangement, sugarcane leaves were bound to the stalk allowing more light for the weeds to grow, so attenuation of radiation might have occurred by the scattering of radiation by the leaves. PAR values at soil level were dramatically lower under canopy shade (Im) as compared to that in between rows (Ir) with extinction values of -70 and -23% respectively; however it is worth noting that when these measurements were taken, crop canopy closure was about 20% of the maximum at full closure when values between rows may be much lower.

Leaf arrangement	Crop canopy		Weeds Underneath Crop Canopy*		Soil Level	
	PAR	Height	PAR		PAR	
	Above Crop (Io)	(cm)	Exposed to Light (Ii)	Shaded (Ia)	Exposed (Ir)	Shaded (Im)
Droopy	642**	66.4	582 (-9.3%)	396 (-38.3%)	494 (-23.1%)	194 (-70%)
Erect	1310	60.2	1150 (-12.2%)	460 (-65%)	1032 (-21.2%)	236 (-82%)

* Weed canopy average height, 30 cm above soil
** Heavy cloud

Table 1. Incident photosynthetically active radiation (PAR, μmol m^{-2} s^{-1}) within a sugarcane canopy (see Figure 1, for Io, Ii, Ia, Ir, and Im, quantum sensor positions). Values in parenthesis are expressed as percentages of Io.

Incident radiation transmitted by the leaves is a function of the extinction coefficient, which quantifies the influence of the arrangement of the leaves in the canopy. According to Tolenaar & Dwyer 1999, in the three-dimensional space above the soil, the leaf area arrangement is determined by (1) plant height, (2) plant spacing (i.e., row width vs. distance between plants in the row), (3) leaf length and width, (4) leaf angle with respect to the horizontal plane, and (5) leaf orientation with respect to north and south (i.e., azimuth angle). The extinction coefficient is relatively constant during the middle of the daytime when close to 90% of the total daily incident photosynthetic photon flux density is received.

Population dynamics (amount and diversity of species) below the canopy is highly influenced by the light environment (shade) and also by sunflecks which play an important role in the germination of new species not seen at the beginning of the crop cycle, as a great number of weed species have small, photoblastic seeds. This is discussed in the second field experiment as part of an strategy of escaping shade at a population level, as new species grown under shade conditions, are more likely to tolerate shade competing successfully with the crop for light. The importance of measurements of light (quantity and quality) in canopies is highlighted by Holt (1995) as a means to improve weed management as many plants possess the ability to adapt quickly to changes in light during the life cycle, species such as *Amaranthus palmeri*, *Crotalaria spectabilis*, *Cyperus rotundus*, *Cyperus esculentus*, *Imperata cylindrica* and *Abutilon theophrasti*, are mentioned as examples of weeds that respond to shade, thus, by understanding physiological and morphological mechanisms of competition for light between weeds and crops will it be possible to manipulate crop canopies to suppress weeds.

3.2 Second field experiment

According to the results of the second field trial, a higher number of species survived after the closure of the canopy of sugarcane plants with erect leaves, as compared to those for droopy leaves, and new species appeared: *Amaranthus* sp, *Bidens pilosa*, *Cyperus rotundus*, *Euphorbia heterophylla*, *Sida* sp, *Aldana dentata*, *Desmodium* sp, *Phyllanthus niruri* and *Eclipta alba*; and the new species (*Aldana sp.*, *Phyllanthus sp.* and *Eclipta sp*) appeared. Under the canopy of sugarcane plants with droopy leaves, *Amaranthus* sp, *Commelina difusa* (new), *Bidens pilosa*, *Mimosa* sp, *Euphorbia heterophylla* and *Desmodium* sp, were observed. In the experimental field of the second experiment, *Rottboellia exaltata* (which is capable of avoiding shade) was not found, and plants of *Leptochloa filiformis*, *Cynodon dactylon*, *Echinochloa colona*, *Ipomoea sp*, *Cucumis melo*, *Ruellia tuberosa* and *Cyperus rotundus*, progressively died under the shade of the canopy with droopy leaves.

It is important to note that some of the species that persisted after canopy closure were located at points within the canopy where light penetration was higher, while others were part of the seed bank (new species that appeared after canopy closure) which, in our opinion, may be an strategy for "agronomic shade avoidance" at a population level. Plants are actually seen growing under the canopy, thus escaping or avoiding shade in some way. In this category, the morning glory group of species (*Ipomoea spp*) referred to as tie-vines, may also be included as they are capable of climbing and forming a dense mat that grows over the crop canopy, escaping shade. Perennial grasses, found under the sugarcane canopy are another example of "agronomic shade avoidance", as sugarcane itself is a grass and conditions able to its development are also conductive to grass weed growth. As seen from the results of this study, perennial grasses such as *Cynodon sp* and *Panicum spp*, persist after canopy closure. In short, during each crop cycle different species are found as part of the biodiversity of the seed bank and the changing environment associated to the quantity (PAR) and quality (wavelength) of light in a closed-in canopy.

Radiation measurements: Energy distribution of visible and near infrared radiation (irradiance and quantum intensity) was measured underneath the sugarcane canopy; as leaf angle with respect to the horizontal plane and leaf length and width, influence the extinction of light within the canopy, and that these variables are associated to variety types, in this second field trial radiation profiles were compared within sugarcane canopies with either planophile (droopy or horizontal leaf arrangement) or erect (vertically inclined) leaves and in selected rows with high and low weed populations.

The energy distribution of radiation above the sugarcane canopy (light profile), is shown in Fig. 5 in terms of Quantum Intensity (QI, in $\mu mol\ m^{-2}s^{-1}nm^{-1}$) of full sunlight on a clear day in Maracay, Aragua ($10^0\ 11'N$, 440 msl). Figures 6 and 7, show energy distribution of visible and near infrared radiation within a canopy of sugarcane leaves with planophile leaves and with low and high populations of weeds growing underneath and Fig. 8 and 9 for a canopy with erect leaves with low or a high populations of weeds respectively.

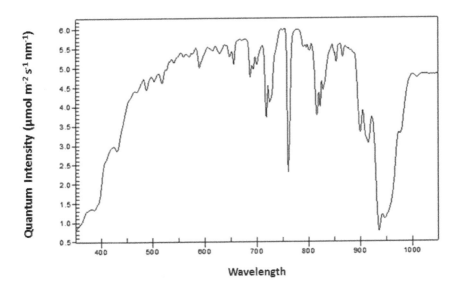

Fig. 5. Quantum energy distribution of full sunlight above a sugarcane canopy in Maracay (10°11' N, 440 msl). Values are given as Quantum Intensity (QI micromol m^{-2} s^{-1} nm^{-1})

The spectroradiometric measurements clearly show that the decreased QI of radiant energy within a sugarcane canopy is not uniform at all wavelengths and that spectral composition of shade light differs from that above the canopy (Fig. 1), because of the selective absorption of PAR (400-700 nm) by the leaves. Therefore, plant responses that may be attributed only to a reduced number of photons, or light intensity of radiant energy, could be confused with responses to a shift in the spectral composition of light received by the shaded leaves. In fact, a decreased R/FR is the most important radiation component within canopies as transmittance of far red radiation (730 nm) is substantial. This is shown in Figs 6 to 11 for the sugarcane canopies of this study. The population of weeds below the crop canopy and also crop architecture, influences the QI distribution of wavelengths reaching the soil at 30 cm from the ground, i.e. above the canopy of weeds. As compared to spectral distribution of sunlight (Fig 1) ,QI increases at wavelengths in the far red (radiation mostly reflected and transmitted by the leaves of the whole plant population crop+weeds) but differences are also found for QI values in canopies with planophile (Fig. 6 and 7) or erect leaves (Figs. 8 and 9) with low or high weed populations growing underneath. In deep shade QI values are lower.

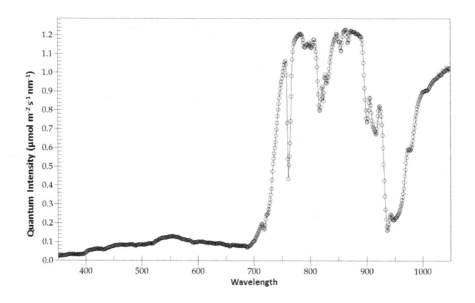

Fig. 6. Quantum energy distribution within and below a closed canopy of field grown sugarcan with planophile (droopy) leaves and a low population of weed growing underneath the canopy. Measurements were taken between adjacent rows at 30 cm from the ground. Values are given as Quantum Intensity (QI micromol m^{-2} s^{-1} nm^{-1})

Shifts in the amount of radiation beneath the canopy with planophile leaves and a low population of weeds, indicates that QI in the visible (400-700 nm) was lower than for a high population of weeds and, as could be expected, a higher QI was observed for wavelengths in the far red when the population of weeds was high (Fig.7). When comparing these results with the radiation profiles, as QI and irradiance, within a canopy of sugarcane leaves with erect leaves (Fig. 8 and 9), radiation reaching the soil in the visible was higher than for a canopy with planophile leaves, but still a higher QI for wavelengths in the near infrared was observed when the population of weeds was high. Except for the near infrared water sensitive portion (940-1040 nm), this is the shade that plants actually "see" and it is perceived at very early stages of development, alters phytochrome photoequilibrium and initiates growth responses to avoid the shade (shade avoidance syndrome).

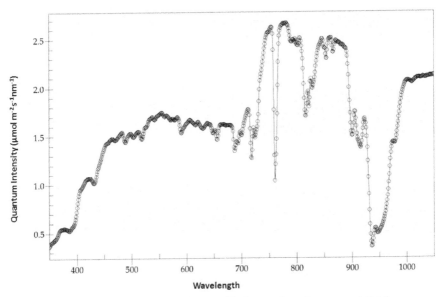

Fig. 7. Quantum energy distribution within and below a closed canopy of field grown sugarcane with planophile leaves and a high population of weeds growing underneath the canopy. Measurements were taken between adjacent rows at 30 cm from the ground. Values are given as Quantum Intensity (QI micromol m^{-2} s^{-1} nm^{-1})

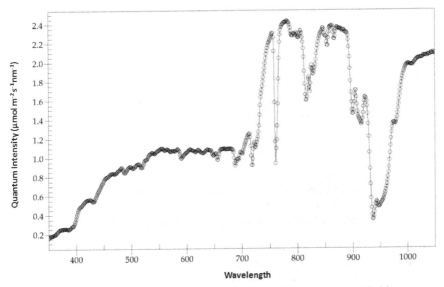

Fig. 8. Quantum energy distribution within and below a closed canopy of field grown sugarcane with erect leaves and a low population of weeds growing underneath the canopy. Measurements were taken between adjacent rows at 30 cm from the ground. Values are given as Quantum Intensity (QI micromol m^{-2} s^{-1} nm^{-1})

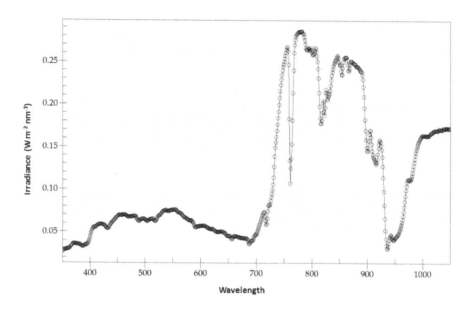

Fig. 9. Spectral irradiance within and below a close canopy of field grown sugarcane with erect leaves and a high population of weeds growing underneath the canopy. Measurements were taken between adjacent rows at 30 cm from the ground. Values are given as irradiance (W m^{-2} nm^{-1})

In Fig 10 an 11 radiation profiles, are shown as spectral irradiance (energy units in Watt m^{-2}nm^{-1}), below and above the **canopy of the weeds** growing underneath the sugarcane canopy with planophille leaves. Under this circumstances, shade conditions are more accentuated and the amount of energy in the visible is almost at limits to sustain growth; in these figures extreme shade conditions are observed within a canopy architecture that favors shade conditions (as for planophile or more horizontal leaf arrangement).

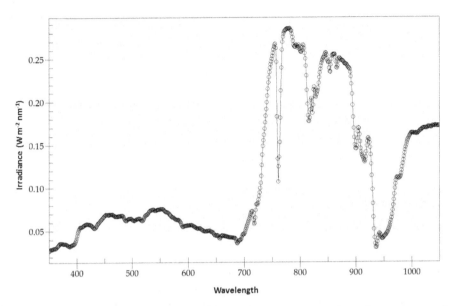

Fig. 10. Spectral irradiance beneath the canopy of the populations of weeds growing under a sugarcane canopy with planophile leaves. (Values are given in W m⁻² nm⁻¹)

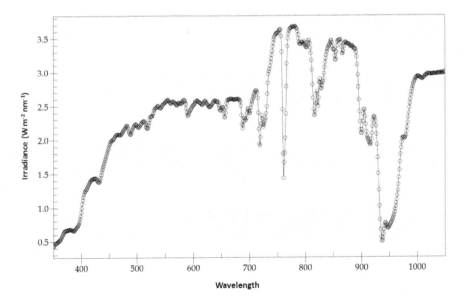

Fig. 11. Spectral irradiance above the canopy of the weed population growing under the sugarcane crop canopy with planophile leaves. (Values are given in W m⁻² nm⁻¹

From the perspectives of this chapter, what is more important is to ascertain which species that were present before crop canopy closure persist. From the establishment of the crop to full canopy closure, different shade intensities are found, and also different species and plant types, then the question arises as to which are capable of developing a syndrome for shade avoidance? the first answer to this question is: those species capable of perceiving, early in their development, that they are being shaded and start building mechanisms or strategies to defend themselves from dying from the shade and to escape the shade; first thing, as seen under field conditions: acceleration of extension growth of stem and petioles. Adaptation takes a little longer.

In the next part of this chapter, experiments under controlled conditions are shown in order to investigate some of the growth strategies most commonly found for three important sugarcane weeds, under an artificial shade.

3.3 Greenhouse experiments

Experiments under controlled conditions are used to simulate and find the causes of plant behavior as seen in the field. The Shade Avoidance Syndrome has been mostly investigated in connection with the ratio of red to far red wavelengths as an indication of neighbor proximity and adaptive plasticity. According to the results obtained by Weing (2000), elongation responses to R:FR are more variable than previously realized and that the observed variability suggests competitive interactions in the natural environment. Other researchers have also stressed on the importance of plant development as influenced by light spectral quality and quantity (Rajcan et al. 2002; Wherley, Gardner & Metzger 2005); tillering dynamics in grasses in relation to R/FR (Evers et al. 2006; Monaco & Briske 2000) and also on the effects of canopy shade on morphological and phenological traits (Brainard, Bellinder & DiTommaso 2005). The effects of reduced irradiance and R/FR on the leaf development of papaya (*Carica papaya*) leaves to simulate canopy shade were studied by Buisson & Lee (1993) using experimental shadehouses; results indicate that although many morphological and anatomical characteristics were affected by reduction in irradiance, some were affected primarily by low R/FR. It is important to note that when vegetation shade is simulated by means of artificial filters in growth cabinets in which R/FR is low but PAR is sufficient to allow for sustained growth, phenological changes are exaggerated (Smith & Whiteham 1997).

In connection with these ideas, the results presented in this chapter were performed with simulated shades of different colors using cellophane paper to grow three of the more severe weeds for the sugarcane crop: *Cyperus rotundus* (purple nutsedge), *Rottboellia cochinchinensis* (ichgrass) and *Leptochloa filiformis*, in order to characterize their growth responses to different light qualities as an expression of the shade avoidance syndrome in these species.

The first species (*Cyperus rotundus*) have been studied mostly due to its susceptibility to canopy and artificial shading, which have been a basis, according to Neeser, Aguero & Swanton (1997) for integrated management under crop cultivation. For the experiment, the plants were grown for 48 days inside plant cabinets with red, blue and green artificial shade, low PFD neutral shade and full exposure to daylight. Results showed that *Cyperus rotundus* plants under low PFD neutral shade had lower values for the number of tillers and corms, dry root and stolon and leaf biomass, which resulted in a lower total dry biomass and leaf area (Lazo & Ascencio 2010). Differences in growth were found between this environment and full exposure to daylight and under the red and blue filters. It is important to emphasize the larger leaf area ratio found under low PFD neutral shade, as resulted from a lower number and corm dry biomass as compared to the aerial part. This species, a C4 plant, is thus highly sensitive to shade as shown by its lower total dry biomass and leaf area under low PFD; this could possibly explain the wide distribution of *Cyperus rotundus* in high light intensity environments, which generally occurs in tropical areas (Bielinski, Morales-paya & Shilling, 1997). However, it is also seen under canopy shade, due to the corms germinating potential and that plants under the canopy flower ("emergency flowering"), to produce seeds that are promptly shed, thus enriching the seed bank. These features may explain its persistence from planting to canopy closure, where sunflecks may play an important role in the maintenance of the plant population seen under cultivation.

Different shade avoidance strategies in biomass production, tillering, leaf area, plant height and flowering, revealing different capacity of acclimation to shade are shown when comparing *Rottboellia exaltata* and *Leptochloa filiformis* (both of the Poaceae family as well as sugarcane). The effect on leaf dry biomass density in *Rottboellia* was similar to full exposure to daylight, under a shade with low PAR and artificial color filters, but a red stimulator effect was observed. In *Leptochloa,* on the other hand, it was observed under blue and red filters, but the differences were not significant when comparing groups among them. The effects on the accumulation of dry biomass of roots, showed higher values for *Rottboellia* in full exposure to daylight and red; in contrast, a remarkable increase in root biomass was observed in *Leptochloa* in full exposure to daylight, which was significantly higher than under low photon flux density shade and blue, red and green filters (Ascencio & Lazo 2009). These results show a higher sensitivity of *Leptochloa* to shade, as a consequence of a lower development of the root system, which did not permit a sustained growth of the plants. This hypothesis is supported by the fact that this species showed "emergency flowering" under conditions of low PAR neutral shade and blue filter, but not total solar exposition and red filters, which are non shade conditions (high R/FR). In contrast, *Rottboellia exaltata* showed shade avoidance responses such as increased petiole length and stem, rendering it capable of competing for a longer time with the sugarcane crop under cultivation. Early flowering was not observed. The effects of shading on *Rottboellia exaltata* were determined under controlled environment conditions by Patterson (1979) under 100, 60, 25 and 2% sunlight; according to the results, this species maintains the capacity for high photosynthetic rates and high growth rates when subsequently exposed to high irradiance, after being shaded, which may explain its competitiveness with crop species.

Besides accelerated shoot growth, decreased tillering and early flowering, and increase in the shoot to root ratio have been part of the responses that have been associated to the SAS; however, this is not always the case as increases or decreases in S/R have been observed. The apparent contradiction is probably due to the complex nature of plant weight as a character. The ability of *Rottboellia* to reduce the effects of shade appears related to increased dry matter partitioning to the shoots. In Table 2, shoot to root ratios on a dry basis are shown for the three species grown exposed to full sunlight, neutral shed of low photon flux density (PAR) and shed plant cabinets in open space under full sunlight and covered with artificial filters made of red, blue and green cellophane paper. The values for S/R or dry matter partitioning are mainly related to biomass allocation to shoots in species capable of escaping shade as *Rottboellia*, however the highest value was found in full sunlight (2.27) but no significant differences were found for S/R with the rest of the light environments. In contrast, a remarkable increase in root biomass as compared to shoots, lowered the S/R in *Leptochloa* in full sunlight, while an increase was observed under shade conditions. The increase in S/A observed for this species and for *Cyperus* are more likely to be related to a lower root biomass allocation under shade conditions.

	Rottboellia exaltata	*Leptochloa filiformis*	*Cyperus rotundus*
	SA/SR	SA/SR	SA/SR
Exposed to Full Sunlight	2.27a	0.15b	0.74c
Neutral Shade (low PAR)	0.93a	0.54b	2.63a
Red cover	0.83a	0.83a,b	1.03c
Blue cover	1.01a	0.48b	0.98c
Green cover	1.58a	1.46a	1.82b

Values with the same letter, within the same column, do not differ according to Tukey HSD test; Homogenous Groups, alpha = .05000

Table 2. Shoot to root ratio (S/R) for *Rotboellia exaltata*, *Leptochloa filiformis* and *Cyperus rotundus* plants, grown: (1) exposed to full sunlight, (2) neutral shed of low PFD, and (3) shed, plant cabinets in open space under full sunlight and covered with red, blue and green cellophane paper.

The partitioning dry biomass to the roots and shoots for *Leptochloa* reflects a higher sensibility of this species to shade impairing the aerial parts to develop strategies to escape or avoid shade. This is shown in lower S/R for *Leptochloa* as compared to the other two species. The sensibility of *Cyperus rotundus* to shade is shown in the lower values of dry biomass for the subterranean parts (roots, corms and stolons) which produces higher values for S/R under low intensity neutral shade (2.63). This plant is totally intolerant to shade and under low intensity light, lower values for the number of tillers and corms, dry root, stolon and leaf biomass were observed (Lazo & Ascencio 2010). Because of the high plasticity of the S/R to environmental conditions, values are not easy to interpret and show less predictable responses to environmental variables such as light.

4. Conclusion

Sugarcane is an important crop in many tropical countries. Under these conditions, the major biotic limiting factor to productivity is the direct interference caused by weeds, especially during sprouting and the first three months of growth (initial stand establishment), when canopy closure has not been completed because crop growth is slow and foliage do not completely cover the area under cultivation. Weeds also serve as host plants of pests and diseases and interfere with crop management practices such as side dressing, sanitary inspection, sampling, maturation, etc., as well as with mechanical harvesting. Weed control is usually performed either by mechanical or the application of chemical herbicides, but the emergence and rapid evolution of weed species resistant to many herbicides, has prompted the search for new alternatives to control, and to consider management strategies based on the knowledge of the dynamics and biology of plants that grow under the canopy of the crop.

When sucrose is the desired sugarcane product, sucrose yield is the ultimate concern, but when the production of ethanol is the main purpose, the accumulation of biomass is the goal to achieve. However, in both cases, the output is determined by the number of stalks, which in turn depends on adequate and timely management of weeds and the good use of agronomic practices. It is important to emphasize that one of the ways to control weeds in this perennial crop, is precisely to take advantage of the intense shading under the sugarcane canopy, which limits the density and biodiversity of plant population grown underneath the crop. At canopy closure, the light environment underneath can exclude a considerable number of plant species, since not all species can tolerate shade, without ruling out the possibility that there are some species that are able to tolerate the shade, and others remain in the seed bank, and become a delayed problem when it is activated by some environmental factor, such as sunflecks. In this connection, it is important to recognize which of the species can tolerate shade and which cannot; therefore, research and studies of the dynamics of weed populations in the sugar cane crop, are required from planting to canopy closure. Light profiles are rarely recorded in field conditions and may be the key to understanding some of the growth responses of different plant species under a crop canopy.

Light quality (wavelength) and quantity (number of photons, irradiance) interact to control growth responses under vegetation canopies and some species underneath sugarcane canopies under field conditions can escape shade competing successfully with the crop for light. Two strategies associated to the Shade Avoidance Syndrome, as seen in the field, were observed: increased internodes length and decreased leaf size, while others species showed a higher sensitivity towards canopy shade and eventually died in this condition. Energy distributions of visible and near infrared radiation above, within, near the soil and above a canopy of weeds underneath a sugarcane canopy, showed that the decreased intensity of radiant energy was not uniform at all wavelengths. Some species are also seen under canopy shade, due to the corms or other subterranean organs and the germinating potential of the seed bank. These features may explain their persistence from time of planting to canopy closure, where sunflecks may play an important role in the maintenance of the plant population seen under crop cultivation. Different shade avoidance strategies in biomass production, tillering, leaf area, plant height and flowering, revealing different capacity of acclimation to shade were shown using experimental shadehouses; experiments under controlled conditions were useful to simulate and find the causes of plant behavior as seen in the field. Even though sugarcane is a crop economically important, research is limited in this area and in our opinion more has to be done on the biology and performance of the population of plants growing underneath the canopy from the beginning of the crop cycle, to improve weed control practices under cultivation.

5. Acknowledgements

We thank Fonacit government grant No. S1-2002000512 for financial support to this research and to Fernando Gil, M.Sc. (Fundacaña) and Jorge Ugarte, M.Sc. (UCV) for technical assistance in field experiments.

6. References

Ascencio, J. & Lazo, J.V. (2009). Respuestas de escape a la sombra en *Rottboellia exaltata* y *Leptochloa filiformis* (Gramineae-Poaceae). *Rev.Fac. Agron. (LUZ)* 26, (December 2009), pp.(490-507). ISSN 1690-9763.

Ascencio, J., Lazo, J.V. & Hernandez, E. (2005). Respuesta a la calidad y cantidad de sombra de *Cyperus rotundus. Revista Saber* 17 supp. (May 2005), pp. (196-198). ISSN 1315-0162.

Azevedo,R.A., Carvalho, R.F., Cia, M.C., Gratao,P.L. (2011). Sugarcane under pressure: An overview of biochemical and physiological studies of abiotic stress. *Tropical Plant Biol.*4,1,(January 2011),pp.(42-51),ISSN 19359756.

Ballare, C.L. & Casal,J.J. (2000). Light signals perceived by crops and weed plants. *Field Crops Research,*67,2,(July 2000),pp.(149-160),ISSN 0378-4290.

Ballaré,C.L., Scopel,A.L. & Sanchez, R.A. (1990). Far-red radiation reflected from adjacent leaves: an early signal of competition in plant canopies. *Science,* 247, 4949,(January 1990),pp.(329-332), ISSN 1095-9203.

Ballaré,C.L., Scopel,A.L. & Sanchez, R.A. (1991). Photocontrol of stem elongation in plant neighborhoods :effects of photon fluence rate under natural conditions of radiation. *Plant, Cell and Environment,*14,1, (January 1991),pp.(57-65). ISSN 1365-3040.

Bielinski, S., Morales-Payan, J.P., Stall, W.M.,Bewick, T.A. & Shilling, D.G. (1997). Effects of shading on the growth of nutsedges (*Cyperus spp.*). *Weed Science,*45, 6, (October 1997), pp. (670-673), ISSN 0043-1745

Brainard,D.C., Bellinder, R.R. & DiTommaso, A. (2005). Effects of canopy shade on the morphology, phenology, and seed characteristics of Powel amaranth (*Amaranthus powellii*). *Weed Science,*53,2, pp. (175-186), ISSN 0043-1745

Buisson, D. & Lee, D.W. (1993). The developmental responses of papaya leaves to simulated canopy shade. *American Journal of Botany*, 80, 8, (June 1993), pp. (947-952), ISSN 0002-9122.

Charles-Edwards,D.A. (1981). *The Mathematics of Photosynthesis and Productivity.* Academic Press,London,127 pp. ISBN 012-170580.

Evers,J.B., Vos,J., Andrieu,B & Struik,P.C. (2006). Cessation of tillereing in spring wheat in relation to light interception and red:far red ratio. *Annals of Botany* 97,4, (March 2006), pp.(649-658). ISSN 1095-8290.

Franklin, K.A. & Whitelam,G.C. (2005). Phytochromes and shade-avoidance responses in plants. *Annals of Botany* 96,2, (August 2005), pp.(169-175). ISSN 1095-8290.

Hatch,M.D. & Slack,C.R. (1966). Photosynthesis by sugarcane leaves. A new carboxylation reaction and the pathway of sugar formation. *Biochem.J.,*101,1,pp.(103-111).ISSN 0264-6021.

Hanlon,D., McMahon,G.C., McGuire,P., Beattie,R:N. & Sringer,J.K. (2000). Managing low sugar preices on farms-short and long term strategies. *Proc. Aust. Soc .Sugarcane Technol.*22,pp.(1-8), ISSN 0726-0822.

Holt,J.S. (1995). Plant responses to light: a potential tool for weed management. *Weed Science,*43,3, (September 1995), pp. (474-482), ISSN 0043-1745.

Lakshmanan,P., Geijskes,R.J., Aitken,K.S., Grof, C.L.P., Bonnett, G.D., & Smith, G.R. (2005). Sugarcane biotechnology: the challenges and opportunities. *In Vitro Cell Dev. Biol. Plant*,41,4,pp.(345-363),ISSN 1054-5476.

Lazo, J.V. & Ascencio, J. (2010). Efecto de diferentes calidades de luz sobre el crecimiento de *Cyperus rotundus*. *Bioagro* 22,2, (December 2010), pp.(153-158). ISSN 1316-3361.

Matsuoka,S., Ferro,J. & Arruda,P. (2009). The Brazilian experience of sugarcane ethanol industry. *In Vitro Cell Dev. Biol. Plant*,45,3,pp.(372-381),ISSN 1054-5476.

Monaco,T.A. & Briske, D.D. (2000). Does resource availability modulate shade avoidance responses to the ratio of red to far red irradiation? an assessment of radiation quantity and soil volume. *New Phytol.*,146,1(April 2000),pp.(37-46),ISSN 0028-646X

Monaco,T.A. & Briske, D.D. (2001).Contrasting shade avoidance responses in two perennial grasses: a field investigation in simulated sparse and dense canopies. *Plant Ecology*, 156,2 pp. (173-182),ISSN 1385-0237

Neeser,C., Aguero, R. & Swanton, C.J. (1997). Incident photosynthetically active radiation as a basis for integrated management of purple nutsedge (*Cyperus rotundus*). *Weed Science*,45, 6, (October 1997), pp. (777-783), ISSN 0043-1745

Patterson, D.T. (1979). The effects of shading on the growth and photo synthetic capacity of ichgrass (*Rottboellia exaltata*). *Weed Science*,27,5, (September 1979), pp. (549-553), ISSN 0043-1745

Pierik, R., Millenaar,F.F., Peeters,A.J.M: & Voesenek,L.A.C.J.(2005). New perspectives in flooding research: the use of shade avoidance and *Arabidopsis thaliana*. *Annals of Botany* 96,2, (August 2005), pp.(533-540). ISSN 1095-8290.

Rajcan,I., AghaAlikhani,M., Swanton, C.J. & Tollenaar,M. (2002). Development of redroot pigweed is influenced by light spectral quality and quantity. *Crop Science*, 42,6,pp. (1930-1936), ISSN 0011-183X

Schmitt,J. (1997). Is photomorphogenic sahde avoidance adaptive? Perspectives from population biology. *Plant, Cell and Environment*, 20,6, (June 1997),pp.(826-830). ISSN 1365-3040.

Smith, H. & Whitelam, G. (1997). The shade avoidance syndrome: multiple responses mediated by multiple phytochromes. *Plant, Cell and Environment*, 20, 6, (June 1997),pp.(840-844),ISSN 1365-3040.

Smith, H. (1982). Light quality, photoperception, and plant strategy. *Ann. Rev. Plant Physiol*, 33 (June 1982), pp.(481-518). ISSN 0066-4294.

Smith,H. Casal, J.J. & Jackson, G.M. (1990). Reflection signals and the perception by phytochrome of the proximity of neighboring vegetation. *Plant, Cell and Environment*, 13, 1,(January 1990), pp.(73-78). ISSN 1365-3040.

Tollenaar,M. & Dwyer, L.M. (1999). Physiology of Maize, In: *Crop Yield physiology and Processes*, Smith, D.L. & Hamel, C., pp. (169-204), Springer-Verlag, ISBN 3-540-64477-6, Berlin.

Weinig,C. (2000). Limits to adaptive plasticity: temperature and photoperiod influence shade-avoidance responses. *American Journal of Botany*, 87,11, (June 2000), pp. (1660-1668), ISSN 0002-9122

Wherley,B.G., Gardner,D.S. & Metzger, J.D. (2005). Tall fescue photomorphogenesis as influenced by changes in the spectral composition and light intensity. *Crop Science* 45,2,(January 2005),pp.(562-568). ISSN 0011-183X

10

Molecular Genetics of Glucosinolate Biosynthesis in *Brassicas*: Genetic Manipulation and Application Aspects

Arvind H. Hirani[1], Genyi Li[1,*], Carla D. Zelmer[1],
Peter B.E. McVetty[1], M. Asif[2] and Aakash Goyal[3]
[1]*Department of Plant Science, University of Manitoba, Winnipeg,*
[2]*Ex-Research Scientist, National Agriculture Research Centre, Islamabad,*
[3]*Lethbridge Research Centre, Agriculture and Agri-Food Canada, Lethbridge,*
[1,3]*Canada*
[2]*Pakistan*

1. Introduction

Glucosinolates are sulphur containing secondary metabolites biosynthesized by many plant species in the order *Brassicales*. Physical tissue or cell injury leads to the breakdown of glucosinolates through the hydrolytic action of the enzyme myrosinase, resulting in the production of compounds including isothiocynates, thiocyanates and nitriles. Derivative compounds of glucosinolates have a wide range of biological functions including anti-carcinogenic properties in humans, anti-nutritional effects of seed meal in animals, insect pest repellent and fungal disease suppression (Mithen et al., 2000; Brader et al., 2006). Glucosinolates play important role in the nutritional qualities of *Brassica* products. *Brassica* products are consumed as oil, meal and as vegetables. Rapeseed (*B. napus*, *B. juncea* and *B. rapa*) is a source of oil and has a protein-rich seed meal. High glucosinolates in the seed meal pose health risks to livestock (Fenwick et al., 1983; Griffiths et al., 1998). Consequently, plant breeders have nearly eliminated erucic acid from the seed oil and have dramatically reduced the level of seed glucosinolates (>100 μmole/g seed to <30 μmole/g seed) via conventional breeding, allowing the nutritious seed meal to be used as an animal feed supplement. There is, however, a significant residual content of glucosinolates in rapeseed/canola seed meal (over 10 μmole/g seed) and further reduction of the total glucosinolate content would be nutritionally beneficial (McVetty et al., 2009). Therefore, to produce healthy seed meal from rapeseed, it is important to genetically manipulate glucosinolate content. *Brassica* vegetables (*B. rapa* and *B. oleracea*) are highly regarded for their nutritional qualities, they are a good source of vitamin A and C, dietary soluble fibres, folic acid, essential micro nutrients and low in calories, fat and health beneficial glucosinolates such as glucoraphanin and sulforaphane. Breeding objectives for these Brassica crops include the enhancement of beneficial glucosinolates and reduction of others. It is, therefore, important to understand the genetic, biosynthetic, transportation and accumulation mechanisms for glucosinolates in *Brassica* species.

---

* Corresponding author

2. Historical background of *Brassica* species

The crops belonging to the genus *Brassica* have been of great importance to humanity. Since ancient times, *Brassica* crops have been used for many purposes, including vegetables, oilseeds, feed, condiments, fodder, green manure and even medical treatments. Early history suggests that rapeseed has been cultivated for several thousand years with its origin in the Mediterranean region although exact time of domestication and the place of origin are still unknown. Sanskrit writings in 2000-1500 BC characterized species identified as *B. rapa* and *B. napus* as oleiferous forms and mustards, respectively. *Brassica juncea* and *B. rapa* are believed to have been crop plants in India long before the Christian era. The Greek, Roman and Chinese literature of 500-200 BC referred *B. rapa* as rapiferous forms and were also described for various medicinal properties (Downey & Röbellen, 1989). In early times, rapeseed oil was used as a lamp oil, which in later centuries led gradually to its use as a valuable cooking oil.

Brassica species are diverse in terms of morphology, agronomy and quality traits. Domestication of rapeseed in Europe seems to have begun in the early Middle Ages. In 1620, *B. rapa* was first recorded in Europe by the Swiss botanist Casper Banhin (Gupta & Pratap, 2007). As a result, *Brassica* crops were adapted and cultivated in many parts of the world (Mehra, 1966). Rapeseed was introduced in Canada before the Second World War (McVetty et al., 2009). Commercial cultivation in Canada began during the Second World War to supply lubricating oil for steamships. Canada's first *B. rapa* rapeseed cultivar, Arlo, with high erucic acid (40 to 45%) and high glucosinolate content (>150 µmole/g seed) was developed in 1958 using selection from open pollinated populations (McVetty et al., 2009). Initially, *B. rapa* was the dominant cultivated species of *Brassica* in western Canada. In late 1980s, a large acreage of *B. rapa* and *B. napus* was grown in the Prairie Provinces. Subsequently, the production area of *B. rapa* declined to about 15 – 20% of its former area in 1990s. The reduction in acreage of *B. rapa* resulted from the introduction of herbicide tolerance canola, which provided the early planting and high yield advantages of *B. napus* cultivars. Currently, *B. rapa* is still grown in small areas in Canada because of its early maturity. Research efforts are underway to develop disease resistant hybrid varieties to increase yield potential of *B. rapa*. *Brassica rapa* are grown as a winter sarson crop in Asian countries such as India, Pakistan, China and Bangladesh. Vegetable forms of *B. rapa* (Chinese cabbage, turnip, pak choi, komatsuna, mizuna green and rapini) are widely cultivated in many parts of the world (Prakash & Hinata, 1980; Takuno et al., 2007).

3. Economic importance of *Brassica* species

The family *Brassicaceae* (syn. *Cruciferae*) is one of the crucial plant families for humans and animals and supplies several products from various plant parts. The little cruciferous weed *A. thaliana* has become an important model organism for the study of plant molecular biology, including the related crop species. The mustard family (*Brassicaceae*) is the fifth largest monophyletic angiosperm family, comprising 338 genera and about 3700 species in 25 tribes (Beilstein et al., 2006). The genus *Brassica* is one of the 51 genera of the tribe *Brassiceae* and includes the economically valuable crop species. *B. napus, B. rapa, B. juncea, B. carinata* and *B. nigra* are grown for edible and industrial oil as well as nutritionally valued seed meal.

Globally, rapeseed and canola oil is being utilized for human consumption, industrial applications and as a feedstock for biodiesel production. Canola oil is considered a healthy edible oil due to its high level of monounsaturated fatty acid (61%), lower level of saturated fatty acid (7%) and moderate amount of polyunsaturated fatty acid (22%) in its overall fatty acid profile (McVetty & Scarth, 2002). Rapeseed that has erucic acid levels greater than 45% also has many industrial applications such as plasticizers, slip agents for fibreglass and oil for the lubrication industry. Additionally, the seed meal is a marketable source of protein rich animal feed supplement.

Rapeseed is the world's third leading oil producing crop after palm and soybean, and it contributes about 15% to the global total vegetable oil production. Canada was the top rapeseed producing country in the world with 12.6 million MT productions in 2008 (FAO 2008). Canola/rapeseed contributes about $14 billion annually to the Canadian economy along with the generation of about 200,000 jobs throughout Canada in the areas of production, transportation, exporting, crushing and refining (Canola Council of Canada, 2010b http://www.canolacouncil.org/canadian_canola_industry.aspx). Canola/rapeseed meal is the second most popular protein feed ingredient in the world after soybean meal. Protein content of canola/rapeseed meal ranges from 36 to 39%, with a good amino acid profile for animal feeding (Newkirk et al., 2003). The major producers and consumers of canola/rapeseed meal are Australia, Canada, China, European Union and India. Along with oil production, *Brassica* species also produce different forms of vegetables and are the most widely cultivated vegetable crops in the world. Most of the production is consumed locally with a small amount of international trade. *B. napus* and *B. juncea* are used as vegetables in Asian countries like China, Japan and India. *B. rapa* is differentiated into seven groups viz., var. *compestris, pekinensis, chinensis, parachinesis, narinosa, japonica* and *rapa*. *Brassica rapa* is cultivated for leafy and root vegetables in the form of Chinese cabbage, pak choi and turnip; *B. oleracea* is cultivated for leafy and floret vegetables in various morph types such as cabbage, cauliflower, kale, collard, kohlrabi, brussels and broccoli.

4. Genomic relationships in *Brassica* species

Genomic relationships between the three diploid and three amphidiploid *Brassica* species were initially established in the 1930s based on various taxonomical and cytogenetic studies (Fig. 1) (Morinaga 1934; U 1935). Three allotetraploid *Brassica* species namely, *B. napus* (AACC, 2n=38), *B. juncea* (AABB, 2n=36) and *B. carinata* (BBCC, 2n=34) have been derived from three diploid elementary species, *B. rapa* (AA, 2n=20), *B. nigra* (BB, 2n=16) and *B. oleracea* (CC, 2n=18).

The genomic relationships of *B. napus* with *B. rapa* and *B. oleracea* have been confirmed by the resynthesis of *B. napus* from *B. rapa* x *B. oleracea* crosses (U 1935; Downey et al., 1975; Olsson & Ellerstrom, 1980). The close relationship between the six *Brassica* species made it feasible to incorporate a trait from one species into others to make the crops more suitable to agricultural systems. Thus, complex traits like glucosinolates can also be manipulated as required through interspecific hybridization. It has been relatively easier to make interspecific crosses among some of these six species (e.g. *B. napus* x *B. rapa*) compared to others (e.g. *B. rapa* x *B. oleracea*). Wide hybridizations are normally performed by the application of embryo rescue techniques. The most recent advances in genome sequencing technology, bioinformatics and data mining have opened an avenue for comparative analysis of ESTs, BACs, genes (families), whole chromosomes and even entire genomes to determine evolutionary relationship between these species and their ancestors (Gao et al., 2004; Gao et al., 2006; Punjabi et al., 2008; Mun et al., 2009; Qiu et al., 2009; Nagoaka et al., 2010).

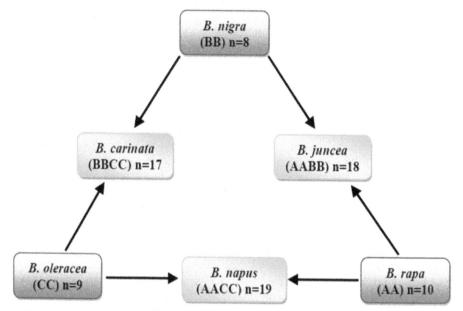

Fig. 1. U-triangle of genomic relationship between diploid and amphidiploid *Brassica* species (U 1935).

4.1 Homoeology between the A, B and C genomes of *Brassica* species

Genome homoeology has been characterized in *Brassica* species by comparative analyses of the genetic and physical maps of *Arabidopsis* with genetic maps of *Brassica* species (Osborn et al., 1997; Lan et al., 2000; Parkin et al., 2002; Lukens et al., 2003; Parkin et al., 2005). These studies indicate that each genomic region has had multiple events of polyploidization and chromosome rearrangements in the *Brassicaceae* lineage after the evolutionary divergence from *Arabidopsis* approximately 14.5 to 20.4 million years ago (MYA) (Yang et al., 1999; Parkin et al., 2002). In the *Brassicaceae*, the *B. nigra* (B) genome separated from the *B. rapa/B. oleracea* (AC) genome lineages about 7.9 MYA (Yang et al., 1999; Lysak et al., 2005).

There are high levels of homoeology among the A- and C-genomes of *B. rapa*, *B. oleracea* and *B. napus* (Parkin et al., 2005; Punjabi et al., 2008). Parkin et al., (2003) reported stretches of collinearity on the linkage groups N1 with N11, N2 with N12 and N3 with N13 of the A- and C-genomes, respectively. Similarly, Osborn et al., (2003) reported reciprocal interstitial translocations of homoeologous regions of linkage groups N7 and N16, and their effects on genome rearrangements and seed yield in *B. napus*. This suggests that inter-genomic translocations and rearrangements have taken place during the evolutionary divergence of *B. oleracea* and *B. rapa* from a polyploid ancestor (Sharpe et al., 1995). As a result of genomic synteny, there have been several reports of homoeologous recombination between the A- and C-genome of *Brassica* species (Udall et al., 2004; Leflon et al., 2006). Cytogenetic and molecular data revealed that small and large collinear genomic regions between the A- and C-genomes of *Brassica* species allow homoeologous recombination-based trait introgression to enhance genetic variability.

5. Plant secondary metabolites and their functions

The sessile nature of plants requires them to produce a large numbers of defence compounds including primary and secondary metabolites. It is believed that the currently discovered plant metabolic compounds account for only about 10% of the actual compounds present naturally within the plant kingdom (Schwab, 2003; Wink, 2003). Plant secondary metabolites are organic biochemical compounds produced in plants during normal growth and development. While they are not directly involved in plant growth, development or reproduction, these secondary metabolites play vital roles in plant defence mechanisms, acting for example, phytoalexins and phytoanticipins. Phytoalexins are antimicrobial defence metabolites synthesized *de novo* in response to biotic and abiotic stresses. Phytoalexins are involved in induced plant defence mechanisms including lytic enzymes, oxidizing agents, cell wall lignifications and pathogenesis-related proteins and transcript stimulation (Pedras et al., 2008). Phytoanticipins are low molecular weight antimicrobial compounds which are constitutively active for defence. Their production may be increased under high biotic or abiotic stresses (Pedras et al., 2007). Certain classes of phytoanticipins require enzymatic modification and derivation in order to become active within the defence systems of the plant. Plant secondary metabolites are broadly categorized into three groups based on their biosynthetic origin

i.   Flavonoids and allied phenolic and polyphenolic compounds
ii. Terpenoid compounds
iii. Nitrogen and sulphur containing alkaloid compounds

5.1 Glucosinolates as secondary metabolites

Glucosinolates are sulphur rich, nitrogen containing anionic natural products, derived from specific amino acids and their precursors (Fenwick et al., 1983). Glucosinolates are reported almost exclusively from the order *Brassicales*, which possesses about 15 families such as *Brassica*ceae, *Capparaceae* and *Caricaceae*. Glucosinolates are also reported in a few members of the family *Euphorbiaceae*, a very distinct family to other glucosinolate containing families (Rodman et al., 1996). Glucosinolates coexist with endogenous thioglucosidases called myrosinases in cruciferous plant species and activate plant defence mechanism against biotic and abiotic stresses. Tissue disruption causes systemic interactions between glucosinolates and myrosinases in the presence of moisture. The interaction produces numerous compounds with diverse biological activities (Bones & Rossiter, 1996; Halkier, 1999). Glucosinolates are some of the most extensively studied plant secondary metabolites; various enzymes and transcription factors involved in biosynthesis have been studied in the model plant *Arabidopsis* and to some extent in *Brassica* crops species. The broad functionality, physiochemical and genetic studies of glucosinolates have led to a model status for research on secondary metabolites (Sønderby et al., 2010).

GSL Name	Trivial name	RF	R Side chain	Chemical Structure	Mol. Wt.[a]
Aliphatic 3C	Sinigrin	1.00	2-Propenyl	$CH_2=CH-CH_2-$	279
	Glucoibervirin	0.80	3-Methylthiopropyl	$CH_3S-CH_2-CH_2-CH_2-$	327
	Glucoiberin	1.07	3-Methylsulfinylpropyl	$CH_3SO-CH_2-CH_2-CH_2-$	343
	Glucocheirolin	1.26	3-Methylsulfonylpropyl	$CH_3SO_2-CH_2-CH_2-CH_2-$	179.26
	Glucoputranjivin	1.00	1-Methylethyl	$CH_3-CH-CH_3$	281
	Glucosisymbrin	1.32	2-hydroxy-1-methylethyl	$OH-CH_2-CH-CH_3$	298
Aliphatic 4C	Gluconapin	1.11	3-Butenyl	$CH_2=CH-CH_2-CH_2-$	293
	Progoitrin	1.09	(2R)-2-Hydroxy-3-butenyl	$CH_2=CH-CH(OH)-CH_2-$	309
	Epi-progoitrin	1.09	(2S)-2-Hydroxy-3-butenyl	$CH_2=CH-CH(OH)-CH_2-$	309
	Glucoerucin	1.04	4-Methylthiobutyl	$CH_3S-CH_2-CH_2-CH_2-CH_2-$	341
	Glucoraphasatin	0.40	4-Methylthio-3-butenyl	$CH_3S-CH=CH-CH_2-CH_2-$	340
	Glucoraphanin	1.07	4-Methylsulfinylbutyl	$CH_3SO-CH_2-(CH_2)_2-CH_2-$	357
	Glucoraphenin	0.90	4-Methylsulfinyl-3-butenyl	$CH_3SO-CH=CH-CH_2-CH_2-$	355
	Glucoconringiin	1.00	2-Hydroxy-2-methylpropyl	$CH_2=CH(OH)-CH_2$	312
Aliphatic 5C	Glucoalyssin	1.07	5-Methylsulfinylpentyl	$CH_3SO-CH_2-CH_2-CH_2-CH_2-CH_2-$	371
	Glucobrassicanapin	1.15	Pent-4-enyl	$CH_2=CH-CH_2-CH_2-CH_2-$	307
	Glucoberteroin	1.05	5-Methylthiopentyl	$CH_3S-CH_2-CH_2-CH_2-CH_2-CH_2-$	354
	Gluconapoleiferin	1.00	2-Hydroxy-pent-4-enyl	$CH_2=CH-CH_2-CH-OH-CH_2-$	323
	Glucocleomin	1.07	2-Hydroxy-2-methylbutyl	$CH_3-CH_2-CH(OH)-CH_2-CH_2-$	326
Aliphatic 6C	Glucolesquerellin	1.00	6-Methylthiohexyl	$CH_3S-CH_2-CH_2-CH_2-CH_2-(CH_2)_2-$	370
	Glucohesperin	1.00	6-Methylsulfinylhexyl	$CH_3SO-CH_2-CH_2-CH_2-CH_2-(CH_2)_2-$	385
Aliphatic 7C	Glucoarabishirsutain	1.00	7-Methylthioheptyl	$CH_3S-CH_2-CH_2-CH_2-CH_2-(CH_2)_3-$	384
	Glucoibarin	1.00	7-Methylsulfinylheptyl	$CH_3SO-CH_2-CH_2-CH_2-CH_2-(CH_2)_3-$	399
Aliphatic 8C	Glucoarabishirsuin	1.10	8-Methylthiooctyl	$CH_3S-CH_2-CH_2-CH_2-CH_2-(CH_2)_4-$	398
	Glucohirsutin	1.10	8-Methylsulfinyloctyl	$CH_3SO-CH_2-CH_2-CH_2-CH_2-(CH_2)_4-$	414

GSL Name	Trivial name	RF	R Side chain	Chemical Structure	Mol. Wt.[a]
Indole	Glucobrassicin	0.29	3-Indolylmethyl		368
	4-Hydroxyglucobrassicin	0.28	4-Hydroxy-3-indolylmethyl		384
	4-Methoxyglucobrassicin	0.25	4-Methoxy-3-indolylmethyl		398
	Neoglucobrassicin	0.20	N-Methoxy-3-indolylmethyl		398
Aromatic	Glucotropaeolin	0.95	Benzyl		329
	Glucosinalbin	0.50	p-Hydroxybenzyl		345
	Gluconasturtiin	0.95	2-Phenethyl		343
	Glucobarbarin	1.09	(2S)-2-Hydroxy-2-phenethyl		360
	Glucomalcomiin	0.40	3-Benzoyloxypropyl		402

Trivial name and chemical formula of R side-chains of glucosinolates identified in *Brassica* species, Mol. Wt.*: molecular weight of desulfoglucosinolates, RF: response factor (Haughn et al., 1991; Griffiths et al., 2000; Brown et al., 2003).

Table 1. Chemical structures of glucosinolates in Brassica species.

5.1.1 Glucosinolates and their biological functions in agriculture and nature

Glucosinolates are a uniform group of thioglucosides with an identical core structure called β-D-glucopyranose bound to a (Z)-N-hydroximinosulfate ester by a sulphur atom with a variable R group. Approximately 120 glucosinolates differing in their R group side chains have been identified (Halkier & Gershenzon, 2006). These glucosinolates are categorized into three classes based on their precursor amino acids and side chain modifications (Table 1). Kliebenstein et al., (2001a) suggested that these three classes of glucosinolates are independently biosynthesized and regulated by different sets of gene families from separate amino acids. Each class is briefly discussed below.

5.1.1.1 Aliphatic glucosinolates

Aliphatic glucosinolates are the major group of glucosinolates in *Brassica* species, contributing about 90% of the total glucosinolate content of the plant. Glucosinolates are constitutively biosynthesized *de novo* in cruciferous plants, although their degradation is highly regulated by spatial and temporal separation of glucosinolates and myrosinases within the plant based on environmental and biotic stresses (Drozdowska et al., 1992). Hydrolysis of glucosinolates produces a large number of biologically active compounds that have a variety of functions. The most common hydrolysis products of aliphatic glucosinolates in many cruciferous species are isothiocyanates that are formed by the rearrangement of aglycone with carbon oxime adjacent to the nitrogen at neutral pH while at acidic pH, nitriles are the predominant products (Fahey et al., 2001). These unstable compounds are cyclised to a class of substances responsible for goiter in animals (Griffiths et al., 1998).

By contrast, sulforaphane is one of the derivatives of glucoraphanin, an aliphatic glucosinolate that has several beneficial properties for humans and animals. It is known as an inducer of phase II enzymes such as glutathione transferases and quinone reductases of the xenobiotic pathway in human prostate cells (Zhang et al., 1992; Faulkner et al., 1998). The phase II enzymes are involved in the detoxification of electrophilic carcinogens that can lead to mutations in DNA and cause different types of cancers (Mithen et al., 2000). Enhanced consumption of cruciferous vegetables appears to reduce the risk of cancers (Nestle, 1997; Talalay 2000; Brooks et al., 2001). The sulforaphane content of these vegetables could be a leading factor in the reduction. Another less documented health benefit of sulforaphane is the inhibition of *Helicobacter pylori*, a pathogen of peptic ulcers and gastric cancer (Fahey et al., 2002). Sulforaphane also protects human retinal cells against severe oxidative stresses (Gao et al., 2001).

Isothiocyanates and other breakdown products of glucosinolates play important roles as repellents of certain insects and pests (Rask et al., 2000; Agrawal & Kurashige, 2003; Barth & Jander, 2006; Benderoth et al., 2006). Leaves of the mutant *myb28myb29* in *Arabidopsis* with low aliphatic glucosinolate content, when fed to the lepidopteran insect *Mamestra brassicae*, enhanced larval weight by 2.6 fold (Beekwilder et al., 2008). Glucosinolates may have specific repellent or anti-nutritional effects on specific classes of insects and microorganisms. Some *in vitro* studies demonstrated that glucosinolate degradation products, isothiocyanates and nitriles, inhibited fungal and bacterial pathogen growth (Brader et al., 2001; Tierens et al., 2001). In *Arabidopsis*, over expression of *CYP79D2* from cassava increased accumulation of isopropyl and methylpropyl aliphatic glucosinolates and transformed plants showed enhanced resistance against a bacterial soft-rot disease (Brader et al., 2006). Birch et al., (1992) reported that biotic stresses such as pest damage in *Brassica* species alters glucosinolate profiles in roots, stems, leaves and flowers. This suggests that a phytoanticipin property of glucosinolates is involved in the plant defence mechanisms of *Brassica*. Glucosinolates and their breakdown products have many biological functions, with a few compounds acting as biopesticides, biofungicides and soil fumigants, while others play roles in attraction of pollinators and provide oviposition cues to certain insects. The attraction of specialized insects could be due to the glucosinolate-sequestering phenomenon of some insects including harlequin bugs, sawflies, and some homoptera including aphids (Bridges et al., 2002; Mewis et al., 2002).

5.1.1.2 Indole glucosinolates

Indole (heterocyclic) glucosinolates in cruciferous plants (including *Arabidopsis*) are derived from tryptophan and possess variable R group side chains. The relatively high content of indole glucosinolates in the model plant *Arabidopsis* has enhanced our knowledge of the biosynthesis, transportation and functional properties of this class of glucosinolates (Petersen et al., 2002; Brown et al., 2003). Side chain modification in indole glucosinolates occurs through hydroxylations and methoxylations catalysed by several enzymes. Indole glucosinolate types and contents in different organs of the plant are strongly affected by environmental conditions. Four main indole glucosinolates have been identified in most cultivated *Brassica* species: glucobrassicin, neoglucobrassicin, 4-methoxyglucobrassicin and 4-hydroxyglucobrassicin. Similar to aliphatic glucosinolates, breakdown products of indole glucosinolates have multiple biological functions. Indole-3-carbinol derived from glucobrassicin has potent anticarcinogenic activity (Hrncirik et al., 2001). The indole

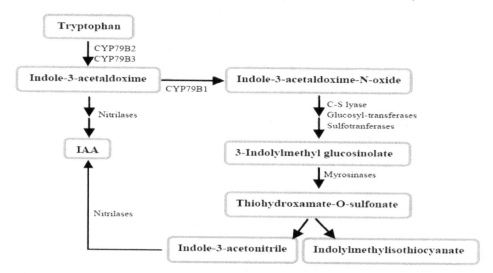

In this pathway, IAA produces from precursors and derivatives of 3-indolylmethyl glucosinolate by various nitrilases.

Fig. 2. Biosynthetic pathway and breakdown products of indole glucosinolates (De Vos et al., 2008).

glucosinolate derived compound 4-methoxyglucobrassicin has strong insect deterrent activity (Kim & Jander, 2007; De Vos et al., 2008). Osbourn, (1996) reported antimicrobial activities of indole glucosinolates and their breakdown products in *Brassica* species. Several studies suggest that there is a metabolic association between indole glucosinolates and the plant hormone indole-3-acetic acid (IAA). In the consecutive reactions, indole glucosinolates are degraded into indole acetonitrile (IAN), which is then hydrolyzed by nitrilases into IAA (Fig. 2). In clubroot infected *Brassica* roots, indole glucosinolate-based induction of IAA was observed to be responsible for gall formation. The IAA production from indole glucosinolates during gall formation is associated with a signalling cascade of IAA and cytokinin complex (Ugajin et al., 2003). Structural similarity data indicates that the indole alkaloid, brassinin, and possibly other cruciferous phytoalexins are derived from glucobrassicin. Studies in rapeseed, mustard and *Arabidopsis* have suggested that methyl jasmonate and wounding induce the biosynthesis of particular indole glucosinolates (Bodnaryk 1992; Brader et al., 2001).

5.1.1.3 Aromatic glucosinolates

The third class of glucosinolates in cruciferous species is aromatic or benzylic glucosinolates, derived from the aromatic parental amino acids phenylalanine and tyrosine. Very limited information is available regarding aromatic glucosinolates at qualitative or quantitative levels. Aromatic glucosinolates are biosynthesized independently from other glucosinolates, which is apparently due to involvement of different amino acid precursors in the biosynthesis of the different classes of glucosinolates (Kliebenstein et al., 2001a). Cloning and functional characterization of the *CYP79A* gene of *Arabidopsis* suggests that cytochrome P450-dependent monooxygenase catalyzes the reaction from phenylalanine to phenylacetaldoxime in aromatic glucosinolate biosynthesis (Wittstock & Halkier, 2000). Five aromatic glucosinolates have been identified in *Brassicaceae*: glucotropaeolin, glucosinalbin, gluconasturtiin, glucobarbarin and glucomalcomiin. The distinctive aroma and spiciness of condiment *Brassica* plant parts, such as the leaves and seeds of white (*Sinapis alba*) and black (*B. nigra*) mustards, is due to the presence of these aromatic glucosinolates (Fenwich et al., 1983).

5.1.2 Biosynthesis of aliphatic glucosinolates

Aliphatic glucosinolates are the most abundant class in *Brassica* species, therefore, the genetic of biosynthesis is described in more detail. Aliphatic glucosinolates are biosynthesized from five amino acids (methionine, alanine, leucine, isoleucine and valine) (Halkier & Gershenzon, 2006). Biosynthesis of aliphatic glucosinolates occurs in three stages at two different locations. The first chain elongation step is catalyzed by BCAT4 in the cytosol (Schuster et al., 2006), whereas development of core structures and secondary side chain modification reactions take place in the chloroplasts (Textor et al., 2007; Sawada et al., 2009). Chain elongation steps produce propyls (3C), butyls (4C), pentyls (5C), hexyls (6C), heptyls (7C) and octyls (8C) aliphatic glucosinolates in cruciferous species including *Arabidopsis*. Glucosinolate side chain modification reactions involve oxygenation, hydroxylation, alkenylation and benzoylation, which are controlled by several gene families. The pattern of glucosinolate biosynthesis varies from organ to organ within the plant; young leaves, buds, flowers and silique walls all have higher rates of glucosinolate biosynthesis than roots, old leaves and presumably seeds (Brown et al., 2003). Various studies also suggest that transportation of glucosinolates and their breakdown products from organ to organ via phloem occurs upon requirement to protect the plant. Seeds, however, are the most important store of total glucosinolates produced by the plants (Brudnell et al., 1999). Seeds contain much higher glucosinolates concentrations than other plant parts and it is thought that leaf glucosinolates are the basis for accumulations of total glucosinolates in seeds (Klienbestein et al., 2001a). This suggests that long distance transportation of glucosinolates from source to sink occurs. A few reports discuss an independent pathway of glucosinolate biosynthesis in seeds, resulting in the high concentration of glucosinolate in seeds (Du & Halkier, 1998; Osbourn, 1996). Experimental evidence, however, is not strong enough to support a separate pathway at this time.

5.1.2.1 Parental amino acid biosynthesis and condensation

Methionine is the main precursor of aliphatic glucosinolates in *Brassica* species. The enzyme BCAT4 catalyzes the initial chain elongation reaction to produce 2-oxo acid from methionine, an analogous process to the formation of the branched chain amino acid valine

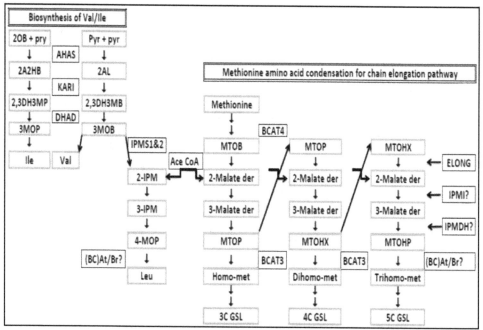

All the reactions are catalyzed by *BCATs, ELONGs, IPMIs* and *IPM-DHs* gene families for 3C, 4C and 5C glucosinolates. Genes shown in gray boxes and derivative products shown in blue boxes. BCAT-branched chain amino transferase, MTOB- 4-methylthio-2-oxobutanoate, MTOP- 6 methylthio-2-oxopentanoate, MTOHX- 4-methylthio-2-oxohexanoate, IPMI- isopropylmalate isomerases, IPMDH-isopropylmalate dehydrogenases, MOB- methyl-2-oxobutanoate, MOP- methyl-2-oxopentanoate, AHAS- acetohydroxyacid synthase, KARI- ketolacid reductoisomerase, DHAD- dihydroxyacid dehydratase, 2AL- 2-acetolactate, 2A2HB- 2-aceto-2-hydroxybutyrate, 2OB- 2-oxobutyrate, 2,3DH3MB-2,3-dihydroxy-3-methylbutyrate, 2,3DH3MP- 2,3-dihydroxy-3-methylpentanoate.

Fig. 3. Methionine amino acid condensation pathway regulated by several gene families (Kroymann et al., 2001; Sawada et al., 2009).

to its chain-elongated homolog leucine (Fig. 3). In *Arabidopsis*, a *bcat4* mutant showed about a 50% reduction in total aliphatic glucosinolates and at the same time increased the level of free methionine and S-methyl-methionine (Schuster et al., 2006). This suggests that the *BCAT4* gene produces an enzyme which is involved in the first deamination reaction. Subsequently, three consecutive reactions of transformations occur. The first is a transamination and condensation reaction with acetyl-CoA catalyzed by *GSL-ELONG* in *Brassica* species (Li & Quiros, 2002). This is homologous to *MAM1* in *Arabidopsis* (Campos de Quirose et al., 2000; Benderoth et al., 2006; Textor et al., 2007). The same reaction occurs for 3C aliphatic glucosinolates which is controlled by isopropylmalate synthase (*IPMS1*, *IPMS2*). Isopropylmalate synthase is homologous to *MAM1* in *Arabidopsis* (Kliebenstein et al., 2001b; Field et al., 2004) and to *GSL-PRO* in *Brassica* species (Li et al., 2003; Gao et al., 2006). The second isomerisation reaction is controlled by isopropylmalate isomerises (*IPMI*) and third reaction is oxidative decarboxylation controlled by isopropylmalate dehydrogenases (*IPM-DH*) (Fig. 3) (Wentzell et al., 2007; Sawada et al., 2009).

These three consecutive reactions produce elongated 2-oxo acids with one or more methylene groups. These compounds are either transaminated by the BCAT enzyme to yield homo-methionine, which can enter into the core glucosinolate skeleton structure formation, or proceed through another round of chain elongation (Fig. 3). Overall, the methionine amino acid condensation pathway produces a range of methionine derivatives such as homo-methionine, dihomo-methionine, and trihomo-methionine, which proceed to the next biosynthesis step called glucosinolate core skeleton formation (Fig. 3).

5.1.2.2 Glucosinolate core skeleton formation

Glucosinolate core skeleton structure formation has been well characterized in *Arabidopsis*, with at least 13 enzymes and five different biochemical reactions, i.e., oxidation, oxidation with conjugation, C-S cleavage, glucosylation and sulfation (Grubb & Abel, 2006; Halkier & Gershenzon, 2006) involved in the formation. The precursors are catalyzed into aldoxime by cytochromes belonging to the *CYP79* gene family (Fig. 4). At least seven *CYP79s* were identified and functionally characterized in *Arabidopsis*. The *CYP79F1* gene converts all short chain methionine derivatives, whereas *CYP79F2* gene is involved in conversions of the long chain methionine derivatives. Similarly, *CYP79B2* and *CYP79B3* catalyze tryptophan derivatives, and *CYP79A2* catalyzes phenylalanine substrates (Fig. 4) (Zang et al., 2008). Subsequently, aldoximes are oxidized into either nitrile oxides or aci-nitro compounds by *CYP83A1* for methionine derivates and *CYP83B1* for tryptophan as well as phenylalanine derivates. This proceeds to a non-enzymatic conjugation to produce S-alkly-thiohydroximates. In this sulphur rich chemical pathway, the next step is C-S cleavage by C-S lyase from S-alkly-thiohydroximate to thiohydroximic acid; C-S lyase forms an enzymatic complex with an S-donating enzyme. The *c-s lyase* mutant of *Arabidopsis* showed complete lack of aliphatic and aromatic glucosinolates in *Arabidopsis*, suggesting that this single gene family has a crucial role in skeleton processing (Mikkelsen et al., 2004).

In the glucosylation step, desulfoglucosinolate is formed by a member of the *UGT74* family. The final reaction of core skeleton formation is accomplished with sulfation of desulfoglucosinolates to produce intact glucosinolates by sulfotransferases *AtST5a*, *AtST5b* and *AtST5C* in *Arabidopsis*. Biochemical characterization of sulfotransferases in *Arabidopsis* revealed that *AtST5a* favour to sulfate phenylalanine and tryptophan derived desulfoglucosinolates, whereas *AtST5b* and *AtST5c* favour to sulfate long chain aliphatic glucosinolates (Piotrowski et al., 2004). In a comparative analysis study between *Arabidopsis* and *B. rapa*, at least 12 paralogs of sulfotransferases were known to be responsible for this reaction (Zang et al., 2008). In glucosinolate skeleton formation reactions, the first four biosynthesis reactions take place in the chloroplast and the last reaction of sulfation occurs in the cytosol. This suggests that shuttle transporters play important roles in the entire biosynthesis process (Klein et al., 2006).

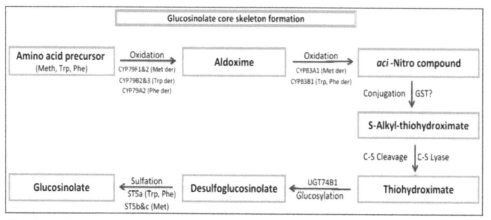

ST- sulfotransferase, UGT- glucuronosyltransferases, GST- glutathione S-transferase.

Fig. 4. Glucosinolate core skeleton structure formation by cytochromes. Methionine amino acid precursors produce aliphatic, tryptophan produces indole and phenylalanine produces aromatic glucosinolate core structures (Grubb & Abel, 2006; Halkier & Gershenzon, 2006).

5.1.2.3 Side chain modification in aliphatic glucosinolates

After glucosinolate core skeleton structure formation, the core skeletons are subjected to a set of reactions known as side chain modification or secondary transformation. Side chain modifications of glucosinolates are the last crucial enzymatic reactions on intact glucosinolates before their transport to sinks or biological degradation by myrosinases occur. Hydrolysis products of individual glucosinolates are recognized based on side chain variation in R groups. A hydrolysis product of glucoraphanin has anticancer properties. The R group modifications of glucoraphanin change their chemical properties, therefore, hydrolysis products have anticarcinogenic functions. Hydrolysis products of progoitrin, however, have anti-nutritional effect in animals, which reduce the palatability of rapeseed meal.

Side chain modification begins with the oxidation of sulphur in the methylthio precursor to produce methylsulfinyl and then methylsulfonyl moieties (Fig. 5). In *Arabidopsis*, this reaction is catalyzed by the flavin monooxygenases, *GSL-FMO$_{OX1-5}$* located within the *GSL-OX1* locus on chromosome I. Phylogenetic analysis revealed a main group of *GSL-FMOs* for cruciferous species, which is further categorized according to subspecies, indicating that functional diversity of S-oxygenation of glucosinolates exists (Hansen et al., 2007; Li et al., 2008). Knockout mutant and over expression studies suggested that *GSL-FMO$_{OX1-4}$* catalyzes the 4-methylthiobutyl to 4-methylsulfinyl reaction and GSL-FMO_{OX5} is involved in the S-oxygenation of long chain glucosinolates in *Arabidopsis* (Li et al., 2008). In *Brassica* vegetables, products of *GSL-FMOs* catalyses are the sources of anticancer compounds from aliphatic glucosinolates. It will be beneficial to identify these genes/loci in *Brassica* species so that they might be further used to manipulate aliphatic glucosinolates towards favourable forms.

A second round of binary side chain modification changes methylsulfinyl to alkenyl- and to hydroxyl- aliphatic glucosinolates (Fig. 5). In *Arabidopsis* these reactions are controlled by a *GSL-ALK/GSL-OHP* locus that has three tandem repeats (*GSL-AOP1*, *GSL-AOP2* and *GSL-AOP3*), which encode 2-oxoglutarate-dependent dioxygenases located on chromosome IV. Functional characterization indicates that *GSL-AOP2* catalyzes the reaction to alkenyl, whereas *GSL-AOP3* controls the reaction toward hydroxyalkenyl. The function of *GSL-AOP1*, however, is not clear in *Arabidopsis*, it might be involved in both reactions (Fig. 5) (Hall et al., 2001; Kliebenstein et al., 2001c; Mithen et al., 1995). The *GSL-ALK* and *GSL-OHP* are either closely linked on the same genomic region or allelic variants of a single genetic locus though they may show variable functions. In *Arabidopsis*, *GSL-OHP* catalyzes the reaction only for 3C aliphatic glucosinolate branches, whereas *GSL-ALK* is involved in 3C, 4C and 5C aliphatic glucosinolate branches. There is no clear functional information available for long chain (6C and so on) aliphatic glucosinolate branches and presumably

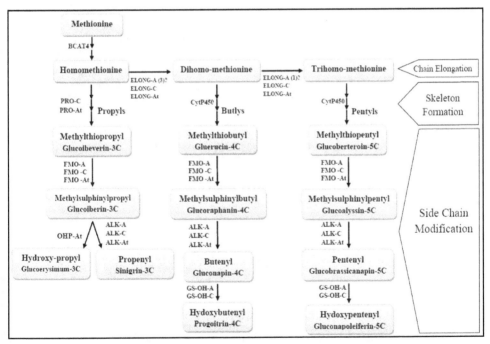

Fig. 5. Glucosinolate core structure and side chain modification pathway for 3C, 4C and 5C aliphatic glucosinolates. In the biosynthesis steps, gene symbols ending with A indicate A genome, C for C genome and At for *A. thaliana* (Magrath et al., 1994; Mithen et al., 1995; Li & Quiros, 2003; Mahmood et al., 2003).

GSL-ALK accomplishes these reactions in *Arabidopsis* (Kliebenstein et al., 2001c; Parkin et al., 1994). In *B. oleracea, GSL-ALK* was inferred by positional cloning and biochemical analysis. The functional allele in collard and the non-functional allele (with 2 bp deletion creating a frame-shift mutation) in broccoli were confirmed. A locus or loci of *GSL-ALK* is also believed to have a role in the catalysis of methylsulfinyl to alkenyl glucosinolates (Li & Quiros, 2003). Hydroxylation changes alkenyl to hydroxy aliphatic glucosinolate (in butyls, pentyls, hexyls and so on) biosynthesis branches in *Arabidopsis* and *Brassica* species; these sets of reactions are controlled by *GSL-OH* dependant on the presence of both *GSL-AOP2* and *GSL-ELONG*. In *Brassica*, the final product of this reaction in 4C glucosinolate biosynthesis is progoitrin and its hydrolytic derivative, oxazolidine-2-thione which causes goiter in animals. These compounds are major obstacles to the use of *Brassica* crops as animal feed (Fenwick et al., 1983).

5.1.3 Diversity of glucosinolates in *Brassicaceae*

Glucosinolates are united by their unique basic skeleton (β-D-glucopyranose) but glucosinolates are diverse in their origins, side chain modifications, degradations and final biological functions. In addition to structural diversity, a diversity of glucosinolates is seen between families, genera, species, subspecies and different accessions of subspecies. This diversity provides insight into glucosinolate biosynthesis at the genomic, physiological, biochemical and host-pathogen interaction levels. The natural variation of glucosinolate profiles between species or different cultivars of same species permits the investigation of the effects of QTL or genes and gene interactions. This can be utilized for advanced breeding applications like MAS, trait introgression and gene pyramiding for beneficial glucosinolates. In *Arabidopsis*, naturally occurring variations in glucosinolates were identified and quantified for 34 types of glucosinolates in the leaves of 39 ecotypes (Hogge et al., 1988; Reichelt et al., 2002). Similarly, different morphotypes of *B. rapa* possess eight different glucosinolates with gluconapin and glucobrassicanapin as predominant aliphatic glucosinolates (He et al., 2000). Padilla et al., (2007) reported 16 different glucosinolates among 116 accessions of turnip greens.

The wide range of variation in glucosinolate profiles provides the opportunity to study individual glucosinolates for their potent biological activities *in planta*. Within different forms of *B. oleracea*, 12 different glucosinolates have been detected. The beneficial glucosinolate glucoraphanin showed significant variation ranging from 44 to 274 μmole/g seed in different genotypes of broccoli (Mithen et al., 2000; Rangkadilok et al., 2002). Furthermore, variation in concentration of individual glucosinolates also exists in cultivars of the same species.

6. Low glucosinolate rapeseed and canola

Early forms of domesticated rapeseed and their cultivars possessed a high concentration of glucosinolates (100 to 180 μmole/g) in their oil-free seed meal. The presence of glucosinolates in rapeseed had hindered the use of rapeseed meal in livestock industries due to anti-nutritional effects of its hydrolysis products in animals. As a result, in the 1970s, plant breeders searched germplasm collections for low glucosinolate contents. A Polish spring rape cultivar, Bronowski, with low glucosinolate content was discovered by The Agriculture Canada Research Station in Saskatoon (Kondra & Stefansson, 1970). This sole genetic source of the low glucosinolate trait has been used to develop all the low glucosinolate cultivars in *B. napus* and *B. rapa* worldwide through conventional plant breeding. *B. napus* and *B. rapa* cultivars with low content of erucic acid and glucosinolate were developed, which ushered in a new era for *Brassica* crop production and its consumption. The world's first double low (low erucic acid and low glucosinolate content) *B. napus* and *B. rapa* cultivars, Tower and Candle, respectively, were developed by pedigree selection in the progenies of interspecific crosses in 1970s (Stefansson & Downey, 1995; McVetty et al., 2009). In Canada, this new type of oilseed rape was designated "Canola". The term "Canola" applies to any rapeseed cultivars with erucic acid content of <2% and glucosinolates content of <30 µmol/g in oil-free seed meal. The Canola term is a registered trademark of the Canadian Canola Association. The name is derived from **Can**adian **O**il **Low** **Ac**id (Canola Council of Canada, 2010a, http://www.canola-council.org/ canola_the_official_definition.aspx). Currently, most rapeseed (high erucic acid) and canola cultivars have glucosinolate levels <15 µmole/g in oil-free seed meal. The development of low erucic acid and low glucosinolate cultivars has also been undertaken for other *Brassica* rapeseed species (e.g. *B. juncea*) and in other parts of the world for the quality improvement of their oils and seed meals.

7. Quantitative trait loci for glucosinolates in major *Brassica* species

Glucosinolate biosynthesis in *Brassica* crops has quantitative inheritance, which is regulated by complex genetic factors and affected by environmental factors. Glucosinolates are functionally diverse and well recognized plant secondary metabolites; so they have been extensively studied in terms of QTL mapping, biosynthesis gene cloning and functional characterization in *Arabidopsis* (Kliebenstein et al., 2001a; Kliebenstein et al., 2001c; Compos de Quiros et al., 2000; Brown et al., 2003; Benderoth et al., 2006; Textor et al., 2007; Li et al., 2008). However, very limited genetic, biochemical and metabolomic information is available on glucosinolate biosynthesis, transport and final product utilization in *Brassica* crops including *B. rapa*. There has been a few QTL mapping studies reported for major *Brassica* crop seed glucosinolates. Uzunova et al., (1995) mapped four QTL for total seed glucosinolate content in a *B. napus* DH population, which accounted 61% total phenotypic variance. Similarly, Toroser et al., (1995), based on a RFLP linkage map, identified two larger and three small effect QTL for total aliphatic glucosinolate content using a DH population in *B. napus*. These QTL explained 70% of the total phenotypic variance. This suggests that several loci with additive or epistatic effect are involved in total seed glucosinolate biosynthesis in different genetic backgrounds. Howell et al., (2003) reported QTL mapping for total seed glucosinolates analyzed by X-ray fluorescence (XRF) and near-infrared reflection spectroscopy (NIRS) in two inter-varietal *B. napus* backcross populations. They identified four QTLs accounting for 76% of the phenotypic variance in the Victor x Tapidor population. These three QTL accounted for 86% of phenotypic variance in this second population. These studies, however, were limited to either total seed glucosinolates or 3C, 4C and 5C aliphatic glucosinolates, and did not infer the genetic loci for individual aliphatic, indole or aromatic glucosinolates. Furthermore, there were no reports of publicly available molecular markers for marker assisted selection of glucosinolates. Such markers, if developed, could be used in breeding to manipulate glucosinolate profiles and contents in *Brassica* crop species.

In another amphidiploid species, *B. juncea,* several studies were conducted for QTL mapping of seed glucosinolates. Cheung et al., (1998) detected two QTL for 2-propenyl and three QTL for 3-butenyl glucosinolates which explained between 89% and 81% of total phenotypic variance. This QTL mapping study was carried out in a DH population derived from the F_1 of two *B. juncea* parental lines, J90-4317 (low glucosinolates) and J90-2733 (high glucosinolates). Mahmood et al., (2003) reported three QTL for 2-propenyl glucosinolate content which explained 78% of the phenotypic variance, while five QTL for total seed aliphatic glucosinolates explained phenotypic variance between 30% and 45%. In this study a DH population and an RFLP linkage map was used. Similarly, Ramchiary et al., (2007) reported six QTL for seed glucosinolate content in the F_1DH and advanced backcross DH (BC$_4$DH) of *B. juncea.* Some of the large effect QTL in advanced backcross (BC$_4$DH) of *B. juncea* were fine mapped using a candidate gene approach and comparative sequence analyses of *Arabidopsis* and *B. oleracea* (Bisht et al., 2009). The results suggested that epistasis and additive effects of glucosinolate genes in different genetic backgrounds in *B. juncea* exist. This study, however, could not explain the homoeologous effects of genes/loci from the A- and B-genomes on the individual or total seed glucosinolate content.

In *B. oleracea, BoGSL-ELONG* a side chain elongation gene was cloned based on the *Arabidopsis* sequence information, and functionally characterized using an RNA interference (RNAi) approach. The results suggested that *BoGSL-ELONG* is involved in 4C and 5C aliphatic glucosinolate biosynthesis in *Brassica* species. The RNAi lines displayed an increased level of propyl glucosinolates suggesting that the precursor homo-methionine concentration enhances the activity of 3C aliphatic glucosinolate biosynthesis in *B. napus* (Li & Quiros, 2002, Liu et al., 2010). A natural mutation in *BoGSL-ELONG* resulting in the failure of excision of the third intron and thus producing a long cDNA fragment has been identified in a white cauliflower genotype (*B. oleracea*) lacking 4C and 5C aliphatic glucosinolates (Li & Quiros, 2002). A molecular marker for this mutation would be useful in *Brassica* breeding programs for modification of glucosinolate profiles. Additionally, a gene *BoGSL-PRO* which control propyl glucosinolate biosynthesis in *B. oleracea* was sequenced using comparative analysis of the *MAM* (*methylthioalkylmalate synthase*) gene family in *Arabidopsis* (Li et al., 2003; Gao et al., 2006).

A glucosinolate side chain modification gene, *BoGSL-ALK,* was cloned using a positional cloning approach based on a closely linked SRAP marker in *B. oleracea* (Li & Quiros, 2003). Functional characterization of *BoGSL-ALK* by overexpression in *Arabidopsis* and RNA interference (RNAi) in *B. napus* suggests that *BoGSL-ALK* is involved in catalyzation of either sulfinylbutyl to butenyl or hydoxybutenyl with high functional redundancy (Li & Quiros, 2003, Liu et al., 2012). Interestingly, a natural frame shift mutation of 2 bp deletions was identified in broccoli, which accumulates sulfinylbutyl glucosinolate by ceasing downstream biosynthesis of other 4C aliphatic glucosinolates.

In *B. rapa,* a single QTL mapping study for leaf glucosinolates has been reported, although it is one of the widely distributed *Brassica* species for oil and vegetable production. Lou et al., (2008) identified six QTL for leaf total aliphatic glucosinolate content, three QTL for total leaf indole glucosinolate conten and three QTL for leaf aromatic glucosinolates in two DH populations of B. *rapa* using an AFLP and SSR based linkage map. There was no information regarding QTL for seed glucosinolates. Glucosinolate content varies greatly between leaves and seeds (Brown et al., 2003). As well, there is variation in the expression patterns of the genetic loci underpinning glucosinolate production in leaves and seeds (Kliebenstein et al., 2001b).

8. Glucosinolate identification and quantification approaches

Early analysis of glucosinolates began with detection of glucosinolates and possible hydrolysis products by paper and thin-layer chromatography. The paper chromatography was applied in combination with high voltage electrophoresis, but it had many complications and low yield (Greer, 1962). Danielak & Borkowski, (1969) analyzed glucosinolates from seeds of 150 different cruciferous species using thin-layer chromatography. Since then, numerous techniques have been employed for quantification of total glucosinolate content with various modifications including steam distillation and titration of isothiocyanates, ELISA, sulfate-release assay, UV spectroscopy and gas chromatography of isothiocyanates. Near infrared reflectance spectroscopy (NIRS) is one of the widely used techniques for seed total glucosinolates quantification, which detects N–H, C–H and O–H groups of total glucosinolates. NIRS is a preferred technique because it can simultaneously quantify oil and protein along with total glucosinolates in canola/rapeseed (Velasco & Becker, 1998). Individual intact glucosinolates can be determined using techniques such as reverse phase HPLC-MS, thermospray LC with tandem MS in the two most common interfaces (ESI or APCI), capillary GC-MS and GC-MS-MS.

Desulfoglucosinolates usually are analyzed by reverse phase HPLC or by X-ray fluorescence spectroscopy (XRF). The reverse-phase HPLC analytical approach has been widely used for quantification of individual intact or desulfo- glucosinolates. The technique was developed in 1984 with UV based detection of either intact glucosinolates or an on-column enzymatic desulfation from plant extracts (Spinks et al., 1984). The photodiode array (PDA) with UV detector can distinguish spectra of aliphatic from indole and aromatic glucosinolates; the indole and aromatic glucosinolates spectra end with a shoulder. This widely applicable method for glucosinolate separation is yet subject to difficulties in interpretation of results because of differences in the time and enzymatic activity for the desulfation reaction, pH effects and mobile phase solvents with an appropriate gradient. Desulfoglucosinolates also have been analyzed by the determination of the sulfur content of the seeds using X-ray fluorescence spectroscopy (XRF) (Schnug & Haneklaus, 1990). The hydrolysis products of glucosinolates, isothiocyanates, nitriles, thiocyanates and benzenedithiol, have been analyzed using techniques including GC or GC-MS and HPLC with or without fluorescent labelling (Kiddle et al., 2001).

9. Molecular markers and their applications for glucosinolates

Molecular markers are efficient, reliable, time saving and cost effective tools that may enhance the capacity of conventional breeding for improvement in agronomy, quality and yield related traits of crop species without adverse effects. Morphological traits such as petal color, leaf shape *etc* were used as markers in classical breeding, where significant time and effort was required to refine crosses. There have been many practical difficulties with the use of morphological markers, including:

i.   a paucity of suitable markers and associations with agriculturally important traits (Ranade et al., 2001),
ii. undesirable pleiotropic effects of many morphological markers on plant phenotypes (Ranade et al., 2001),
iii. high linkage drag (Ranade et al., 2001), and
iv. trait of interest easily can be lost in a breeding cycle if there is no strong linkage between marker and traits (Ranade et al., 2001).

Advancements in molecular biology tools and techniques have overcome some of the difficulties of classical breeding. Different types of DNA molecular markers (hybridization based e.g. RFLP; PCR based e.g. SSR, RAPD, SCAR, and SRAP) have been used for gene/QTL mapping, cloning, genetic map construction and marker assisted selection in plant breeding. Most recently, the conversion of various molecular markers (RFLP, RAPD, SRAP, AFLP, SSR, SNP etc.) to simple PCR based SCAR markers for marker assisted selection has overcome the difficulties of other markers. It is feasible and cost effective to use SCAR markers for marker assisted selection of populations.

Marker assisted selection in plant breeding is well supported by the availability of molecular maps developed using various marker systems in different mapping populations. The use of molecular markers has facilitated introgression of important traits through intra or interspecific as well as inter-generic crosses. Similar to agronomic, disease resistance and yield related traits, seed quality traits such as glucosinolates can be genetically manipulated using interspecific hybridization followed by marker assisted selection for introgression or replacement of a native gene with the allied gene. Natural mutations for glucosinolate biosynthesis genes have been identified in accessions of *B. oleracea* (Li & Quiros, 2002; 2003) and molecular markers have been developed. These molecular markers have been employed for the manipulation of glucosinolate profiles in *Brassica* through interspecific hybridization and marker assisted selection. In our QTL mapping study in *B. rapa* RIL mapping population, we identified single major QTL for 5C aliphatic glucosinolates (glucobrassicanapin, glucoalyssin and gluconapoleiferin) on chromosome A3 and gene specific SCAR molecular markers were developed and utilized that markers for marker assisted selection in other *Brassica* interspecific crosses (unpublished). Hasan et al., (2008) reported linkage of SSR markers to candidate genetic loci of glucosinolate biosynthesis genes in *Brassica napus* through structure-based allele-trait association studies, and found potential application of these markers in marker assisted selection for glucosinolates.

On the other hands, Niu (2008) attempted to replace the functional *GSL-ALK* gene of *B. rapa* by the null allele from *B. oleracea* (broccoli) using a gene specific SCAR marker. However, introgression of the *GSL-ALK* null allele or replacement of a single locus with small effect did not change the glucosinolate profile of the *B. rapa* in this study. This suggests that multiple loci with functional redundancy play important roles in glucosinolate biosynthesis in *Brassica* species. This approach has met with very little or no success. This might be due to many reasons, such as:

i.   duplicated or triplicated genomic regions may mask the effect of the single locus being replaced for a quantitative traits like glucosinolate profile and concentration

ii. lack of similarity of gene and spacer sequences between alien and host chromosomes in monosomic or disomic alien chromosome addition lines

iii. presence of active homoeologous recombination regulator genes during meiosis

iv. directional exchange of genetic materials in trivalent formations during meiosis because of distinct chromosome behaviour

v.   host genome chromosome numbers and amount of homology between host and alien chromosomes

Several traits in *Brassica* species have been improved through introgression of functional genes from allied species through interspecific or inter-generic crosses such as *B. rapa* x *B. oleracea* and *B. rapa* x *B. oxyrrhina* (Srinivasan et al., 1998). In near future, development of molecular markers using sequenced genome information of *B. rapa* and *Arabidopsis* will hasten marker assisted selection of glucosinolates to increase beneficial glucosinolates such as glucoraphanin and glucoerucin in *Brassica* vegetables and to reduce total glucosinolates in rapeseed meal.

10. Acknowledgments

The authors are grateful to the Genome Canada/Genome Alberta/Genome Prairie and Manitoba Provincial Government for financial support. The authors also extend their thanks to Dr. Habibur Rahman, Department of Agricultural, Food & Nutritional Science, University of Alberta, Canada and Dr. Anne Worley, Department of Biological Sciences, University of Manitoba, Canada for reviewing this chapter.

11. References

Agrawal, A. A. & Kurashige, N. S. (2003) A role for isothiocyanates in plant resistance against the specialist herbivore *Pieris rapae*. *J Chem Ecol.* 29: 1403-1415.

Barth, C. & Jander, G. (2006) *Arabidopsis* myrosinases TGG1 and TGG2 have redundant function in glucosinolate breakdown and insect defense. *Plant J.* 46: 549-562.

Beekwilder, J., Leeuwen, W., Dam, N. M., Bertossi, M., Grandi,V., Mizzi, L., Soloviev, M., Szabados, L., Molthoff, J. W., Schipper, B., Verbocht, H., Vos, R. C. H., Morandini, P., Aarts, M. G. M. & Bovy, A. (2008) The impact of the absence of aliphatic glucosinolates on insect herbivory in *Arabidopsis*. *PLoS One.* 3(4): e2068.

Beilstein, M. A., Al-Shehbaz, I. A. & Kellogg, E. A. (2006) *Brassicaceae* phylogeny and trichome evolution. *Am J Bot.* 93: 607-619.

Benderoth, M., Textor, S., Windsor, A. J., Mitchell-Olds, T., Gershenzon, J. & Kroymann, J. (2006) Positive selection driving diversification in plant secondary metabolism. *Proc Natl Acad Sci USA* 103(24): 9118-9123.

Birch, A. N. E., Griffiths, D. W., Hopkins, R. J., Smith, W. H. M. & McKinlay, R. G. (1992) Glucosinolate responses of swede, kale, forage and oilseed rape to root damage by turnip root fly (*Delia floralis*) larvae. *J Sci Food Agric.* 60: 1-9.

Bisht, N. C., Gupta, V., Ramchiary, N., Sodhi, Y. S., Mukhopadhyay, A., Arumugam, N., Pental, D. & Pradhan, A. K. (2009) Fine mapping of loci involved with glucosinolate biosynthesis in oilseed mustard (*Brassica juncea*) using genomic information from allied species. *Theor Appl Genet.* 118: 413-421.

Bodnaryk, R. P. (1992) Effects of wounding on glucosinolates in the cotyledons of oilseed rape and mustard. *Phytochem.* 31: 2671-2677.

Bones, A. M. & Rossiter, J. T. (1996) The myrosinase-glucosinolate system, its organization and biochemistry. *Physiol Plant.* 97: 194-208.

Brader, G., Mikkelsen, M. D., Halkier, B. A. & Palva, E. T. (2006) Altering glucosinolate profiles modulates disease resistance in plants. *Plant J.* 46: 758-767.

Bridges, M., Jines, A. M. E., Bones, A. M., Hodgson, C., Cole, R., Bartlet, E., Wallsgrove, R., Karapapa, V. K., Watts, N. & Rossiter, J. T. (2002) Spatial organization of the glucosinolate-myrosinase system in *Brassica* specialist aphids is similar to that of the host plant. *Proc Biol Sci.* 269(1487): 187-191.

Brooks, J., Paton, V. & Vidanes, G. (2001) Potent induction of phase 2 enzymes in human prostate cells by sulforaphane. *Cancer Epidemiol Biomarkers Prev.* 10: 949-954.

Brown, P. D., Tokuhisa, J. G., Reichelt, M. & Gershenzon, J. (2003) Variation of glucosinolate accumulation among different organs and developmental stages of *Arabidopsis thaliana*. *Phytochem.* 62: 471-481.

Brudenell, A. J. P., Griffiths, H., Rossiter, J.T. & Baker, D. A. (1999) The phloem mobility of glucosinolates. *J Exp Botany.* 50(335): 745-756.

Cheung, W. Y., Landry, B. S., Raney, P. & Rakow, G. F. W. (1998) Molecular mapping of seed quality traits in *Brassica juncea* L. Czern., and Coss. *Acta Hort.* 459: 139-147.

Compos de Quiros, H., Magrath, R., McCallum, D., Kroymann, J., Schnabelrauch, D., Mitchell-Olds, T. & Mithen, R. (2000) α-Keto acid elongation and glucosinolate biosynthesis in *Arabidopsis thaliana*. *Theor Appl Genet.* 101: 429-437.

Danielak, R. & Borkowski, B. (1969) Biologically active compounds in seeds of crucifers Part III. Chromatographical search for glucosinolates. *Dissert in Pharm and Pharmacol.* 21: 563-575.

De Vos, M., Kriksunov, K. L. & Jander, G. (2008) Indole-3-Acetonitrile production from indole glucosinolates deters oviposition by *Pieris rapae*. *Plant Physiol* 146: 916-926.

Downey, R. K. & Röbellen, G. (1989) *Brassica* species. In G. Röbellen, R. K. Downey and A. Ashri, (eds) Oil crops of the world, McGraw Hill Publishing Company, New York, USA, pp. 339-362.

Downey, R. K., Stringam, G. R., McGregor, D. I. & Stefansson, B. R. (1975) Breeding rapeseed and mustard crops. In: J.T. Harapiak (eds) Oilseed and pulse crops in western Canada. Western Cooperative Fertilizers Ltd., Calgary, Alberta, Canada, pp 157-183.

Drozdowska, L., Thangstad, O. P., Beisvaag, T., Evjen, K., Bones, A. & Iversen, T. H. (1992) Myrosinase and myrosin cell development during embryogenesis and seed maturation. *Israel J. B.* 41: 213-223.

Du, L. & Halkier, B. A. (1998) Biosynthesis of glucosinolates in the developing silique walls and seeds of *Sinapis alba*. *Phytochem.* 48(7): 1145-1150.

Fahey, J. W., Haristoy, X., Dolan, P., Kensler, T., Scholtus, I., Stephenson, K., Talalay, P. & Lozniewski, A. (2002) Sulforaphane inhibits extracellular, intracellular, and antibiotic-resistant strains of *Helicobacter pylori* and prevents benzo[α]pyrene-induced stomach ulcers. *Proc Natl Acad Sci USA*. 99: 7610-7615.

Fahey, J. W., Zalcmann, A. T. & Talalay, P. (2001) The chemical diversity and distribution of glucosinolates and isothiocyanates among plants. *Phytochem*. 56: 5-51.

Faulkner, K., Mithen, R. & Williamson, G. (1998) Selective increase of the potential anticarcinogen 4-methylsulphinylbutyl glucosinolate in broccoli. *Carcinogenesis*. 19: 605-609.

Fenwick, G. R., Heaney, R. K. & Mullin, W. J. (1983) Glucosinolates and their breakdown products in food and food plants. *Crit Rev Food Sci Nutr*. 18: 123-201.

Field, B., Cardon, G., Traka, M., Botterman, J., Vancammeyt, G. & Mithen, R. (2004) Glucosinolate and amino acid biosynthesis in *Arabidopsis*. *Plant Physiol*. 135: 828-839.

Gao, M., Li, G., Potter, D., McCombie, W. R. & Quiros, C. F. (2006) Comparative analysis of *methylthioalkylmalate synthase* (*MAM*) gene family and flanking DNA sequences in *Brassica oleracea* and *Arabidopsis thaliana*. *Plant Cell Rep*. 25: 592-598.

Gao, M., Li, G., Yang, B., McCombie, W.R. & Quiros, C.F. (2004) Comparative analysis of a *Brassica* BAC clones containing several major aliphatic glucosinolate genes with its corresponding *Arabidopsis* sequence. *Genome*. 47: 666-679.

Gao, X., Dinkova-Kostova, A. & Talalay, P. (2001) Powerful and prolonged protection of human retinal pigment epithelial cells, keratinocytes, and mouse leukemia cells against oxidative damage: the indirect antioxidant effects of sulforaphane. *Proc Natl Acad Sci USA*. 98: 15221-15226.

Greer, M. A. (1962) The isolation and identification of progoitrin from *Brassica* seed. *Arch of Biochem and Biophy*. 99: 369-371.

Griffiths, D. W., Bain, H., Deighton, N., Botting, N.P. & Robertson, A. A. B. (2000) Evaluation of liquid chromatography-atmospheric pressure chemical ionisation-mass spectrometry for the identification and quantification of desulphoglucosinolates. *Phytochem Anal*. 11: 216-225.

Griffiths, D. W., Birch, A. N. E. & Hillman, J. R. (1998) Anti-nutritional compounds in the *Brassicaceae* – analysis, biosynthesis, chemistry and dietary effects. *J Hort Sci Biotech*. 73: 1-18.

Grubb, C. D. & Abel, S. (2006) Glucosinolate metabolism and its control. *Trends Plant Sci*. 11(2): 89-100.

Gupta, S. K. & Pratap, A. (2007) History, origin and evolution. In: Gupta SK (eds) Advances in botanical research, rapeseed breeding. Academic Press, Elsevier, San Diego, CA, USA, pp 2-17.

Halkier, B. A. (1999) Glucosinolates. In: Ikan R (eds) Naturally occurring glycosides: chemistry, distribution and biological properties. John Wiley and Sons Ltd, London, UK, pp. 193-223.

Halkier, B. A. & Gershenzon, J. (2006) Biology and biochemistry of glucosinolates. *Annu Rev Plant Biol*. 57: 303-333.

Hall, C., McCallum, D., Prescott, A. & Mithen, R. (2001) Biochemical genetics of glucosinolate modification in *Arabidopsis* and *Brassica*. *Theor Appl Genet*. 102: 369-374.

Hansen, B.G., Kliebenstein, D. J. & Halkier, B. A. (2007) Identification of a *flavin-monooxygenase* as the S-oxygenating enzyme in aliphatic glucosinolate biosynthesis in *Arabidopsis*. *Plant J*. 50: 902-910.

Hasan, M., Friedt, W., Pons-Kuhnemann, J., Freitag, N. M., Link, K. & Snowdon, R. J. (2008) Association of gene-linked SSR markers to seed glucosinolate content in oilseed rape (*Brassica napus* spp. napus). *Theor Appl Genet*. 116(8): 1035-1049.

Haughn, G. W., Davin, L., Giblin, M. & Underhill, E. W. (1991) Biochemical genetics of plant secondary metabolites in *Arabidopsis thaliana*. *Plant Physiol*. 97: 217-226.

He, H., Fingerling, G. & Schnitzler, W. H. (2000) Glucosinolate contents and patterns in different organs of Chinese cabbages, Chinese kale (*Brassica alboglabra* bailey) and Choy sum (*Brassica campestris* L. spp chinensis Var. Utilis tsen et lee). *J App Bot*.74: 21-25.

Hogge, L. R., Reed, D. W., Underhill, E. W. & Haughn, G. W. (1988) HPLC separation of glucosinolates from leaves and seeds of *Arabidopsis thaliana* and their identification using thermospray liquid chromatography-mass spectrometry. *J Chromatgr Sci*. 26: 551-556.

Howell, P. M., Sharpe, A. G. & Lydiate, D. J. (2003) Homoeologous loci control the accumulation of seed glucosinolates in oilseed rape (*Brassica napus*). *Genome*. 46: 454-460.

Hrncirik, K., Valusek, J. & Velisek, J. (2001) Investigation of ascorbigen as a breakdown product of glucobrassicin autolysis in *Brassica* vegetables. *Eur Food Res Technol*. 212: 576-581.

Kiddle, G., Bennett, R. N., Botting, N. P., Davidson, N. E., Robertson, A. A. B. & Wallsgrove, R. M. (2001) High-performance liquid chromatographic separation of natural and synthetic desulfoglucosinolates and their chemical validation by UV, NMR and chemical ionisatioin-MS methods. *Phytochem Anal*. 12: 226-242.

Kim, J. H. & Jander, G. (2007) Myzus persicae (green peach aphid) feeding on *Arabidopsis* induces the formation of a deterrent indole glucosinolates. *Plant J*. 49: 1008-1019.

Klein, M., Reichelt, M., Gershenzon, J. & Papenbrock, J. (2006) The three desulfoglucosinolate sulfotransferase proteins in *Arabidopsis* have different substrate specificities and are differentially expressed. *FEBS J*. 273(1): 122-136.

Kliebenstein, D. J., Gershenzon, J. & Mitchell-Olds, T. (2001a) Comparative quantitative trait loci mapping of aliphatic, indole and benzylic glucosinolate production in *Arabidopsis thaliana* leaves and seeds. *Genetics*. 159: 359-370.

Kliebenstein, D. J., Kroymann, J., Brown, P., Figuth, A., Pedersen, D., Gershenzon, J. & Mitchell-Olds, T. (2001b) Genetic control of natural variation in *Arabidopsis* glucosinolate accumulation. *Plant Physiol*. 126: 811-825.

Kliebenstein, D. J., Lambrix, V. M., Reichelt, M., Gershenzon, J. & Mitchell-Olds, T. (2001c) Gene duplication in the diversification of secondary metabolism: tandem 2-oxoglutarate-dependent dioxygenases control glucosinolate biosynthesis in *Arabidopsis*. *Plant Cell*. 13: 681-693.

Kondra, Z. P. & Stefannson, B. R. (1970) Inheritance of major glucosinolates in rapeseed (*Brassica napus*) meal. *Can J Plant Sci.* 50: 643-647.

Kroymann, J., Textor, S., Tokuhisa, J.G., Falk, K. L., Bartram, S., Gershenzon, J. & Mitchell-Olds, T. (2001) A gene controlling variation in *Arabidopsis* glucosinolate composition is part of the methionine chain elongation pathway. *Plant Physiol.* 127: 1077-1088.

Lan, T. H., DelMonte, T. A., Reischmann, K. P., Hyman, J., Kowalski, S. P., McFerson, J., Kresovich, S. & Paterson, A. H. (2000) An EST-enriched comparative map of *Brassica oleracea* and *Arabidopsis thaliana*. *Genome Res.* 10: 776-788.

Leflon, M., Eber, F., Letanneur, J. C., Chelysheva, L., Coriton, O., Huteau, V., Ryder, C. D., Barker, G., Jenezewski, E. & Chevre, A. M. (2006) Pairing and recombination at meiosis of *Brassica rapa* (AA) X *Brassica napus* (AACC) hybrids. *Theor Appl Genet.* 113: 1467-1480.

Li, G., Gao, M., Yang, B. & Quiros, C. F. (2003) Gene for gene alignment between the *Brassica* and *Arabidopsis* genomes by direct transcriptome mapping. *Theor Appl Genet.* 107: 168-180.

Li, G. & Quiros, C. F. (2002) Genetic analysis, expression and molecular characterization of *Bo-GSL-ELONG*, a major gene involved in the aliphatic glucosinolate pathway of *Brassica* species. *Genetics.* 162: 1937-1943

Li, G. & Quiros, C. F. (2003) In planta side-chain glucosinolate modification in *Arabidopsis* by introduction of dioxygenase *Brassica* homolog *BoGSL-ALK*. *Theor Appl Genet.* 106: 1116-1121.

Li, J., Hansen, B. G., Ober, J. A., Kliebenstein, D. J. & Halkier, B. A. (2008) Subclade of flavin-monooxygenases involved in aliphatic glucosinolate biosynthesis. *Plant Physiol.* 148: 1721-1733.

Liu, Z., Hammerlindl, J., Keller, W., McVetty, P. B. E., Daayf, F., Quiros, C. F. & Li, G. (2010) *MAM* gene silencing leads to the induction of C3 and reduction of C4 and C5 side-chain aliphatic glucosinolates in *Brassica napus*. *Mol Breed.* 27(4): 467-478.

Liu, Z., Hirani, A. H., McVetty, P. B. E., Daayf, F., Quiros, C. F. & Li, G. (2012) Reducing progoitrin and enriching glucoraphanin in *B. napus* seeds through silencing of the *GSL-ALK* gene family. *Plant Mol Biol.* DOI 10.1007/s11103-012-9905-2.

Lou, P., Zhao, J., He, H., Hanhart, C., Del Carpio, D. P., Verkerk, R., Custers, J., Koornneef, M. & Bonnema, G. (2008) Quantitative trait loci for glucosinolate accumulation in *Brassica rapa* leaves. *New Phytol.* 179: 1017-1032.

Lukens, L., Zou, F., Lydiate, D., Parkin, I. A. P. & Osborn, T. (2003) Comparison of a *Brassica oleracea* genetic map with the genome of *Arabidopsis thaliana*. *Genetics.* 164: 359-372.

Lysak, M. A., Koch, M., Pecinka, A. & Schubert, I. (2005) Chromosome triplication found across the tribe *Brassiceae*. *Genome Res.* 15: 516-525.

Magrath, R., Bano, F., Morgner, M., Parkin, I., Sharpe, A., Lister, C., Dean, C., Turner, J., Lydiate, D. & Mithen, R. (1994) Genetics of aliphatic glucosinolates. I. Side chain elongation in *Brassica napus* and *Arabidopsis thaliana*. *Heridity.* 72: 290-299.

Mahmood, T., Ekuere, U., Yeh, F., Good, A. G. & Stringam, G. R. (2003) Molecular mapping of seed aliphatic glucosinolates in *Brassica juncea*. *Genome.* 46: 753-760.

McVetty, P. B. E., Fernando, D., Li, G., Tahir, M. & Zelmer, C. (2009) High-erucic acid, and low-glucosinolate rapeseed (HEAR) cultivar development in Canada. In: Hou CT, Shaw JF (eds) Biocatalysis and agricultural biotechnology. CRC, Boca Raton, FL, USA pp 43-61.

McVetty, P. B. E. & Scarth, R. (2002) Breeding for improved oil quality in *Brassica* oilseed species. *J Crop Prod.* 5: 345-369.

Mehra, K. L. (1966) History and ethnobotany of mustard in India. *Advancing Frontiers of Plant Sciences.* 19: 51-59.

Mewis, I. Z., Ulrich, C. & Schnitzler, W. H. (2002) The role of glucosinolates and their hydrolysis products in oviposition and host-plant finding by cabbage webworm, *Hellula undalis. Entomol Exp Appl.* 105: 129-139.

Mikkelsen, M. D., Naur, P. & Halkier, B. A. (2004) *Arabidopsis* mutants in the C-S lyase of glucosinolate biosynthesis establish a critical role for indole-3-acetaldoxime in auxin homeostasis. *Plant J.* 37: 770-777.

Mithen, R. F., Clarke, J., Lister, C. & Dean, C. (1995) Genetics of aliphatic glucosinolates. III. Side chain structure of aliphatic glucosinolates in *Arabidopsis thaliana. Heredity.* 74: 210-215.

Mithen, R. F., Dekker, M., Verkerk, R., Rabot, S. & Johnson, I. T. (2000) The nutritional significance, biosynthesis and bioavailability of glucosinolates in human foods. *J Sci Food Agri.* 80: 967-984.

Morinaga, T. (1934) Interspecific hybridization in *Brassica.* VI. Then cytology of F_1 hybrid of *B. juncea* and B. nigra. *Cytologia.* 6: 62-67.

Mun, J. H., Kwon, S. J., Yang, T. J., Seol, Y. J., Jin, M., Kim, J. A., Lim, M. H., Kim, J. S., Baek, S., Choi, B. S., Yu, H. J., Kim, D. S., Kim, N., Lim, K. B., Lee, S. I., Hahn, J. H., Lim, Y. P., Bancroft, I. & Park, B. S. (2009) Genome-wide comparative analysis of the *Brassica rapa* gene space reveals genome shrinkage and differential loss of duplicated genes after whole genome triplication. *Genome Biol.* 10(10): R111.

Nagoaka, T., Daullah, M. A. U., Matsumoto, S., Kawasaki, S., Ishikawa, T., Hir, H., Okazaki, K. (2010) Identification of QTLs that control resistance in *Brassica oleracea* and comparative analysis of clubroot resistance genes between *B. rapa* and *B. oleracea. Theor Appl Genet.* 120(7): 1335-1346.

Nestle, M. (1997) Broccoli sprouts as inducers of carcinogen-detoxifying enzyme systems; clinical, dietary, and policy implications. *Proc Natl Acad Sci USA.* 94: 11149-11151.

Newkirk, R. W., Classen, H. L., Scott, T. A. & Edney, M. J. (2003) The availability and content of amino acids in toasted and non-toasted canola meals. *Can J Anim Sci.* 83: 131-139.

Niu, Z. (2008) Manipulation of biosynthesis of aliphatic glucosinolates in brassica crops and *Arabidopsis* through gene replacement and RNA interference. Ph.D. thesis, Depart of Plant Science, Uni of Manitoba, Winnipeg, Canada.

Olsson, G. & Ellerstrom, S. (1980) Polyploidy breeding in Europe. In: S. Tsunoda, K. Hinata, and C. Gomez-Campo (eds) *Brassica* crops and wild allies. Japan Science Society Press, Tokyo, Japan, pp 167-190.

Osborn, T. C., Butrulle, D. V., Sharpe, A. G., Pickering, K. J., Parkin, I. A. P., Parker, J. S. & Lydiate, D. J. (2003) Detection and effects of a homeologous reciprocal transposition in *Brassica napus*. *Genetics*. 165: 1569-1577.

Osborn, T. C., Kale, C., Parkin, I. A. P., Sharpe, A. G., Kuiper, M., Lydiate, D. J. & Tricj, M. (1997) Comparison of flowering time genes in *Brassica rapa, B. napus* and *Arabidopsis thaliana*. *Genetics*. 146: 1123-1129.

Osbourn, A. E. (1996) Performed antimicrobial compounds and plant defense against fungal attack. *Plant Cell*. 8: 1821-1831.

Padilla, G., Cartea, M. E., Velasco, P., De Haro, A. & Ordas, A. (2007) Variation of glucosinolates in vegetable crops of *Brassica rapa*. *Phytochem*. 68: 536-545.

Parkin, I., Magrath, R., Keith, D., Sharpe, A., Mithen, R. & Lydiate, D. (1994) Genetics of aliphatic glucosinolates. II. Hydroxylation of alkenyl glucosinolates in *Brassica napus*. *Heredity*. 72: 594-598.

Parkin, I. A. P., Lydiate, D. J. & Trick, M. (2002) Assessing the level of collinearity between *Arabidopsis thaliana* and *Brassica napus* for *A. thaliana* chromosome 5. *Genome*. 45: 356-366.

Parkin, I. A. P., Sharpe, A. G. & Lydiate, D. J. (2003) Patterns of genome duplication within the *Brassica napus* genome. *Genome*. 46: 291-303.

Pedras, M. S. C., Zheng, Q. A. & Sarna-Manillapalle, V. K. (2007) The phytoalexins from *Brassicaceae*: structure, biological activity, synthesis and biosynthesis. *Nat Prod Comm*. 2: 319-330.

Pedras, M. S. C., Zheng, Q. A. & Strelkov, S. (2008) Metabolic changes in roots of the oilseed canola infected with the biotroph *Plasmodiophora brassicae*: Phytoalexins and phytoanticipins. *J Agric Food Chem*. 56: 9949-9961.

Petersen, B. L., Chen, S., Hansen, C. H., Olsen, C. E. & Halkier, B. A. (2002) Composition and content of glucosinolates in developing *Arabidopsis thaliana*. *Planta*. 214: 562-571.

Piotrowski, M., Schemenewitz, A., Lopukhina, A., Müller, A., Janowitz, T., Weiler, E. W. & Oecking, C. (2004) Desulfoglucosinolate sulfotransferases from *Arabidopsis thaliana* catalyze the final step in the biosynthesis of the glucosinolate core structure. *J Biol Chem*. 279(49): 50717-50725.

Prakash, S. & Hinata, K. (1980) Taxonomy, cytogenetics and origin of crop *Brassica*, a review. *Opera Bot*. 55: 1-57.

Punjabi, P., Jagannath, A., Bisht, N.C., Padmaja, K. L., Sharma, S., Gupta, V., Pradhan, A. K. & Pental, D. (2008) Comparative mapping of *Brassica juncea* and *Arabidopsis thaliana* using intron polymorphism (IP) markers: homoeologous relationships, diversification and evolution of the A, B and C *Brassica* genomes. *BMC Genomics*. 9: 113.

Qiu, D., Gao, M., Li, G. & Quiros, C. F. (2009) Comparative sequence analysis for *Brassica oleracea* with similar sequences in *B. rapa* and *Arabidopsis thaliana*. *Plant Cell Rep*. 28: 649-661.

Ramchiary, N., Bisht, N. C., Gupta, V., Mukhopadhyay, A., Arumugam, N., Sodhi, Y. S., Pental, D. & Pradhan, A. K. (2007) QTL analysis reveals context-dependent loci for seed glucosinolate trait in the oilseed *Brassica juncea*: importance of recurrent selection backcross scheme for the identification of 'true' QTL. *Theor App Genet.* 116: 77-85.

Ranade, S. A., Farooqui, N., Bhattacharya, E. & Verma, A. (2001) Gene tagging with random amplified polymorphic DNA (RAPD) markers for molecular breeding in plants. *Crit Reviews in Plant Sci.* 20(3): 251-275.

Rangkadilok, N., Nicolas, M. E., Bennett, R. N., Premier, R. R., Eagling, D. R. & Taylor, P. W. J. (2002) Developmental changes of sinigrin and glucoraphanin in three *Brassica* species (*Brassica nigra, Brassica juncea* and *Brassica oleracea* var. italic). *Scientia Hort.* 96: 11-26.

Rask, L., Andreasson, E., Ekbom, B., Eriksson, S., Pontoppidan, B. & Meijer, J. (2000) Myrosinase: gene family evolution and herbivore defense in *Brassicaceae. Plant Mol Biol.* 42: 93-113.

Reichelt, M., Brown, P. D., Schneider, B., Oldham, N. J., Stauber, E., Tokuhisa, J., Kliebenstein, D. J., Mitchell-Olds, T. & Gershenzon, J. (2002) Benzoic acid glucosinolate esters and other glucosinolates from *Arabidopsis thaliana. Phytochem.* 59: 663-671.

Rodman, J. E., Karol, K. G., Price, R. A.& Sytsma, K. J. (1996) Molecules, morphology, and Dahlgrens expanded order Capparales. *Syst Bot.* 21: 289-307.

Sawada, Y., Kuwahara, A., Nagano, M., Narisawa, T., Sakata, A., Saito, K. & Hirai, M. Y. (2009) Omics-based approaches to methionine side chain elongation in *Arabidopsis*: characterization of gene encoding methylthioalkylmalate isomerase and methylthioalkylmalate dehydrogenase. *Plant Cell Physiol.* 50(7): 1180-1190.

Schnug, E. & Haneklaus, S. (1990) Quantitative glucosinolate analysis in *Brassica* seeds by X-ray fluorescence spectroscopy. *Phytochem Anal.* 1: 40-43.

Schuster, J., Knill, T., Reichelt, M., Gershenzon, J. & Binder, S. (2006) *BRANCED-CHAIN AMINOTRANSFERASE4* is part of the chain elongation pathway in the biosynthesis of methionine-derived glucosinolates in *Arabidopsis. Plant Cell.* 18(10): 2664-2679.

Schwab, W. (2003) Metabolome diversity: too few genes, too many metabolites? *Phytochem.* 62(6): 837-837.

Sharpe, A. G., Parkin, I. A. P., Keith, D. J. & Lydiate, D. J. (1995) Frequent non-reciprocal translocations in the amphidiploid genome of oilseed rape. *Genome.* 38: 1112-1121.

Sønderby, I. C., Geu-Flores, F. & Halkier, B. A. (2010) Biosynthesis of glucosinolates-gene discovery and beyond. *Trends Plant Sci.* 15(5): 283-290.

Spinks, E. A., Sones, K. & Fenwick, G. R. (1984) The quantitative analysis of glucosinolates in cruciferous vegetables, oilseed and forages using high performance liquid chromatography. *Fette Seifen Anstrichmittel.* 86: 228-231.

Srinivasan, K., Malathi, V. G., Kirti, P. B., Prakash, S. & Chopra, V. I. (1998) Generation and characterization of monosomic chromosome addition lines of *Brassica campestris-B. oxyrrhina. Theor Appl Genet.* 97: 976-981.

Stefansson, B. R. & Downey, R. K. (1995) Rapeseed. In: Slinkard AE and DR Knott (eds) Harvest of gold. University Extension Press, University of Saskatoon, SK, Canada, pp 140-152.

Takuno, S., Kawahara, T. & Ohnishi, O. (2007) Phylogenetic relationships among cultivated types of *Brassica rapa* L. em. Metzg. as revealed by AFLP analysis. *Gen Resour and Crop Evol.* 54: 279-285.

Talalay, P. (2000) Chemoprotection against cancer by induction of phase 2 enzymes. *Bio Factors.* 12: 5-11.

Textor, S., De Kraker, J. W., Hause, B., Gershenzon, J., Tokuhisa, J. G. (2007) *MAM3* catalyzes the formation of all aliphatic glucosinolate chain lengths in *Arabidopsis. Plant Physiol.* 144(1): 60-71.

Tierens, K. F., Thomma, B. P., Brouwer, M., Schmidt, J., Kistner, K., Porzel, A., Mauch-Mani, B., Cammue, B. P., Broekaert, W. F. (2001) Study of the role of antimicrobial glucosinolate-derived isothiocyanates in resistance of *Arabidopsis* to microbial pathogens. *Plant Physiol.* 125: 1688-1699.

Toroser, D., Thormann, C. E., Osborn, T. C. & Mithen, R. (1995) RFLP mapping of quantitative trait loci controlling seed aliphatic glucosinolate content in oilseed rape (*Brassica napus* L.) *Theor Appl Genet.* 91: 802-808.

U. N. (1935) Genome analysis in *Brassica* with special reference to the experimental formation of *B. napus* and peculiar mode of fertilization. *Japanese J Bot.* 7: 389-452.

Udall, J., Quijada, P. & Osborn, T. C. (2004) Detection of chromosomal rearrangements derived from homologous recombination in four mapping populations of *Brassica napus* L. *Genetics.* 169: 967-979.

Ugajin, T., Takita, K., Takahashi, H., Muraoka, S., Tada, T., Mitsui, T., Hayakawa, T., Ohyama, T. & Hori, H. (2003) Increase in indole 3-acetic acid (IAA) level and nitrilase activity in turnips induced by *Plasmodiophora brassicae* infection. *Plant Biotech.* 20: 215-220.

Uzunova, M., Ecke, W., Weissleder, K., Robbelen, G. (1995) Mapping the genome of rapeseed (*Brassica napus* L.). I. Construction of an RFLP linkage map and localization of QTLs for seed glucosinolate content. *Theor Appl Genet.* 90: 194-204.

Velasco, L. & Becker, H. C. (1998) Analysis of total glucosinolate content and individual glucosinolates in *Brassica* spp. by near-infrared reflectance spectroscopy. *Plant Breed.* 117: 97-102

Wentzell, A. M., Rowe, H. C., Hansen, B. G., Ticconi, C., Halkier, B. A. & Kliebenstein, D. J. (2007) Linking metabolic QTLs with network and cis-eQTLs controlling biosynthetic pathways. *PLos Genet.* 3(9): 1687-1701.

Wink, M. (2003) Evolution of secondary metabolites from an ecological and molecular phylogenetic perspective. *Phytochem.* 64: 3-19.

Wittstock, U. & Halkier, B. A. (2000) Cytochrome P450 *CYP79A2* from *Arabidopsis thaliana* L. catalyzes the conversion of L-phenyl alanine to phenylacetaldoxime in the biosynthesis of the benzylglucosinolate. *J Biol Chem.* 275: 14659-14666.

Yang, Y. W., Lai, K. N., Tai, P. Y. & Li, W. H. (1999) Rates of nucleotide substitution in angiosperm mitochondrial DNA sequences and dates of divergence between *Brassica* and other angiosperm lineages. *J Mol Evol.* 48: 597-604.

Zang, Y. X., Lim, M. H., Park, B. S., Hong, S. B. & Kim, D. H. (2008) Metabolic engineering of indole glucosinolates in Chinese cabbage plants by expression of *Arabidopsis CYP79B2, CYP79B3,* and *CYP83B1. Mol Cells.* 25(2): 231-241.

Zhang, Y., Talalay, P., Cho, C. G. & Posner, G. H. (1992) A major inducer of anticarcinogenic protective enzymes from broccoli: isolation and elucidation of structure. *Proc Natl Acad Sci USA*. 89: 2399-2403.

Permissions

The contributors of this book come from diverse backgrounds, making this book a truly international effort. This book will bring forth new frontiers with its revolutionizing research information and detailed analysis of the nascent developments around the world.

We would like to thank Aakash Goyal, for lending his expertise to make the book truly unique. He has played a crucial role in the development of this book. Without his invaluable contribution this book wouldn't have been possible. He has made vital efforts to compile up to date information on the varied aspects of this subject to make this book a valuable addition to the collection of many professionals and students.

This book was conceptualized with the vision of imparting up-to-date information and advanced data in this field. To ensure the same, a matchless editorial board was set up. Every individual on the board went through rigorous rounds of assessment to prove their worth. After which they invested a large part of their time researching and compiling the most relevant data for our readers. Conferences and sessions were held from time to time between the editorial board and the contributing authors to present the data in the most comprehensible form. The editorial team has worked tirelessly to provide valuable and valid information to help people across the globe.

Every chapter published in this book has been scrutinized by our experts. Their significance has been extensively debated. The topics covered herein carry significant findings which will fuel the growth of the discipline. They may even be implemented as practical applications or may be referred to as a beginning point for another development. Chapters in this book were first published by InTech; hereby published with permission under the Creative Commons Attribution License or equivalent.

The editorial board has been involved in producing this book since its inception. They have spent rigorous hours researching and exploring the diverse topics which have resulted in the successful publishing of this book. They have passed on their knowledge of decades through this book. To expedite this challenging task, the publisher supported the team at every step. A small team of assistant editors was also appointed to further simplify the editing procedure and attain best results for the readers.

Our editorial team has been hand-picked from every corner of the world. Their multi-ethnicity adds dynamic inputs to the discussions which result in innovative outcomes. These outcomes are then further discussed with the researchers and contributors who give their valuable feedback and opinion regarding the same. The feedback is then collaborated with the researches and they are edited in a comprehensive manner to aid the understanding of the subject.

Apart from the editorial board, the designing team has also invested a significant amount of their time in understanding the subject and creating the most relevant covers. They scrutinized every image to scout for the most suitable representation of the subject and create an appropriate cover for the book.

The publishing team has been involved in this book since its early stages. They were actively engaged in every process, be it collecting the data, connecting with the contributors or procuring relevant information. The team has been an ardent support to the editorial, designing and production team. Their endless efforts to recruit the best for this project, has resulted in the accomplishment of this book. They are a veteran in the field of academics and their pool of knowledge is as vast as their experience in printing. Their expertise and guidance has proved useful at every step. Their uncompromising quality standards have made this book an exceptional effort. Their encouragement from time to time has been an inspiration for everyone.

The publisher and the editorial board hope that this book will prove to be a valuable piece of knowledge for researchers, students, practitioners and scholars across the globe.

List of Contributors

Herry S. Utomo, Ida Wenefrida and Steve D. Linscombe
Louisiana State University Agricultural Center, USA

Abhijit Sarkar
CSIR-SRF, Laboratory of Air Pollution and Global Climate Change, Ecology Research Circle, Department of Botany, Banaras Hindu University, Varanasi, Uttar Pradesh, India

Ganesh Kumar Agrawal
Research Laboratory for Biotechnology and Biochemistry (RLABB), Kathmandu, Nepal

Kyoungwon Cho
KRFC Research Fellow, Seoul Center, Korea Basic Science Institute, Seoul, South Korea

Junko Shibato
Department of Anatomy I, School of Medicine, Showa University, Tokyo, Japan

Randeep Rakwal
Research Laboratory for Biotechnology and Biochemistry (RLABB), Kathmandu, Nepal
Department of Anatomy I, School of Medicine, Showa University, Tokyo, Japan
Graduate School of Life and Environmental Sciences, University of Tsukuba, Ibaraki, Japan

Pallavi Mittal
ITS Paramedical College, Ghaziabad, India

Rashmi Yadav
All India Institute of Medical Science, Delhi, India

Ruma Devi
PAU Regional Station, Gurdaspur, India

Shubhangini Sharma
Aptara (Techbook International), Delhi, India

Aakash Goyal
Bayer Crop Science Saskatoon, Canada

Mukhtar Ahmed
Department of Agronomy, PMAS Arid Agriculture University Rawalpindi, Pakistan

Muhammad Asif
Agricultural, Food and Nutritional Science, 4-10 Agriculture/Forestry Centre, Univ. of Alberta, Edmonton, AB, Canada

Aakash Goyal
Bayer Crop Science, Saskatoon, Canada

Waqar Ahmad
Faculty of Agriculture, Food, and Natural Resources, The University of Sydney, Australia

Munir H. Zia
Research & Development Section, Fauji Fertilizer Company Ltd, Rawalpindi, Pakistan

Sukhdev S. Malhi
Agriculture and Agri-Food Canada, Melfort, Saskatchewan, Canada

Abid Niaz
Soil Chemistry Section, Institute of Soil Chemistry & Environmental Sciences, Ayub Agricultural Research Institute, Faisalabad, Pakistan

Saifullah
Institute of Soil and Environmental Sciences, University of Agriculture Faisalabad, Pakistan
School of Earth and Environment, Faculty of Natural and Agricultural Sciences, The University of Western Australia, Crawley, Perth, Australia

Kanika Narula, Eman Elagamey, Asis Datta, Niranjan Chakraborty and Subhra Chakraborty
National Institute of Plant Genome Research, Aruna Asaf Ali Marg, New Delhi, India

José Henrique Cattanio
Federal University of Pará – UFPA, Brazil

María A. Morel and Victoria Braña
Molecular Microbiology, Biological Sciences Institute Clemente Estable, Uruguay

Susana Castro-Sowinski
Molecular Microbiology, Biological Sciences Institute Clemente Estable, Uruguay
Biochemistry and Molecular Biology, Faculty of Science, Montevideo, Uruguay

Jocelyne Ascencio and Jose Vicente Lazo
Universidad Central de Venezuela, Facultad de Agronomía, Maracay, Venezuela

Arvind H. Hirani, Genyi Li, Carla D. Zelmer and Peter B.E. McVetty
Department of Plant Science, University of Manitoba, Winnipeg, Canada

M. Asif
Ex-Research Scientist, National Agriculture Research Centre, Islamabad, Pakistan

Aakash Goyal
Lethbridge Research Centre, Agriculture and Agri-Food Canada, Lethbridge, Canada

Printed in the USA
CPSIA information can be obtained
at www.ICGtesting.com
JSHW011433221024
72173JS00004B/780